So Many Snakes, So Little Time

So Many Snakes, So Little Time

Uncovering the Secret Lives of Australia's Serpents

Rick Shine

CRC Press
Taylor & Francis Group
Boca Raton London

CRC Press is an imprint of the
Taylor & Francis Group, an **informa** business

Photo credit: Diamond Pythons gather in large mating aggregations in springtime, so that radio-tracking one of these snakes can lead the researcher to a spectacular sight. Photograph by Terri Shine.

First edition published 2022
by CRC Press
6000 Broken Sound Parkway NW, Suite 300, Boca Raton, FL 33487–2742

and by CRC Press
4 Park Square, Milton Park, Abingdon, Oxon, OX14 4RN

CRC Press is an imprint of Taylor & Francis Group, LLC

Library of Congress Cataloging-in-Publication Data
Names: Shine, Richard, author.
Title: So many snakes, so little time : uncovering the secret lives of Australia's serpents / Rick Shine.
Description: First edition. | Boca Raton : CRC Press, 2022. | Includes bibliographical references and index.
Identifiers: LCCN 2021060471 (print) | LCCN 2021060472 (ebook) | ISBN 9781032232928 (paperback) | ISBN 9781032234618 (hardback) | ISBN 9781003277712 (ebook)
Subjects: LCSH: Serpents—Behavior—Australia. | Serpents—Ecology—Australia.
Classification: LCC QL666.O6 S424 2022 (print) | LCC QL666.O6 (ebook) | DDC 597.96/0994—dc23/eng/20220106
LC record available at https://lccn.loc.gov/2021060471
LC ebook record available at https://lccn.loc.gov/2021060472

ISBN: 978-1-032-23461-8 (hbk)
ISBN: 978-1-032-23292-8 (pbk)
ISBN: 978-1-003-27771-2 (ebk)

DOI: 10.1201/b22815

Typeset in Times
by Apex CoVantage, LLC

Contents

Preface vii
About the Author ix

Chapter 1 Boyhood and Adolescence 1

Chapter 2 Serpents in the Sheep Paddock 19

Chapter 3 Peering into the Love Lives of Blacksnakes 51

Chapter 4 Long-Dead Snakes at the Museum 81

Chapter 5 A Plethora of Pythons 101

Chapter 6 Between a Rock and a Hard Place 127

Chapter 7 Snakes in Need of a Defamation Lawyer 151

Chapter 8 Rough Characters in the Billabong 173

Chapter 9 Snakes, Rats, and Rainfall 203

Chapter 10 Science on the Floodplain 227

Chapter 11 Our Evolving Relationship with Serpents 247

Acknowledgments 257

Bibliography 259

Index 273

Preface

Observing snakes is not a popular aspect of field zoology. Indeed to the uninterested person it appears more like insanity than reason, especially when such an aggressive form as the Tiger Snake is taken into consideration.

—Percy Gilbert, 1935

I was given an encyclopedia of the natural world on my 10th birthday. Delirious with excitement, I turned straight to the chapter on my favorite animals—reptiles. But I was crushed. After devoting 200 pages to the glories of mammals and birds, the author dismissed reptiles and amphibians in a 10-page chapter entitled "Backwaters of Evolution". My beloved snakes were scorned as ugly and primitive creatures.

And that attitude had been the mainstream view for a century or more. In a short story written in 1894, famed Australian author Henry Lawson praised feral cats for "dragging in . . . ugly, loathsome, crawling abortions . . . [like the] long, wriggling, horrid black snake". In 1953, the Chamber of Commerce in the Queensland city of Mackay wrote to the Australian government to request that scientists develop ways to exterminate all snakes from Australia. Snakes, they said, "don't serve any useful purpose". That attitude still dominated Australian culture when I was an undergraduate student in the late 1960s. Ecological perspectives on the other native inhabitants of my home country were equally negative. Australia was a life raft for third-class animals, surviving to the modern day only because their isolation had protected them from contact with "more advanced" species.

Today, we have a different view of Australian wildlife in general, and reptiles in particular. We understand that there's nothing "inferior" about the cold-blooded Aussies. Modern textbooks of evolution and ecology don't force the diversity of life into a simple linear scale with humans at the top, pushing reptiles (and anything Australian) to the bottom of the list. As we learned about the biology of reptiles and of Australia, we discarded that stereotyped view of their inferiority.

There's a lot we still don't know. Mystery still surrounds basic questions like how many species of snakes occur in Australia, let alone the details of their private lives. But we have learnt a lot since the 1970s, and I wrote this book to illuminate our increased understanding as well as our continuing ignorance. Unlike most "snake books", this one focuses on adventures rather than misadventures. Snakes spend most of their time doing things that are far more interesting than retaliating against aggressive primates. Looking behind the smokescreen of exaggerated deadliness, venom is only a small part of the story.

About 100 million years ago, a small group of legless lizards evolved into snakes. Those first wormlike burrowers spawned an incredible success story. Today, the planet contains more than 3,000 species of snakes—under the ground, on the ground, in the water, up in the trees. Snakes are amazing. From tiny wormlike burrowers that eat ant eggs to giant pythons that swallow kangaroos; from agile tree snakes in the rainforest through to sea snakes in the open ocean. One of the great success stories in the history of life on Earth. Over the course of a long career, I've had the privilege to study fascinating twigs of that mighty evolutionary tree. An archeologist might get the same thrill from unravelling the history of a mighty empire, or a sports journalist from documenting the history of a world-beating team.

Fate gave me a front-row seat to this global rethink about snakes and about Australia. I carried out the first detailed ecological research on Australian snakes, and these fantastic animals have been the focus of my professional life. I have witnessed a revolution in perspectives about the ecology of snakes, and about the functioning of Australian ecosystems. So although this book is partly about my own career, it is mostly a song of praise—a celebration of magnificent creatures in an incredible continent.

About the Author

Rick Shine has conducted pioneering research on the ecology and conservation of snakes. He has studied many types of snakes in many types of ecosystems, and has revolutionized our understanding and appreciation of these mysterious creatures. His work has resulted in more than a thousand scientific publications, and has attracted numerous national and international awards.

1 Boyhood and Adolescence

Three months after my 15th birthday, I almost died. Death tapped my nose, but didn't manage to sink his fangs in.

The scene is etched into my brain. Sheep pastures running down to the Molonglo River, near the city of Canberra. Willow trees along the edge of the river, and steep banks where floods had eroded the sandy soils. Between the banks and the water, a narrow sandy beach. And on that beach, as I walked along the banks above, lay the biggest Brownsnake I had ever seen.

The Eastern Brownsnake is the great success story of snakedom in southern Australia. As its name suggests, an adult Brownsnake is drab brown above, beautifully camouflaged against the gray-brown Australian bush—just another dead branch amid the thousands littering the paddocks. Juvenile Brownsnakes feed mostly on frogs and lizards, but when they reach a meter (3 feet) in length, these slender-bodied lightning-fast speedsters switch across to eat house-mice. And that's a good tactic, because there's a lot of mice out there to eat. Clearing the eucalypt forests to create pastures for sheep and cattle has provided a smorgasbord for rodents. As a result, Brownsnakes thrived. Their numbers boomed.

Eastern Brownsnakes possess one of the deadliest venoms of any snake species, worldwide. Combine that toxicity with high abundance, especially around farms, and you have a recipe for disaster. And there are other factors as well. These elongate Australian mambas are some of the fastest-moving snakes I've ever dealt with; and they are among a small minority of snake species that, if harassed, will opt for vigorous retaliation. Sometimes, they even chase their tormentor rather than trying to escape. As a result, Brownsnakes kill more people than do any other snake species in Australia.

Even a tiny Brownsnake—as thick as a pencil—has enough venom to kill an adult human. And the snake that lay below me on the banks of the Molonglo was at least 2 meters (6 feet) long, and thicker than my scrawny forearms. And I was miles from anywhere. Mobile phones had yet to be invented, nobody knew where I was, and there was no way to send a distress call if anything went wrong. If I were bitten, it would take hours for me to walk out to the nearest road and flag down a passing car. And I wouldn't have hours to spare. A bite would be fatal.

Only an idiot would have tackled that snake. But I was 15 years old, and this was the biggest snake I'd ever seen. Without pausing to look around for a stick with which to pin the snake's head down, I took a flying leap off the bank, landed beside the snake, seized it by the tail, and lifted that tail into the air.

Very quickly, several things became apparent. First, the snake was huge. Even when I stood on tip-toes and held its tail as high as I could, its head still reached the

DOI: 10.1201/b22815-1

ground. Second, the snake wasn't happy. It exploded into action, lunging at me with an athleticism that belied its bulk. Third, I realized—belatedly—that I had no way out. The bank was too steep to scramble back up: I was trapped on a tiny beach with an enraged snake. It was too heavy to fling away—if I tried, it was likely to come straight back for revenge. And fourth, not grabbing a stout stick beforehand was a serious error—because as far as I could see (in between ducking and weaving as the snake's head shot past my shoulder), there were no suitable sticks on the beach.

It's amazing what adrenalin can do. That snake was bouncing around, mouth open, hurling itself in every direction. It was an important lesson about the power of a large venomous snake. And a lesson also in my own fallibility—I discovered that it was very tiring to stand on tip-toes holding a monster snake by the tail. My muscles were quivering and cramping, and something had to happen very soon. I frantically looked around for a stick to pin down the reptilian whirling dervish so that I could grab it behind the head—and saw an old branch, a foot long, by the water's edge. Looking around for a stick meant I wasn't focused on the snake—but I felt its weight shift as it threw itself upward. I looked up to see an open mouth inches from my face and coming down fast. As I drew back, the snake's snout tapped my nose, but he didn't get his fangs into me.

It was now or never. Unless I got my hands around that snake's neck very soon, I would lose the contest. So I dropped to my knees, grabbed that short stick, and pushed it down across the snake's neck when it next hit the ground. The stick broke immediately, but this was my only chance. I slammed my hand down on the gyrating head, pressing the business end of the massive reptile into the sandy beach. And by good fortune, it worked.

Panting, trembling, I lay on top of the writhing snake until my body was able to obey the signals from my brain. Somehow-or-other, I managed to stuff the snake into a bag and take it home. But even with my 15-year-old's enthusiasm, I could understand my parent's reluctance to have this giant snake added to my backyard menagerie.

I've caught a lot of snakes since then, but none have frightened me as much as that first huge Brownsnake. That magnificent animal taught me to think about what I'm doing before *the battle is joined, not afterward. It was an important education for someone destined to spend the next 50 years working with snakes. By scaring the hell out of me, that huge Brownsnake helped me survive a thousand similar situations in the decades to come.*

At some stage, every child worries his or her parents by exhibiting peculiar behavior. But some behaviors are more peculiar than others, and a fetish for snakes must rank highly among "Behaviors That Induce Parental Anxiety". All over the world, a small minority of young boys and girls manifest a passion for snakes. Most of those children later shift their interest to dinosaurs or video games, but a few stubbornly retain their scaly obsession. We pester our parents to take us to the Reptile House at the zoo, and we spend hours poring over books and websites that feature snake stories. And eventually, we decide that life isn't complete without a pet snake.

In most parts of the world, that's not a major problem. The snake that young Jack or Amelia encounters in the backyard is harmless, so catching it won't entail more than a few scratches. And if the backyard lacks snakes, you can buy a python at the pet store. But those options weren't available for people growing up where I did (in Australia) when I did (in the 1950s and 1960s). Uniquely among the world's continents, Australia has more venomous than non-venomous snake species. That's especially true in the southern part of the continent, where I lived as a boy. And the commercial trade in snakes as pets didn't begin until much later. When I was a youngster, the only way to obtain a pet snake was to catch one yourself. It was legal—regulations about wildlife protection didn't include reptiles until the 1970s—but catching your own snake was a daunting challenge.

With snakes in the "too hard" basket, my childhood activities centered on pet lizards instead of pet snakes. Serpents were deadly and elusive, whereas lizards were harmless and common. And I had plenty of opportunities to look for lizards in this halcyon era, before fear of strangers convinced parents to drive their kids to school. I walked from my home to the train station every morning to catch the train to school, and back again in the afternoon. The walk took 20 minutes if I took the straightest path, but instead I embarked on circuitous routes to look for Bluetongue Skinks. Up to 60 cm (24 inches) long, these reptilian rouseabouts handle urbanization better than most other reptiles. They are common in suburban gardens, but fiendishly difficult to catch. Blueys plan their escape routes in advance, and they don't stray far from a safe haven. I rarely caught a lizard the first time I saw it, but I was persistent. After I located the home of a local lizard, I would creep up to the appropriate garden every day. If the lizard was slower or sleepier than usual, I could leap onto it before it reached the safety of an impregnable drainage pipe. My success rates were low, but once in a while I came home triumphant, with a large lizard to be added to the outdoor pen in our backyard. I was better at catching lizards than at building secure cages, so the rate of escapees exceeded the rate of new arrivals. I never ran out of room.

Snakes fascinated me but they scared me as well, because everyone I talked to regarded snakes as malevolent monsters. The common catch-cry was "the only good snake is a dead snake". In today's world, the internet makes information easily accessible—but facts about snakes were hard to come by when I was a youngster. Even the basic issues—what snakes occur where, and how can you tell the difference between one species and another?—were shrouded in mystery. The only books on Australian snakes were dry academic tomes. Even if a schoolkid was prepared to fight his or her way through technical terms (and I was), the information was limited and often wrong. So, my ignorance about snakes was matched by my fear of them.

There's nothing like unadulterated terror to make a meeting memorable. The first snakes I saw in the bush were Red-Bellied Blacksnakes, a species that later became a focus of my research. Blacksnakes are magnificent. Awe-inspiring. Big, heavy-bodied, glossy black above, scarlet along the sides, pink on the belly . . . rippling muscles . . . and an amiable personality and relatively mild venom to boot. But in the days of my childhood, the sight of a jet-black body glistening in the sunlight filled me with terror rather than reverence. I fled in fear, expecting these demons to hurl themselves at me with venom dripping from their fangs. In retrospect, that's like expecting Mother Teresa to whip out a switchblade and eviscerate me. But snakes were an unknown entity, and everything I heard about them was negative.

FIGURE 1.1 The Eastern Brownsnake combines agility and attitude with highly toxic venom. Photo by Rob Valentic.

Early high school years in the northern suburbs of Sydney liberated me to wander further from home, and I discovered a tract of bushland with sandstone cliffs that might house reptiles. And sure enough, it wasn't long before I turned over a large rock and saw a slender olive-green snake beneath. Fortunately for me, it was a harmless species—a Green Tree Snake. I managed to capture the snake, albeit clumsily. And I brought it home in triumph.

My fascination with reptiles surprised my parents. They expected me to develop other interests as I grew up, but in the meantime they were willing to let me indulge my passion. My parents' generation had a far harder life than my own. Mum was brilliant but never finished high school; in the society she grew up in, "education is wasted on girls". Dad completed high school and started a university course, but he couldn't afford the fees. He dropped out and sold encyclopedias door-to-door during the Great Depression. My mother and father were married during the Second World War, just days before he left for Canada for his military training—and from there he was sent off to an English airstrip. As navigator in a Lancaster Bomber, his job was to bomb the Germans. It was incredibly dangerous; only a quarter of the Lancaster airmen survived the war. In 1944, Mum received a telegram from the War Office informing her that Dad's plane had been shot down over enemy territory, and he was believed dead. Months later, his name appeared on a list of prisoners-of-war, and Molly Shine knew that she wasn't yet a widow. Compared to that, my life has been a bed of roses.

FIGURE 1.2 Resilient to the challenges of suburban living, large Bluetongue Lizards can survive even in densely populated cities like Sydney. For reptile-obsessed youngsters growing up in eastern Australia, Blueys were the "big game" we all tried to catch. Photograph by Jules Farquar.

FIGURE 1.3 The first snake I ever caught in the wild was a Green Tree Snake, a harmless species that remains common even on the outskirts of suburban Sydney. Photograph by Stephen Mahony.

In hindsight, my parents' forbearance was perilously close to deserving sainthood. With no special interest in animals themselves, they were bemused by their youngest son's obsession with scaly creatures. But they supported me, and Dad made running repairs to my rickety backyard lizard pen so that it offered at least a temporary home to my pet reptiles. Snakes, though? Surely that was going too far? I'm sure that there were some long and difficult parental conversations (out of my hearing) about young Ricky's peculiar passions. Bless their hearts, they let me keep that first snake I caught. I proudly housed it in an aquarium in my bedroom.

That snake was soon followed by others of the same species. And fossicking around the same area a few weeks later, I discovered rock crevices that housed Brown Tree Snakes, a mildly venomous and delightfully pugnacious species. They joined the collection. I even had a Yellow-Faced Whipsnake briefly—a small slender olive-green venomous species with a bite like a bee-sting. My parents' tolerance was testimony to their unwavering support for their children. My older brother John's passion was chemistry, and the result of one unanticipated chemical reaction was a lung-full of chlorine gas that sent John gasping and wheezing from the garage. My sister Judith never manifested a passion for venomous snakes or dangerous explosions, but any teenage girl gives her parents plenty to worry about. So, snakes may have seemed like a low-risk option.

I have very few photographs of myself as a youngster, with or without a snake in hand. Film was expensive, and the Shine family wasn't rolling in cash. A twist of fate, however, has given me several photographs from a snake-catching trip that took

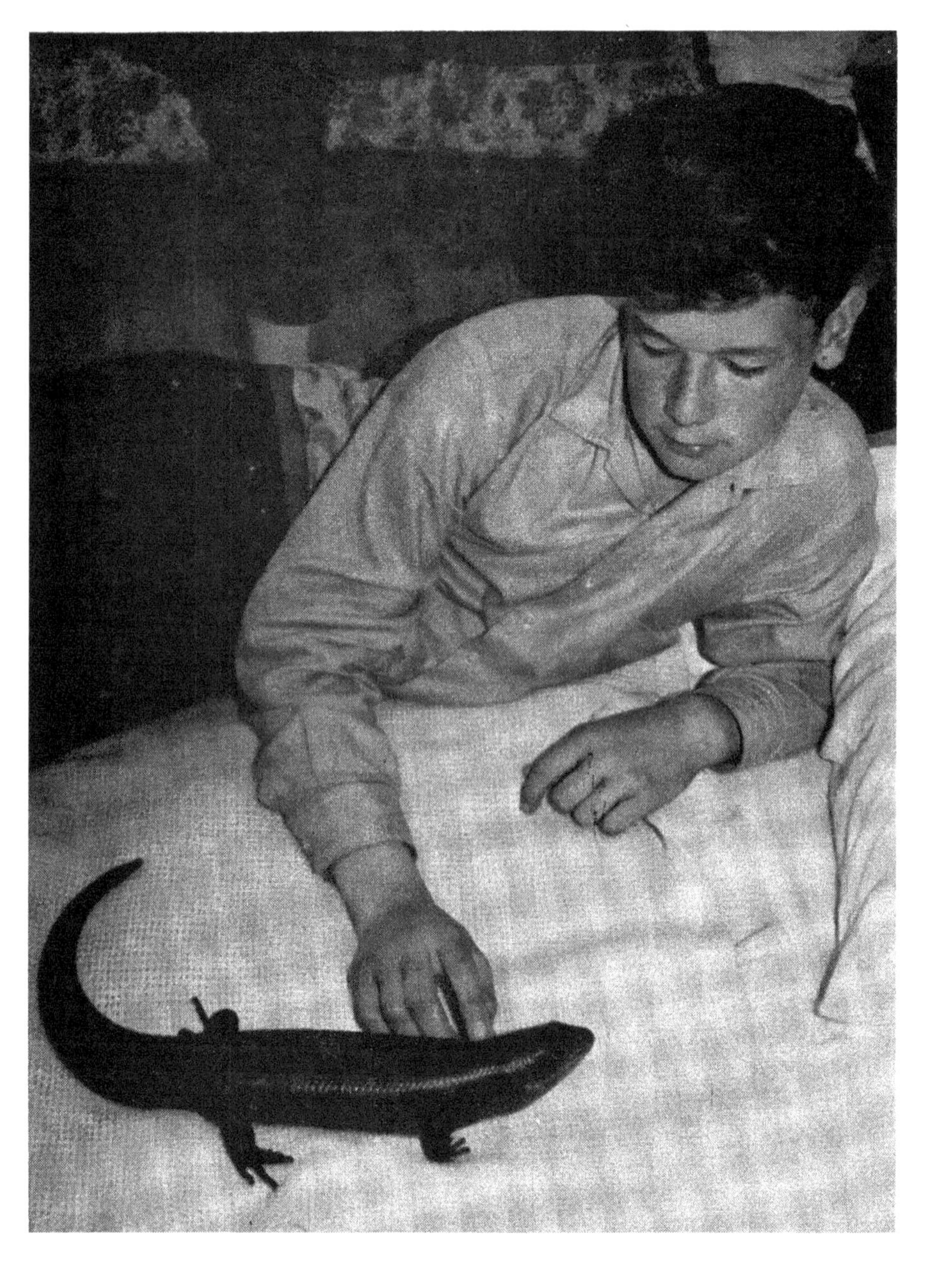

FIGURE 1.4 The author as a teenage reptile enthusiast, sharing a bed with a Land Mullet. Photograph by Michael J. Strachan.

place soon after I began high school. With the idea of producing a small children's book on the topic, an aspiring author offered to drive me and two of my school-friends to somewhere that we could find snakes. Transport to a snake-hunting site was too good an opportunity to refuse, so we leapt at the opportunity to visit the fabled golf course at La Perouse. I caught two Whipsnakes that day and my picture adorns the cover of a slim volume entitled *The Snake Catchers*. Photographic evidence of my youthful obsession!

The next Great Leap Forward occurred a couple of years later. As soon as my brother John (the garage-chemist, four years older than me) obtained his driving license, I had a way to travel further afield. And one of those trips took me to a patch of bushland where I had my first encounter with the rockstar of snake life in southern Australia. As its name suggests, the Diamond Python is a living jewel. This is no shoelace-thin tiny serpent in olive-drab colors. It's a huge snake—up to 3 meters (9 feet) long, chunky and muscular, and splendidly adorned with fluorescent yellow spots against a jet-black background. It makes most other snakes look like earthworms.

Pythons were the stuff of dreams for any young snake enthusiast, but I never expected to see one in the flesh. When a 2-meter vision of beauty slithered across the bush track in front of me, I was shocked. I was already keeping a field notebook when I was 14 years old, so I can nominate the exact date of this great adventure. My friend Tim and I were wandering along on the 22nd of August 1964, when the giant python materialized in front of us. Neither of us believed it was real, and the snake had almost slithered away into the bush before I came to my senses and grabbed its tail. The python was by far the largest snake I had ever touched, so I expected fireworks—and possibly, near-lethal wounds—but the snake just quietly hissed its displeasure. I'd read somewhere that Diamond Pythons are mellow snakes, reluctant to bite. Foolishly, I believed it. On the car ride back to my house, I cradled that snake in my lap.

The next challenge was to convince my parents to let me keep this monster. Dad wasn't home, and Mum was horrified. I was in the front yard, with my armful of giant snake, when she came out of the house to block my way. Irresistible force meeting immovable object. I had no intention of giving up the snake of my dreams, and Mum was irrevocably opposed to housing this huge reptile in the backyard. The high rates of escape of my captive Bluetongues and Green Tree Snakes may have reinforced her opinion. But I was equally fervent. Believing in the Diamond Python's mellow nature, I supported a few coils in my arms, letting the head move about freely . . . and as I was pointing out the angelic attitude of Diamond Pythons to my skeptical mother, the snake finally ran out of patience. It struck out hard and fast, fortunately at my leg rather than my face. Pythons have big heads, with dental equipment that evolved to handle formidable prey like Brushtail Possums. That snake hit my leg like a battering ram.

When the snake hammered into my leg, I was knocked to the ground. But it was a blessing in disguise. My continued defense of the python, despite the inconvenient fact that its teeth were embedded in my leg, made my mother realize that her opposition was futile. She said she'd talk to Dad about it when he got home, and she wandered back into the house. After she left, I carefully disengaged those long recurved teeth from my calf, and reverently transferred the new family member to a cage in the backyard. It escaped a week later.

But the snakes of the Sydney area were soon to get a reprieve. My father was offered a more senior job in Canberra, so the Shine family moved from the green forests of the coast to the dry sheep paddocks of the inland. I was partway through high school, but the move wasn't a surprise. We rarely lived in one place for more than a few years. By the time we headed to Canberra, I had already lived in three cities: Brisbane, Melbourne, and Sydney. After the trauma of his wartime years, Dad couldn't settle down.

My first reaction to a change of domicile was: what new species of snakes might I encounter? Within an hour of Dad breaking the news about our move, I took out my dog-eared copy of David McPhee's little pocket-book *Some Common Snakes and Lizards of Australia*, and made a list of all the species that might be found around our new home. The list was a long one, full of exciting animals like Tigersnakes and Copperheads, and I went to bed that night with dreams of all the new snakes I would meet. But I was daunted as well, because most of those serpents were highly venomous. I would need to change my approach. Leaping onto a snake and wrestling it into submission was OK for a Green Tree Snake, and even a Diamond Python. I had relied on enthusiasm, with no clear tactics. But around Canberra, that devil-may-care approach might land me in terminal trouble. I needed to learn about snake-handling from somebody who had the skills that I lacked. But I didn't know anybody with those skills.

****

Lady Luck was on my side. On my first day at Narrabundah High, my new school in Canberra, one of the other new boys I met was Russell Wombey, who lived in a suburb close to mine. And hearing of my reptilian interests, Russell volunteered an astonishing fact: his older brother John was a snake enthusiast. Indeed, John was a reptile fanatic; his passion for snakes rivalled mine, and his expertise was a thousand times greater. In later years, I got to know John much better, through our mutual attendance at reptile conferences. Fiercely proud of his Scottish heritage, John enlivened the conference proceedings by marching through the room with his red hair flying, his kilt waving, and his bagpipe thundering. And that enthusiastic approach to life and reptiles was evident from the outset. As soon as he learned that there was a fellow snake-lover at his brother's high school, John invited me on a trip to catch snakes at a riverside location half an hour's drive west of Canberra. So, that same weekend (my notebook tells me it was the 4th of April 1965), I had my first experience of catching venomous snakes.

The environment around Canberra, Australia's capital city, is far from spectacular. Unless, that is, you enjoy looking at sheep. Most of the eucalypt woodland has been transformed into pastures, with an occasional tree retained to provide shade for the stock. But the streams that crisscross that unexciting landscape are bordered by vegetation—including spectacular river gums—and the native fauna that once roamed freely across the hills is still to be found along the river's edge. We walked quietly along the banks of the Murrumbidgee River until we found a large Blacksnake basking in the sunlight. It headed for the water, but John sprinted across and—marvel of marvels—seized it by the tail and held it aloft. The snake writhed and snapped, but soon tired. John pinned down its head with a stick, seized it by the neck, and carefully transferred it into a cloth bag.

It isn't easy to pick up a venomous snake by the tail without being bitten. That first wild explosion of calisthenics has the snake's head perilously close to the hand that's holding the tail, and to the body and face of the would-be snake-catcher. Heavy-bodied species like Blacksnakes have the muscle to launch a well-aimed strike back toward their assailant's body; and a light-bodied Brownsnake can throw the front half of its body further and faster than seems permissible under the laws of physics. Held by the tail and unceremoniously lifted into the air, a hot snake transforms into a whirlwind. It's not a game for the faint-hearted.

I didn't catch any snakes myself that first day, but I learned a great deal. "Tailing" was the traditional way to catch snakes in Australia, whereas (as I later discovered) snake enthusiasts from other countries rely on purpose-made snake-tongs. Most venomous snakes in the Northern Hemisphere are heavy-bodied vipers and rattlesnakes, slow to move but fast to strike. Easy to manhandle from a distance, but a catastrophe if you enter the strike zone. In contrast, most Australian venomous snakes are long and slender. Sliding a hook under a rattlesnake to lift it off the ground works fine, but don't try it on a Taipan. Within the next few seconds, it will either be far away in the bush or else chewing on your leg. So, pioneer Aussie snake-catchers like "Pambo" Eades developed the simple but risky method of "tailing".

These days tailing large venomous snakes has fallen out of favor. Even the Alpha Males of Australia use snake-tongs, except when under the influence of a surfeit of testosterone and/or alcohol—but it's still a fallible human being against a deadly snake. Any mistake may be your last. Most of the "top ten" deadliest snakes in the world live in Australia.

But tailing was the way that John Wombey caught snakes, so I happily agreed to give it a try when John offered to hand that first snake over to me. Grasping the snake's tail and keeping away from its fangs were just the first steps in a lengthy process. After my opponent tired, I needed to get it into a cloth bag. There are some tricks to doing that—for example, it should be done alone. Your helpful assistant is likely to move where and when you don't expect him/her to. Best to keep it simple. My rule when handling deadly snakes is that only one person—the snake-handler—is allowed to move. Everyone else stands stock still.

Unfortunately, the snake won't cooperate. John showed me how to pin the snake's head against the ground with a stick, and to seize it firmly by the neck (keeping my fingers away from those fangs that are slashing around frantically, a few millimeters away). Then I could slip a cloth bag inside-out over my other hand like a glove, and transfer my grip of the snake to that hand. Then turn the bag right-side-out again and hey presto, the snake is inside and the hand holding it is outside. So, then I just had to stuff the body in and tie the bag shut. Nothing to it, really—except that it's difficult to keep a firm grip on a snake's head through a cloth bag; and forcing the snake's body into the bag and tying up the bag with one hand requires a cool head and good coordination.

By now, I was gazing at John with a reverence rarely seen outside a cathedral. But just to demonstrate human fallibility, the next lesson came from a mistake. Just after John tied the bulging snake-bag to his belt, the Blacksnake found a small hole in the bag and began to force its way out. The snake got its head and neck through the narrow opening but then stuck fast and flailed around wildly, bumping against

FIGURE 1.5 Don't try this at home. My first experience of "tailing" a Red-Bellied Blacksnake. Photograph by Michael J. Strahan.

John's leg. Miraculously, it didn't seize the opportunity to grab a mouthful of flesh. Memo to aspiring snake-catchers: check every bag carefully before you put a snake into it, and never allow a bagful of venomous snakes to bump against your genitals.

The lessons kept on coming. John caught another two Blacksnakes. He realized that I was passionate about snakes but incredibly naive, and he tolerated my questions with good humor. It was heady stuff. And although his methods were risky, John's commonsense approach gave me a great start in my own snake-catching career. That day was my baptism in snake-catching. After that I was on my own. I never had another lesson.

If they knew what I was doing, my parents would have been horrified. But I painted a rosy picture of the sweet-natured, "mildly venomous" snakes that were the focus of my activities every weekend during warmer months of the year. And even in winter, when puddles on the road were frozen as I walked to the school bus stop in the morning, I ranged far and wide looking for snakes under rocks by the riverbank. Most days I didn't find anything, but an occasional scaly bonanza kept me motivated.

Predictably, my first near-fatal highlight involved a Brownsnake. One cold winter morning, I was shocked to find a medium-sized Brownsnake (60 cm [2 feet] long) under a large boulder I turned over beside the Molonglo River. It was the first Brownsnake I'd ever seen. I stumbled backward in fright—I had heard of the awesome venom and rapid movements of these lithe speedsters, coupled with exaggerated tales of their ferocity. But the snake was so cold that it could barely move, so my courage returned. It was easy to pin the snake down and transfer it to a cloth bag. Encouraged by this spectacular find, I turned many more rocks over the next couple of hours but without success. The day was warm, despite the overnight frost, so I stopped to rest and rehydrate. And of course, I just *had* to take a look at that magnificent snake I had captured earlier. It didn't occur to me that 2 hours in my backpack would heat up the snake, transforming it from a sluggish snake-sicle to a fearsome warrior. So, like an idiot, I sat down on a convenient rock, placed the bag in my lap, and opened it. My plan was to peer inside, just to take a look, but the snake had other ideas. As soon as the bag opened, it shot out onto my lap, raised itself into a cobra-like threat posture, and looked around deciding where to bite me.

Any movement on my part would precipitate a strike, faster than I could evade. Fortunately, terror had me rooted to the spot. I don't know how long I sat there—it seemed like eternity. Eventually, the snake calmed down, convinced that it was the only living thing in the vicinity, and crawled away off my lap. Foolishly, I jumped up and grabbed it by the tail and managed to get it back into its bag. I kept it for a few months but finally took it back to the river and released it. If ever a snake deserved freedom, it was that Brownsnake. I realized that stories of ferocious snakes were nonsense. The same is true for almost all wildlife. When asked what kinds of animals were most dangerous, the 18th-century explorer Paul Du Chaillu (the inspiration for the movie character Tarzan) replied "those that defend themselves when attacked". That's an accurate summary for snakes as well.

During my teenage years, I spent a lot of time looking for snakes, catching them, and keeping them at home. But my menagerie came under threat after the Canberra newspaper ran a headline story about a fatal snakebite. Belatedly realizing the dangers involved, my parents decided that a garage-full of venomous snakes was too big a risk. That evening, we had a long and awful family conference—just Mum, Dad, and me around the dinner table. Only one topic on the agenda: I had to get rid of my snakes. Immediately. I was horrified, but my pleas fell on deaf ears. Their minds were made up. I went to bed in tears. But catastrophe was averted the very next day, when a letter arrived to say that I had been awarded an academic scholarship for the final two years of my high school career. Those funds would make it a lot easier to balance the family budget.

My parents were kindly souls, already distressed at breaking my heart in that prolonged argument the evening before. They had been shocked at my intense reaction to their decision. Faced with the evidence that I had been working hard at school, my parents relented. The Brownsnakes had to go, but I could keep my beloved Blacksnakes. And deep in my heart, I agreed with them. Teenage boys and deadly snakes are a poor combination.

I experimented with various ways of catching snakes, but none of them worked as well as John Wombey's "creep up on 'em, chase 'em down, tail 'em" technique. My attempt to catch snakes with a butterfly net, for example, was a total disaster. And my attempts at keeping snakes in captivity weren't much better. An animal in a cage doesn't thrill me, and I was a poor snake-keeper. I'd prefer a quick glimpse of a snake in the wild to an hour of staring at it in a cage. The visceral fear of wrestling that giant Brownsnake is still with me, half a century after the riverside encounter with which I started this chapter—but when I later looked at that snake in a cage, it seemed less impressive. Before long, I stopped keeping snakes. I still spent a lot of time and effort trying to catch them, but then I just let them go.

As my school years concluded and I became a university student, snakes faded a little in my pantheon of hobbies. I developed other passions—for playing rugby; for trout-fishing; for motorbikes; and for girls. I still leapt upon snakes whenever I found them, but most of those encounters were accidental, when a snake appeared in front of me as I crept along the riverbank toward a rising trout.

So, in summary, snakes played a central role in my youth and adolescence. Why? Most people don't find snakes especially fascinating. Why were these animals such a lightning rod for my own enthusiasms? I have no sensible answer. I didn't have any weird uncles or aunts with a penchant for reptiles. Nor even any snake-keeping neighbors. It's just that snakes enthrall me, and they always have. It's a visceral reaction—partly driven by fear, partly by wonder, partly by aesthetics.

What would my life have been like if my ancestors hadn't emigrated from Ireland to New Zealand in the 1800s? Or if a New Zealand doctor hadn't sent my grandparents to Australia in 1909 so that the warmth could cure a persistent lung problem? Neither Ireland nor New Zealand has snakes. If I'd grown up in one of those cold little countries, would I have fixated on something else? Football? Fishing? Poetry? Fast cars? Or would I have fled to a warmer country to fulfill my passion for contact with serpents? It's difficult for me to imagine a life without snakes.

My zealotry about snakes is simply part of me. My awe of snakes defies rational explanation, but has always been strong enough to withstand opposition. Especially

FIGURE 1.6 Eastern Brownsnakes had several opportunities to terminate my existence during my schoolboy years; I'm grateful for their forbearance. Photograph by Kris Bell.

FIGURE 1.7 The author during his high school years, holding a pet Carpet Python in the backyard of his Canberra home.

in my younger days, many people questioned the wisdom of my fixation on serpents. Sometimes that opposition came from mates who didn't understand why I wanted to squander my career on such peculiar creatures. But as friends, they usually abandoned the argument when they saw the depth of my passion. Far more confronting were my discussions with senior biologists. None of them studied snakes, and they were deeply disparaging about my choice of study animals. Several people tried to make me see reason and switch across to studying "real animals" like birds or mammals.

One of the most confronting conversations about the idiocy of snake ecology as a career path happened one evening in a pub in Canberra, partway through one of the first scientific conferences I ever attended. I was still an undergraduate student, already determined to become a snake researcher but without any clear thinking as to why that made sense. My companion was an erudite but somewhat arrogant Englishman, temporarily in Australia for postdoctoral research in the colonies before inevitably heading back to Cambridge or Oxford for a comfortable academic life in the Old World. I'm not sure if he was trying to offer helpful advice to a neophyte, or simply enjoyed inflicting his deeply held prejudices on a naive young enthusiast. But regardless, it was a frontal attack on my deeply held beliefs.

He adopted a lofty position, befitting someone who understood the realities of science far better than I did, and carefully spelt out reasons why studying the ecology of snakes was a foolish idea. "Useless bloody animals", he declaimed. Ugly, aggressive, and ecologically unimportant. Research on snakes would be a scientific equivalent of the board game Trivial Pursuit. There were too few types of snakes, and most of them were too scarce to study, and they were all pretty much the same anyway (long tubes, no legs). Minor players, at best, in evolutionary and ecological issues. Unlikely to shed light on any fundamental questions. Simple and primitive creatures, uninteresting in their own right. He kept repeating the phrase "boring inflexible cold-blooded robots". As the drinks flowed, his vehemence increased.

I was horrified. Everything he said made sense at one level, but it was utterly contrary to my own views and my plans. I had no good answers, and was feeling seriously deflated. But when he began hurling the same criticisms at Australia-based research ("if someone ever decides to give Earth an enema, they'll insert the tube into Canberra"), my mood shifted from frustration to irritation. It was bad enough having my favorite animals disparaged by a drunken Englishman. Hearing him revile my country was a step too far. Increasingly, his smug diatribe revealed a worldview confined to his own academic interests. He was flailing away at species and places outside his own experience. I was annoyed—deeply annoyed—but it was a valuable lesson. A professional career in snake ecology had always seemed like the right path for me, but that decision had been driven by emotion not intellect. If my obnoxious English colleague was correct, I was heading at full speed down a one-way street, with nothing in front of me but a brick wall.

After I recovered from the assault, I took stock. Setting aside the prejudices on view, there were still some cogent arguments that I needed to confront. Was it professional suicide to focus my research on snakes? Would it lead me into an intellectual backwater, describing trivial variations on an obscure theme? And finally, after weeks wrestling with the challenge, I made my decision. Life's not a dress rehearsal. The possibility of a research career focusing on the ecology of snakes was more exciting than any other option I could imagine. At the very least, I would test the validity of the anti-snake sentiments held by so many scientists, and so cogently expressed by that acerbic Englishman. And with luck, I would be able to prove to myself that the naysayers were wrong.

2 Serpents in the Sheep Paddock

Midnight, in a small country hotel, with a splitting headache. The sleep I craved was out of the question. As I lay on a billiard table, looking up at a "Dungowan Pub" sign, I could hear my fellow Ph.D. students talking and laughing, and the clatter of pool cues hitting balls on the adjacent table. And as I waited for my friends to finish their game and their beers, I reflected on a wonderful day that had turned ugly.

The morning had been terrific. With three other reptile-fanatics enrolled at the same university as myself, I had left the small rural town of Armidale at first light to spend the day collecting snakes and lizards. Our supervisor wanted to document the geographic distributions of all the local reptile species, but the data were sparse—many locations had never been checked. So wherever the map showed an unexplored area, it was a great excuse for us to spend a day roaming around the bush catching reptiles. Most of the land was dominated by farms, but there were plenty of small patches of woodland where we could search for animals.

I had arrived in Armidale only recently to begin my Ph.D., and Hal Heatwole's other students were still making up their minds about me. Was I just a stuck-up city boy? I was lonely, and desperate to make friends with my fellow-students. It was daunting, and the conversation in the car sometimes felt like a job interview as the other students checked out the new guy. But although I was shy and socially clumsy, I had one ace up my sleeve. I was experienced at finding and catching reptiles, so maybe I could impress my new peers with those skills, even if I couldn't manage sparkling conversation. The day was turning out well. I had spotted the first snake of the trip—a big Blacksnake by the side of the road. Only a few inches of its body were visible, and nobody else in the car had seen it. They were dubious when I called out "snake!" But they stopped . . . and when I returned to the car holding a Blacksnake by the tail, they were impressed. At every stop, I found as many lizards and frogs as anybody else, and I felt that I was establishing some credibility with the locals.

Pride comes before a fall. The morning was warming up when I turned over a log to reveal a slender meter-long (3-foot) olive-brown snake coiled up beneath. It was a Yellow-Faced Whipsnake, a species that hadn't been found previously in the area. So, a stellar find. Assuming that the snake was still in its cold overnight retreat, and hence as athletic as a slug, I reached down to pick it up behind the head. But I was wrong. The snake had been basking in the sunlight and had retreated to its shelter as I approached. As one might guess from their common name, Whipsnakes are slender—and like most slender snakes, they move like greased lightning when they are hot. The snake shot away as I reached down, and the only bit I could reach was its fast-disappearing tail. I grabbed it, and the snake wheeled, struck, and danced around faster than the eye could follow. When it tired, I slipped it into a

DOI: 10.1201/b22815-2

cloth bag—and then noticed blood dripping from two small puncture wounds on my forearm. The snake had bitten me during its first furious calisthenics, but I hadn't felt it at the time.

I tried to recall everything I had read about the venom of Whipsnakes. My snake was small, but it had bitten me in a place where the venom could easily get into my bloodstream. Was that a problem? I wasn't sure, because very little was known about the venoms of small species of snakes. Snakebite researchers were drawn to large and deadly species like Taipans, Brownsnakes, and Tigersnakes. Nobody had ever measured the toxicity of a Yellow-Faced Whipsnake's venom. I doubted that such a small snake could inflict a lethal bite—but on the other hand, I had heard of people dying after bites from small supposedly harmless species.

With the stupidity of youth, I played the tough guy. I showed off my bite wounds to my new mates, and we kept on collecting. And foolishly, we punctuated our reptile hunts with social drinking, stopping at local hotels for a beer between each collecting site. It was a hot day, and we stopped at 20 different spots. By mid-afternoon I was feeling terrible—perhaps because of dehydration and alcohol consumption rather than snakebite, but there was no way to be sure.

As dusk fell, we stopped at "one more pub" on the way home—in the town of Dungowan. By then, I was in bad shape. And our "quick drink" turned into dinner, several beers, and games of pool. My fellow students were more used to alcohol than I was, plus they hadn't been bitten by a venomous snake. As they became more raucous, I sunk into the private hell that is known only to migraine sufferers. I hoped that passing out on the unoccupied billiards table might have aroused some sympathy, but my friends were having too much fun to call an end to the evening.

We didn't get back to the university dormitory where I lived until 2 A.M. What to do now? What if I developed more serious symptoms? I scribbled a note saying "I have been bitten by an adult Yellow-Faced Whipsnake, on my right forearm", and laid it beside my bed. That way, I thought, anyone finding me unconscious in the morning could work out what was wrong and get me to hospital. The problem was that nobody was likely to find me. So reluctantly I knocked on the door of my neighbor, a serious-minded Chinese student who was studying economics, and woke him up. I asked Eddie to check on me in the morning because I might be seriously ill with snakebite by then. Having arranged everything, I then fell into a dreamless sleep, awakening mid-morning with a clear head and a hearty appetite. Unfortunately, things weren't quite so rosy for my neighbor. Terrified that he would find a lifeless corpse in the morning, Eddie came back into my room and sat beside me all night, checking every few minutes to see if I was still breathing. When I woke up, I was in much better shape than he was.

After I finished high school, I stayed in Canberra to do my undergraduate degree in science at the Australian National University. I was mentored by Dick Barwick, a gentlemanly New Zealand–born academic. Dick was genial, and a man of diverse talents ranging from scientific illustration to wine appreciation (in his spare time, he was Master of the University Cellars). He was also a charismatic raconteur, full of exciting stories about research in the Antarctic earlier in his career. In pride of place in his office, he displayed an old black-and-white photograph of himself and

FIGURE 2.1 The Yellow-Faced Whipsnake is an agile speedster, and gave me my first unpleasant experience of snakebite. Photograph by Matt Greenlees.

four colleagues with an immense penis (recently detached from a sperm whale) lying across their laps. I was impressed, as any 17-year-old male would be.

There were many times during my undergraduate days in Canberra when my life might have taken off in a different direction. For example, I bought a motorbike and (betrayed by an awesomely poor sense of balance) fell off it frequently—but fortunately those accidents were minor. Lots of blood, but no broken bones. One critical fork in the road came when I was 20 years old. Unable to attract enough foot-soldiers to fight in the massively unpopular Vietnam War, the Australian government introduced conscription to fill the ranks. An annual lottery was held, and anyone with their birthday falling on one of the randomly selected dates was sent off to war. I missed the lottery date by a single day, so was able to stay in Canberra rather than head off to the jungles of Southeast Asia.

University was fun, and my grades improved. I had been a mediocre student at high school, but I flourished under the open-ended challenges of tertiary education. Like most university students, I learned more about myself than I did about any other subject. Ironically, one of the greatest learning experiences came from a catastrophe. At the end of my first year's university studies, I landed a job for the summer holiday season as a fieldwork research assistant for a government agency, the Commonwealth Scientific Research Organisation (universally abbreviated to "CSIRO"). It was a dream come true. The work involved frequent checking of reproductive state in kangaroos and wallabies held in extensive outdoor pens—and because the marsupials didn't enjoy being captured and examined, it was great exercise as well as fascinating scientifically. But I had always suffered from terrible hay-fever, and the pollen from grass in the kangaroo yards set off a massive fit of sneezing every time I entered a pen. No medicine I found could control my allergy, and it was simply impossible to continue. Broken-hearted, I resigned after a week.

With my dreams shattered, I looked around for another job that would enable me to earn some money over the long summer break. And I was lucky—a job came up in the Library of Parliament House, the seat of Australia's federal government. The work was repetitive—mostly just gluing information cards into books, and returning books to the shelves after they had been borrowed by parliamentarians. But at least I could work inside, away from grass pollen. The Library was a busy workplace, with around 20 employees. I was the lowest-ranked person on the office totem pole and importantly, I was also the only male. Up until that point, I had suffered from an affliction that was almost universal among the young men of my generation—I was uncomfortable with women outside those in my own family. At least until the end of high school, men and women lived fairly separate lives in 1960s Australia. So this new job was a crash course in spending time with females of my own species. I loved it. By the time I finished the job a couple of months later, I saw women as fellow human beings rather than as terrifying aliens.

That lesson was reinforced during my later undergraduate years, when I began volunteering for unpaid fieldwork with academics at the University. I joined a team led by Professor Hugh Tyndale-Biscoe, partly because he was a superb lecturer and partly because his graduate students were conducting exciting field-based research on Greater Gliding Possums at a site not far from Canberra. It was the kind of thing I had always wanted to do—cruising along forest roads at night in a four-wheel-drive vehicle, searching for and collecting native animals so that we could find out how best to conserve the species. The leader of that research program was a Ph.D. student named Marilyn Renfree: energetic, forthright, talented, and willing to educate an enthusiastic undergraduate about what we were doing, and why. Marilyn took me under her wing, and gave me my first experience of the creativity, forethought, ambition, hard work, and stoicism needed to be a professional researcher. She remains a good friend.

As I learned more about biology, I discovered that snakes weren't the only interesting animals on the planet—for example, marsupials were pretty exciting as well—but I never wavered in my conviction that snakes were the *most* interesting. In my final undergraduate year, I conducted my Honors research program under Dick Barwick's benevolent supervision. Snakes were not an option, because the idea of bringing venomous animals into the Department elicited an instant and absolute veto. So instead of snakes, I studied lizards—specifically, the Cunningham's Rock Skink. These large and heavily armored lizards were abundant among rock outcrops scattered in agricultural pastures close to the city. Dick took a hands-off approach, allowing me to come up with the original idea for the project, and to attack it on my own. As a result, I made a lot of mistakes—but I learned a great deal in the process. I was excited to be finally conducting my own research on reptile behavior, but it gradually dawned on me that I had neglected a critical assumption—that the lizards would behave. I spent weeks crouching in a metal-sided hide, peering intently through binoculars at lizards in a nearby outcrop. For hour after hour, day after day, I waited for the lizards to do something interesting. They never did. Most of the time, they just basked in the sunlight, ignoring each other and everything else. It wasn't the lizards' fault, but every time I see a Cunningham's Skink, I still feel a strong urge to thump it on the head for revenge.

So, despite my outstanding undergraduate record, my first solo attempt at research was mediocre. Nonetheless, my overall academic credentials were good enough for me to be offered Ph.D. scholarships at several places, and I chose the University of New England. It's a small institution, in the country town of Armidale, in the New England highlands of northern New South Wales—an area of rolling hills, quaint farming towns, cold winters, hot summers, and an enormous number of sheep.

Why choose Armidale rather than a more prestigious institution? I was offered scholarships at "better" universities, but they wouldn't have allowed me to pursue my dream. Back in the 1970s, very few scientists studied the ecology of reptiles—especially in Australia. Reptiles were second-class denizens of the natural world, unworthy of serious research. Real biologists studied marsupials or birds, leaving the "primitive" cold-blooded animals for showmen or zealots. In all of Australia, only one group of academics was doing the kind of research I wanted to do. That group was based in Armidale and led by Hal Heatwole, an American with superhuman energy—a kindred spirit, with broad scientific interests and a special passion for reptiles; the ideal supervisor for my Ph.D.

FIGURE 2.2 Cunningham's Skinks are large rock-dwelling lizards, common in rocky outcrops among sheep paddocks near Canberra. By doing nothing for hours on end, these lizards exhausted my patience and convinced me to seek a career as an ecologist and not a behavioral biologist. Photograph by Kris Bell.

By any standards, Hal Heatwole is a remarkable man. He was born in 1934, into a small Mennonite community in rural Virginia, an unlikely background for an evolutionary biologist. Hal's passion for education and adventure so horrified the local religious fundamentalists that he was expelled from high school for misconduct. But his energy and thirst for knowledge, combined with a powerful intellect, saw him survive the pressures of a rigid system and graduate from the prestigious University of Michigan in 1960. Most university students focus primarily on their studies, with no time to spare for hobbies—but Hal marches to a different drummer. During his college career, Hal was intramural boxing champion at Goshen College in 1954, and a member of the gymnastics team, the volleyball team, the Audubon Society, the Literary Society, the Spanish Club, the Science Club, was Head Usher for the Music Series, and was a member of the Committee for Equal Housing for Negroes—to mention just a few of his roles.

And although Hal was mid-career when I joined his research group in Armidale, he showed no signs of slowing down. His evenings were taken up with competitive dancing on roller skates, with square dancing, and with teaching Spanish at a community hall. Hal was a perpetual-motion machine, constantly leaping from one activity to another. Lightly built to the point of near-emaciation, red-haired, and with a nautically trimmed red beard, Hal hurtled through life with more energy than a dingo pack in hot pursuit of a wallaby. And although he was usually doing several things at once, Hal managed to bring almost all of his scientific projects to fruition. Most academics look back on their Ph.D. thesis as the greatest accomplishment of their professional career. They are proud to have survived the pressure. But Hal is the only person I know who has written not one, but *three*, Ph.D. theses—the first on amphibians in 1959, the second on island vegetation in 1986, and the third on sea snakes in 2012. And somehow or other, along the way, Hal has joined or conducted research expeditions to almost every country in the world. And that's no exaggeration. Last time I talked to Hal, there were only three or four countries that he hasn't visited, and he hopes to get to them as well.

Although he was far and away the best choice for me to have as a supervisor, Hal's focus was different from mine: he worked mostly on lizards not snakes, and on physiology not ecology. But Hal was happy to leap in where angels fear to tread; issues that threatened cardiac arrest for most professors didn't faze Hal in the slightest. Taking on a young student who wanted to go out and catch hundreds of deadly snakes, for example, wasn't an issue. A risk-taker himself, Hal let people make their own decisions. His Reptile Empire comprised half-a-dozen students who were delving into the private lives of scaly creatures from crocodiles through to turtles. The passion was palpable. It seemed like the perfect place for me.

Sadly, my timing was poor. I arrived at the end of Hal's golden era in Armidale, just as he was preparing to leave. He still had extraordinary energy—sleeping only a few hours per night and running many projects simultaneously—but he was increasingly frustrated with university politics. His research output exceeded that of the rest of the Zoology Department combined, but the conservative academic hierarchy didn't take kindly to a brash American. Hal decided to return to the country of his birth, and he began applying for jobs in the United States shortly before I arrived. So, he was friendly but not deeply involved—he expected to leave Armidale long before

I finished my thesis. Added to that, Hal loves to travel. He flew out for fieldwork in distant parts of the world the day after he concluded his undergraduate lectures each term, and he returned just in time to give the first lecture of next term. In Hal's absence, I was officially supervised by a delightfully old-school academic, a kind and gentle man named John De Bavay—but in practice, I had free rein.

Without pressure from micromanaging supervisors, I designed a Ph.D. that followed on from my Honors-year research in Canberra. Although serpents were my passion, everyone told me that snakes were too scarce around Armidale for me to work on them. OK, I thought, I'll work on lizards instead (which were far more common) and look for ways to do a bit of research on snakes on the side. Perhaps I was still rattled by that Englishman in the pub, and his unshakeable belief that studying snakes was idiocy. The project I chose was to measure energy flow at an ecological level, looking at how energy in the food consumed by a population of lizards was allocated into growth, reproduction, and metabolism. It was a sensible project, but not exciting. I found a population of Golden Water Skinks at nearby Dangars Falls, and I set to work.

Water Skinks are widely distributed across eastern Australia, and I already knew them well. Like most skinks, they are glossy in appearance—shimmering in the sunlight—and a very attractive beast. Olive-brown on top, with scattered dark spots along the side and a pale belly. And by international standards, they are huge—often around 30 cm (12 inches) long. Australia is the home of giant lizards, so no Aussie reptile enthusiast gets too excited at this common species—but few other countries contain lizards that are so large and so bold. My international reptile-biology visitors squeal with excitement when they see their first Water Skink.

Dangars Falls is a reserve because of a spectacular waterfall that provides a photographic opportunity for tourists; but my study area was a few hundred meters upstream of the falls, where the creek flows through eucalypt woodland before plunging off the cliffs. Few tourists visited the upstream area, so my days in the field were tranquil. Colorful parrots in the gum trees, swamp wallabies lazing amidst the streamside bushes, and enough lizards to keep me busy. It was an idyllic location, and the research went well. I spent my days catching lizards, measuring and releasing them, and obtaining the information I needed to build a clear picture of energy flow.

If I'd stayed with that project, I would have obtained my Ph.D. But I was bored. The site was beautiful, the lizards were elegant, and the theme was interesting enough. But somehow or other, it didn't get my juices flowing. The highlight of my time in the field was an occasional encounter with my favorite animal—the Red-Bellied Blacksnake. Once every couple of days, I would stumble across a snake as I crept up on one of my lizards. Needing a justification (no matter how feeble) for incorporating these wonderful animals into my research program, I hit upon the idea that they must be major predators of my lizards. Hence, I could add snakes into my lizard project. All I would need to do was estimate the rates at which the snakes ate my study lizards, what kinds of lizards (sexes, ages) they ate, and so forth. That way, I could track energy flow not only through the lizard population, but up through the next level in the food web. And it wouldn't require anything complicated—I just needed to collect snakes, humanely kill them, and look inside their stomachs to work

FIGURE 2.3 The Golden Water Skink is an alert intelligent lizard, common over much of eastern Australia and the subject of my first (ultimately unsuccessful) Ph.D. project. Photo by Sylvain Dubey.

out what they were eating. It was flimsy stuff, but I convinced my supervisors to let me add snake-collecting to my regular routine.

When I dissected those Blacksnakes the following winter, my rationale for expanding the project proved to be complete nonsense. I rarely found a Water Skink inside a Blacksnake's stomach: only 6 of them, out of 300 prey items. Instead, Blacksnakes ate frogs, frogs, and more frogs, plus occasional slow-moving lizards of other species. The super-agile, hyper-smart Water Skinks are out of the Blacksnake's league. Tortoises can't catch gazelles.

But I didn't discover that error until I had been in Armidale for a year. Winters in Armidale can be bitterly cold, at least by Australian standards; we even had light snowfall occasionally. As a result, fieldwork on reptiles was out of the question during the colder months. After my first summer field season, I spent the winter hauling snakes out of the freezer, defrosting them, and dissecting them. It was a revelation. Learning that Water Skinks were rarely on the Blacksnake's menu was only one of a thousand discoveries. I began to grasp the basics of Blacksnake biology. For example, I discovered that male Blacksnakes grow much larger than females; and the testicles of males are enlarged only for a brief period in spring, corresponding with when I'd found male–female pairs together (clearly, this was the mating season). Females produce about a dozen live young in autumn, and so forth. Basic natural history. Nothing astonishing. But when I went to the University library to compare my results to previous studies, I made an even more startling discovery. There *weren't* any previous studies on snake ecology in Australia. My project was the first.

Red-Bellied Blacksnakes are an iconic species, common around the largest cities in Australia. I had assumed that their natural history was well known, and my contribution would be a modest one: comparing my local snakes to those in other parts of the country (studied in more detail, by someone more expert than me). But to my astonishment, no such studies had ever been done. In his 1967 book on Australian reptiles, the renowned authority Hal Cogger wrote "*no species of elapid snake has ever been studied intensely enough to provide an accurate picture of its biology and life history*". I typed that quote onto a piece of paper, and I stuck it up on my office wall. My studies were providing the first detailed scientific information on the ecology of Australian snakes. I was a pioneer.

That was mind-blowing. I could blaze a new trail, by conducting research on the most fascinating animals in the world! And I could focus on the topics that most interested me—diets, reproduction, growth, habitat use. And so, a momentous idea occurred to me. I could discard the lizard project. Switch across to snakes, and still get my degree. A Ph.D. is a test of endurance—work away at any topic for a few years and you'll emerge with a doctorate. But what then? Jobs are hard to come by. Switching to snakes might be academic suicide. No university was likely to employ someone who conducted natural history studies on snakes. My original project on lizard energetics was trendier, and more likely to get me a job one day.

My passion for snakes overcame my common sense. The thought of a Ph.D. on snake ecology excited me, whereas the prospect of three years measuring the energy content of lizard poo bored me to tears. If my supervisors had been heavily involved, things may have been different. They might have convinced me to stay on the straight and narrow rather than tread the perilous path of snake natural history. But Hal and John let me determine my own destiny, and so the die was cast. At the age of 22, I became Australia's first full-time snake ecologist.

As a pioneer, I could set my own agenda. I could decide what to measure, and everything I learned was new. From the time I arrived in Armidale, I had collected every snake I could find, so I had information about Tigersnakes and Copperheads as well as Blacksnakes—and a few Brownsnakes also, plus cryptic smaller species. It wasn't a huge assortment, but that's all there was in the cold climate of the New England highlands. And all of those species were venomous snakes of the family Elapidae—the group that contains Black Mambas in Africa, King Cobras in Asia, and Coral Snakes in South America. The Elapidae is one of the main evolutionary lineages of venomous snakes worldwide. I would have preferred a greater diversity of snakes (a few pythons and so forth), but what the hell—nobody knew much about *any* Australian snake. The only downside to working with large elapids was that a bite could kill me. But on the other hand, the danger added zest. As an introverted person, uncomfortable with other people, I felt depressingly "normal" and boring. Capturing venomous snakes gave me a claim to be slightly interesting. I didn't think about the danger; 22-year-old males believe that they are immortal.

Without any clear idea as to why I was doing it, I just went out and caught snakes. From spring to autumn, I spent my days out in the bush. For Blacksnakes, any of the small streams running through sheep pastures were fine. For Tigersnakes, the small lakes near Uralla, 20 km south of Armidale, were surefire spots. Especially in the early morning or late afternoon, a quiet stroll around the lake edge was likely

FIGURE 2.4 Catching a Tigersnake beside Uralla Lagoon, during my Ph.D. studies. By this time, I had abandoned "tailing" snakes, and opted for a safer method using tongs and a hoop bag. You can tell the picture is from the 1970s: long hair, partial nudity, flared jeans.

to reveal a glistening coil—a thickset olive-brown snake, banded in yellow—and although Tigersnakes are famously deadly, they are also relatively slow. I rarely left the lagoons without half-a-dozen snakes in the bag. Copperheads are about the same size as Tigersnakes—averaging a bit over a meter (3 feet) long, and equally heavy-bodied. Their colors are subtle but elegant—dark above, shading to paler on the sides and yellow on the belly—with an iridescent bronze sheen across the body and distinctive white bars on the lips. They prefer colder climates than the other elapid species, so for Copperheads I headed up to higher colder areas—to swamps with delightful names like Llangothlin Lagoon and Mother of Ducks Lagoon. The snakes weren't abundant—I averaged one capture an hour—but I caught hundreds of snakes over the next three years.

In the early days, I went collecting by myself. Before long, though, a few near-misses made me realize that I was treading a fine line between enthusiasm and recklessness. The project would be a lot safer if I had a companion, to drive me to hospital if I was bitten. I decided to keep looking for Blacksnakes on my own, but to have someone else with me when I was after the deadlier Tigersnakes and Copperheads. Fortunately, I lived in a university dormitory with 200 resident students, and many of them could be cajoled into spending a day in the bush looking for snakes.

And there was another attraction—half of those students were young women. As a single man in my early 20s, the possibility of a romantic tryst was never far from my mind (or what passes for a mind, in a young male). Fascinated with snakes since childhood, I saw snake-catching as the ultimate Adventure Sport, so I hoped that young women would swoon with admiration at my fearless athleticism. They didn't. It was an important lesson about life and love. Those young women enjoyed watching a testosterone-soaked halfwit risk his life, but none of them had the slightest interest in romance with such a moron. Women, I realized, were far more sophisticated than men. My male friends were impressed by my snake-catching skills, but on the other hand they were also impressed by heavy drinking and flatulence.

As time passed, I learned more about snakes too. My reflexes were so fine-tuned that I could hurl myself at a fast-disappearing tail as soon as I saw it. I caught snakes that, a year earlier, I wouldn't have got close to. But while wandering slowly along the edges of creeks and swamps, I also pondered what the animals were doing. Finding a snake is much more difficult than catching it—and thinking like a snake was the best way to predict where and when my quarry would be lurking. That not only helped me to find them, but it suggested all kinds of novel ideas about snake ecology. A typical morning jaunt might yield three animals. That gave me time to think through all kinds of issues, ranging from abstruse mathematical models of sex-ratio evolution through to the ethics and philosophy of science. And of course, why I wasn't finding more snakes. Many scientists are too busy doing science to dwell on the bigger issues. I benefitted enormously from those hours of quiet reflection.

One clear lesson was that stories about "aggressive" snakes were hogwash. The snakes I met were shy creatures, far more interested in escaping than in biting me. In the 1950s, the American biologist Clifford Pope wrote "Snakes are first cowards, then bluffers, and last of all warriors". Truer words were never written. An aroused Tigersnake or Brownsnake is a formidable beast. If its escape route is blocked, the snake will flatten its body, hiss loudly, and strike out at its enemy. And that scaly body

FIGURE 2.5 Like many other kinds of snakes in cool climates, Copperheads are attracted to farm buildings that provide warm shelters where corrugated iron sheets are exposed to the sun. Photograph by Rob Valentic.

can hurtle across the ground very quickly, especially if the snake is hot. I sometimes had snakes fling themselves into the air, straight at me, but they always fell short. And given a route to safety, that animal will switch instantly from fight to flight. I was lucky that the snakes were so forgiving. Handling hundreds of deadly snakes in remote bushland, I had many moments that could have proven disastrous. Twice I seized the wrong end of a half-seen snake in thick grass, grasping the head instead of the tail. Fortunately, both times I realized my error before the snake reacted.

Some of my scariest moments came when snakes were hidden up in trees or down in holes. For example, one day a bird-watcher told me of seeing a huge Tigersnake crawling into a Magpie nest, meters up a large gum tree. I found the tree, and I clambered up it. But when I peered into the nest, I was staring straight at the snake's head from a few inches away. It had heard me coming and was ready to retaliate if I attacked it—but fortunately, it was in a forgiving mood. The same thing happened when a massive Blacksnake disappeared into a small hole in the ground. I was determined to get that snake, so I returned the next day with a shovel, and excavated a huge amount of soil. No sign of the snake; the burrow system just kept going deep into the bowels of the earth. I lay on the ground and plunged my head and shoulders inside the hole I'd made, to peer down into the hole . . . but then I felt tongue-tips flicking the back of my neck. The snake had moved into a shallow side-branch, and was now tongue-flicking me to identify the strange creature that was disturbing its peace. Both times, I was a sitting duck. If those snakes had wanted to bite me, I would have been in deep trouble—bites to the neck or face are a nightmare to treat. But each time, the tolerant serpents allowed themselves to be captured without making a fuss.

FIGURE 2.6 Despite their fearsome reputation, I have always found Tigersnakes to be inquisitive rather than aggressive. Photograph by Rob Valentic.

As I spent more time with venomous snakes, the fear factor was replaced by respect and affection. Like many young men, I was initially attracted to snakes by the adrenalin rush of handling an animal capable of killing me. As our American cousins might say, venomous snakes scared the living crap out of me. My obsessive desire to catch every snake that crossed my path was due, at least in part, to my adolescent view of snake-catching as ritual combat with an animal that would kill me if I gave it even the slightest opportunity to do so. But as time wore on, and I became more comfortable with snakes, I realized that they are inoffensive creatures that bear no hatred toward the human race. I could tell a thousand stories about their forbearance.

In the end, the only bites I ever received during my Ph.D. project were in the lab not in the field. Both times, I was bitten while trying out new (and unsuccessful) ways to handle Blacksnakes. Neither bite caused any symptoms, because no venom was injected. Each time, the snake only managed to slip a single fang into my finger, and my rapid reaction—pulling my hand away—meant that very little venom was injected.

I did have two nasty accidents during my Ph.D., but snakes weren't the villains in either case. One incident occurred in the lab, as I was preserving a road-killed King Brown Snake that a fellow student had found in western New South Wales. The snake was more than 2 meters (6 feet) long, but it was in terrible condition. It had been lying on the road for a few days after being run over by a truck, so it was putrid, and crawling with maggots. But it was a useful specimen, since this is an uncommon species. To stop the flesh from rotting any further, I filled a large syringe with concentrated formalin, and injected the snake with vast amounts of full-strength preservative. My fellow students evacuated the lab as soon as the pungent odor of decomposing snake began to spread, so it was a solitary as well as smelly process. I had to keep refilling the syringe and plunging it into the snake's body. And after a few dozen times, I became careless. After filling the syringe with formalin, I reinserted the needle—but holding the base of the needle in my fingertips, with the sharp end pointed at the palm of my hand. And with my fingers slippery with rancid snake-fat, the syringe slipped through my fingers when I pushed. The needle plunged straight through my hand. It took all my strength to haul the needle back out; and all my self-control not to scream as the formalin spread through my system. By good luck, not much had been injected. Otherwise, I would have faced the future as a one-handed snake-catcher.

The other accident was far more painful. I began a side-project catching, marking and releasing snakes on the Little Bellingen River near Coffs Harbour, a coastal town, 3-hour drive from Armidale. The aim of the project was to recapture those snakes, and so learn about their rates of growth and movement—but that never happened, because the project came to a premature end. The Little Belligen River was a beautiful clear stream running through rainforest, a refreshing change from the agricultural landscapes of Canberra and Armidale. Blacksnakes and Carpet Pythons were common, and I hoped to turn the project into a major study. But it all came crashing down one day when on my way back to Armidale from Bellingen, I stopped to drop off a drum of formalin for the rangers in New England National Park. In those dim and distant days, ignorant rednecks weren't the only people killing snakes. Even the National Park rangers—the people whose job was to conserve wildlife—killed every snake they saw, "to protect the public". I couldn't convince them to stop the massacre, so I asked them to save the carcasses for me. But fate played an unkind trick. The only way into the Park was a dirt road that had recently been regraded. And the work had been done incompetently, with little regard for safety. On one gravel-covered bend where the road sloped the wrong way, my car slid across the road and toward a cliff. I hit the brakes—exactly the worst thing to do, because the tires locked and I lost all control. As I slid inexorably toward the edge of the road, time passed slowly—it felt like a slow-motion movie. But a few seconds later I hit a tree on the edge of the cliff rather than toppling over.

Although I was only travelling at around 50 km (30 miles) per hour, the car was a write-off, and its passengers didn't fare much better. My girlfriend in the passenger seat had the worst luck—her seatbelt broke, and she hurtled forward to hit the dashboard, breaking her collarbone and knocking her unconscious. In the era before airbags, much of the carnage from car accidents occurred when bodies smashed into the dashboard or the steering wheel. My seatbelt held, but it didn't stop my head from lurching forward and downward into the steering wheel. The wheel hit my face just above my upper lip, almost tearing off my nose as my head slammed downward. As I bounced back upright, the lower part of my nose was ripped away from my face; the rest was sticking straight out, and a geyser of blood spurted out in a spectacular fountain. My hat fell into my lap, and my first memory after regaining my senses was of looking down to see blood sloshing around inside my hat.

I turned off the engine, jumped out of the car, and dragged my unconscious girlfriend up the road to the shade of a tree, in case the car burst into flames. After she recovered consciousness, I set off to walk back to the ranger station. You might expect me to have felt downbeat, but in fact I was euphoric—a reaction to the shock. Despite having most of my nose detached from my face, I was keenly aware of the sun shining, the birds singing, and the fact that I and my girlfriend were still alive. It wasn't until several hours later that I went into shock, and started trembling uncontrollably. Until then, I was a happy camper. When a car appeared on the road coming toward me, I cheerfully waved them down, and was surprised to see a passenger in the front seat pass out in shock. It took a while to revive her. When you go for a drive in the local park, you don't expect to see someone drenched in blood, with a horizontally protruding nose, and surrounded by a cloud of flies. She recovered but then one of her friends also passed out when she got a close look at my ex-nose, so I decided it was best hidden from view. I tied a large handkerchief over my bloodied proboscis to prevent further episodes of fainting. The ambulance arrived 2 hours later, and off we went to the hospital.

I thought that our problems were over, but I hadn't counted on the chaos in small country hospitals. There's only one doctor on duty so if more than one urgent case comes in, that doctor has to make difficult decisions. My girlfriend was soon X-rayed and patched up, but my own treatment was delayed when a sick child was brought in. The doctor injected local anesthetic into my face so he could put my nose back in place and sew me up—but each time he started stitching, the child suffered a relapse and the doctor had to leave. In the end, he had to sew up my face without anesthetic.

The earlier injections had worn off, and regulations forbade multiple doses. Those stitches, and the doctor's fingers inside my face as he squeezed my nose back into shape, hurt more than the car accident.

Compared to those incidents, my interactions with snakes were mostly mundane. But one memorable event occurred early on, while I was still pursuing the project on lizards at Dangars Falls. Water Skinks are smart animals, which is why Blacksnakes have trouble catching them. (I've never run IQ tests on snakes, but Blacksnakes must be near the bottom of the class.) Anyway, it took full concentration to sneak up on these lizards with a noosing pole. If they had been caught before, those lizards knew exactly what I had in mind. I needed to focus intensely on the job to have any chance of success. And to my frustration, a Magpie living in the study site took an intense dislike to me. These big black-and-white birds (related to crows and ravens) are common in the suburbs as well as the bushland, and have achieved iconic status in Australian culture. They are the bird that fights back against humans, terrorizing pedestrians and pushbike-riders. Magpies are notorious for swooping down silently toward the heads of people they don't like and clicking their beaks loudly beside their target's ear. Or even, removing a tuft of hair or a slice of skin with their sharp bills. Individual birds take strong dislikes to particular people, presumably as a result of some unpleasant experience. They are smart enough to distinguish between different people, and they direct their aggression toward specific intruders. I was blameless in this case—I never harassed that Magpie—but the bird must have had a prior nasty encounter with someone who looked like me. So, the winged avenger decided that I was Public Enemy Number One.

Having the bird startle me as I was trying to catch a lizard was frustrating, but one sunny afternoon he took it to another level. I was creeping up on a Water Skink in thick grass when a large Blacksnake materialized in front of me and hurtled away. My only chance of capturing it was to act immediately. I threw myself forward full-length and got my fingers around the tip of the meter-long snake's tail. It instantly darted back to defend itself, creating a tricky problem since I had to evade the oncoming ophidian missile while lying stretched out on the ground. I dodged that first strike and hastily scrambled upright, with the snake's tail still in my hand, only to receive a stunning blow to the back of my head. My nemesis the Magpie had noticed that I was fully occupied doing something in the grass and had decided it was a fine time to give me a scare. Unfortunately, because he arrived as I leapt to my feet to avoid the snake's lunges, the Magpie thundered into me instead of skimming above me. The solid thump to the back of my head caused me to drop to my knees again, releasing the snake as I did so, and an observer might have found some humor in the sight of a human, a snake, and a large black-and-white bird all thrashing around in the grass beside each other, none of them quite knowing what was going on. More by luck than anything else, I was able to get hold of the snake again before it took off. The Magpie flew away while I was getting the snake into a bag.

I brought a rifle with me to the field site on my next field trip, planning vengeance. But all the Magpies looked the same, so I couldn't identify my attacker. And that bird never attacked me again.

****

For someone who loved being alone in the bush, and loved catching snakes, my Ph.D. research was wonderful. Unlike the academic system in many other countries, Australian universities don't require Ph.D. students to take classes or teach in labs, so I was free to focus on my research project. There weren't many academic distractions in the Department of Zoology. It was a pleasant workplace, but intellectually it was a sleepy hollow. Well away from the mainstream research world, we had occasional academic visitors but no regular schedule of seminars. In the tearoom, we talked about football not research.

Although the Armidale scene was disappointing in terms of science, it offered a vibrant social life. Most students lived on campus, in dormitories that offered the chance to meet a diverse array of people, discuss every topic under the sun, play sport in inter-college rivalries, and learn about life. Like many students, I played several sports, and threw myself into the social whirl with enthusiasm. Over the space of a couple of years, I developed close friendships and took on more responsible roles within my college, Drummond. I didn't develop into a raging extrovert, but I began to feel more comfortable in social situations. That evolution of a more confident persona was critical in my later career, as well as making my life a lot more enjoyable. So in the end, my choice of Armidale for my Ph.D. institution turned out well. If I had selected a busier and less socially active campus, I might have been pushed into more fashionable science, but I wouldn't have been free to develop my own approach—and just as importantly, I might not have developed an ability to interact comfortably

FIGURE 2.7 Even after being harassed, a Tigersnake generally tries to bluff its way out of trouble by flattening its neck and hissing rather than launching an all-out attack. Photograph by Kris Bell.

with my own species. Without supervisors looking over my shoulder, my winters were about football and parties, and cold evenings inside a warm pub. But somehow or other, in between the parties and the football games, I found time to dissect the snakes I had collected the previous summer. And that gave me more than enough information for my Ph.D.

There's a large Elephant in the Room here, and I might as well confront it head-on. I've been talking about my love of snakes, and yet my Ph.D. was based on slaughtering them. In my defense, I tried to minimize the numbers I killed. Many of the snakes that I dissected were dead before I obtained them—run over by vehicles, or killed by farmers or fishermen. I went to a lot of trouble to get my hands on those carcasses. I erected signs, left drums of preservative with Park rangers, and gave talks at fishing clubs where I asked the members to save the bodies of any snakes they killed. I pleaded with my colleagues to pick up any roadkill they found, no matter how stinky, and give it to me. But the main basis of my research was the 300 snakes I caught and killed myself. As a snake-lover, how could I do it?

I hated killing the animals, even though it was a quick and painless death. Mostly I just put the labeled bag into the fridge and then, a few hours later, into the freezer. The snakes cooled down, went to sleep, and never woke up. Others were killed by an injection of alcohol straight into the brain, causing death in seconds. But there was no escaping the fact that I was turning magnificent living creatures into corpses. I justified the massacre by reflecting on the eventual benefits to the species involved—if we understand their biology, we can work out how best to conserve them. And I didn't know any other way to get the information that I needed. I extracted every shred of information I could from my dissections, and then preserved the bodies and lodged them in the Australian Museum for other researchers to use in future. And I trolled the university to find other scientists who might be able to use tissues from the animals I was killing. That worked well, so we got a great deal of scientific knowledge (about topics like blood chemistry) from those dead snakes. But I won't pretend it was guilt-free. At the end of my Ph.D., I vowed to never again conduct a research program that depended on killing snakes. Henceforth, I would study snakes without harming them. I've killed very few snakes in the 40 years since I left Armidale. Snakes are beautiful creatures, and I would rather see one crawl away into the bush than kill it to get a data point. That was a luxury I couldn't afford as a graduate student, but it was gut-wrenching to end up as a mass-murderer of organisms I had cherished from childhood.

When I eventually returned to Armidale more than 20 years later, my guilt was lessened—but at a terrible price. Most of my study areas had been "developed" for agriculture or urbanization. The bush was trashed, and the streams were polluted. The snakes were gone. Many of the snakes that I killed wouldn't have lived much longer anyway. Their descendants faced a far worse death than the one I meted out. That didn't help the snakes I had slaughtered—but it put scientific collecting into a broader perspective.

All around the world, natural habitats are being destroyed. Ecosystems are collapsing, and we can't conserve our wildlife if we don't understand it. Warm fuzzy feelings aren't enough. We need to value our wildlife, and care about wild places.

But when it comes down to the pointy end of management—setting priorities for conservation spending—we have to know how ecosystems function. To maintain natural places, we need to understand them. Sometimes, that will involve stressing—and even killing—the species we want to conserve. It's a dilemma that no wildlife ecologist can escape. The collateral damage we inflict can't be denied, but it's trivial compared to pressures wrought by human overpopulation, habitat destruction, pollution, and climate change. Science must address those issues. In the process of discovering how to buffer those threats, researchers will hurt and sometimes kill their study animals.

My angst about killing snakes wasn't shared by the people I met around Armidale. I knocked on the doors of many country homes, asking farmers for permission to look for snakes on their properties. I heard many tales about ferocious attacks by snakes, but most had the hallmarks of myth—vagueness about when, where, who was involved, and so on. Often, I heard the same Tall Tale from different people. Actual case-histories ("Uncle Billy was bitten on the bum by a Blacksnake") were rare. And to my surprise, farmers were as ignorant about the local snakes as were city-dwellers. Often, I was told when I arrived "You're welcome to have a look, but I haven't seen a snake on my property for years". When I stopped by a few hours later on the way out, they were horrified to hear that I had found several snakes. One man became abusive, claiming that I must have brought the snakes with me because there were none on his farm. And even when people knew there were snakes about, they often didn't know what species were involved. One farmer near Guyra proudly told me about the long history of his family in the region, but he didn't know that the snakes on his farm were Copperheads. He called them "Yellow-Bellied Blacksnakes", a commonly used name (it even features in a song by the 1980s rock group, Midnight Oil) that is scientifically meaningless. You won't find "Yellow-Bellied Blacksnakes" in any snake book. Snakes were shadowy creatures for most people, even in the bush—rarely seen, unidentified, the stuff of legends.

To replace those myths with hard evidence, I needed to document the basic biology of the snakes. For example, why were different species found in different habitats? I found Copperheads in the highland swamps, Tigersnakes in the lowland swamps, and Blacksnakes along the banks of streams. Brownsnakes were in the dry places, especially in farmland. Why? It's an easy question to ask, but a hard one to answer. Each snake species eats different kinds of prey, but do those dietary differences *cause* the habitat preference (snakes live where they can find their favorite food) or are they a *result* of it (snakes eat whatever food occurs in their preferred habitats)? We still don't know. And food isn't the only important difference between habitat types—perhaps some snakes prefer warmer areas, or habitats that lack a predator or disease that threatens them? Or if different species of snakes eat each other, a good place for a Blacksnake may be a dangerous place for a Tigersnake? All I could do was document the patterns, and we still don't really understand much about why snakes (or indeed, other animals) live in the places they do, and not in other places.

Some of my most memorable "wow!" moments while dissecting snakes came from looking inside their stomachs. Snake biologists have a huge advantage here. If you look inside the stomach of a lizard, for example, all you find are mashed-up fragments of insects. Lizards eat small things, and lots of them, and their powerful jaws and strong teeth reduce even a large grasshopper to a shredded mess before it travels down into the stomach. A challenge to identify. With a couple of exceptions (species that eat lizard-tails or crab-legs), snakes feed on large animals and swallow them whole. A snake biologist doesn't have to assemble a jigsaw of blended frogs to work out what has been eaten—the intact prey items inside a snake's stomach are easy to identify.

As an aside, snakes sometimes don't even bother to kill their prey before they eat it. A small frog can't harm a large snake, so the snake just gobbles it down. That creates an interesting situation when a snake-catcher comes along and lifts the snake up by the tail: a snake with a full belly is likely to regurgitate its prey. And every once in a while, one or two of those prey items will blink, then slowly hop away. They have a hell of a story to tell their friends.

Especially if the snake is venomous, however—like the Armidale serpents—most prey items are defunct by the time they reach the stomach. Some of the snakes I captured contained huge prey, like adult Bluetongue Lizards. It must take hours for a snake to overpower and eat such a formidable meal. Most often, though, the snakes had snacked on small frogs and lizards—sometimes dozens in a single gluttonous binge. Many Blacksnakes had wall-to-wall frogs along the entire length of their stomachs, all the way up to their throats.

Identifying a snake's latest meal by looking inside its stomach isn't rocket science, and I didn't need any advice about how to do it. But internal organs were trickier. It's easy to recognize a lung or a liver, but not as simple to distinguish kidneys, spleens, and testicles. The task is made more difficult by the snake's long slender body plan, which has forced evolution to change the sizes, shapes, and positions of organs to fit within a long thin tube. As a result, it's hard for a novice to know what organ is what. Adding to the challenge, paired organs (like kidneys, ovaries, and testicles) are arranged one-above-the-other not side-by-side; that way, they fit better into the body without creating bulges that would interfere with the snake's mobility. Sometimes, one half of that pair is far smaller than the other—or missing entirely. Most snakes have only a single lung. And of course, each species is different. Snakes with slender bodies show the most dramatic rearrangements of their internal structures. For example, lithe tree-dwellers hold the eggs from their two ovaries one-above-the-other rather than side-by-side.

As a naive young student, how could I navigate my way around inside a dead snake's body, recognize the important bits, and know what kinds of information to record? There weren't any websites—or even, books—to teach me snake anatomy. And here, I had a tremendous piece of luck—one that changed my entire scientific career. My official supervisor (Hal Heatwole) could have helped me learn about snake anatomy, but he was in Tunisia studying ants. Fortunately, however, a sabbatical visitor from the United States had arrived in our department to investigate evolutionary relationships in Australian turtles. John Legler was a gruff, cantankerous field biologist, unwilling to waste his time on anyone who wasn't worthy of the effort. And John felt that most of the Armidale students fell into that category, so he was more likely to express withering scorn than praise. But he and I got along well, perhaps because we were fascinated by the same questions. And when I plucked up the courage to ask him to show me around the insides of a dead snake, he readily agreed. John was a superb anatomist, so it was a master class. I instantly forgot most of the detail, but John gave me an incredible gift—he taught me how to recognize the body parts inside a snake, and how to record information about those innards. Those skills were exactly what I needed in order to conduct a detailed study of reproductive cycles.

Anyone with a successful career can look back and identify minor events that made everything else possible. For me, those few hours in the lab with John Legler, looking inside defrosted Blacksnakes, set me up for the next phase of my professional development. And John did something else for me as well. He brought with him a new graduate student, who was doing his Ph.D. on turtle chromosomes. Jim Bull was to become my first scientific collaborator, and a lifelong friend.

Jim is tall, thin, intense, and brilliant. Mercurial, intellectual, uninterested in idle chit-chat. His idea of a pleasant weekend is a couple of days of uninterrupted thinking about complex eco-evolutionary dynamics. If Jim was an actor, he would be ideally suited to playing an old-school version of Sherlock Holmes. Jim's career has prospered since those days in Armidale, and he's now one of the world's top evolutionary biologists. He still enjoys snakes, but his research is on much smaller animals—bacteria. He studies rapid evolutionary change in these tiny life-forms in the lab, builds mathematical models of evolutionary processes, and says things so terrifyingly intelligent that they make me feel like a lower life-form. Jim marches to a different drummer than most of us, but he and I have remained firm friends over many years.

FIGURE 2.8 The author with long-time friend and colleague Jim Bull, with a snake we found while cruising the roads at night.

Jim had spent his undergrad days at Texas Tech in Lubbock, a small hot dry Texas town (one of my favorite lines from a country and western song is "Happiness was Lubbock Texas in the rear-view mirror"). One thing that Lubbock does have in its favor, though, is its proximity to the Big Bend region of Texas—a desolate, cactus-covered, snake-catcher's nirvana. Deserts are harsh environments, well-suited to the tough and adaptable nature of snakes. And soon after Jim and I met in Armidale, I was listening spellbound as he told me stories of road-cruising at night through the Texan desert, watching in the headlights' gleam for glorious Black-Tailed Rattlesnakes, Trans-Pecos Ratsnakes, and the like. Cruising the roads at night with his mate Greg Mengden—later to become a close friend of mine also—Jim had accumulated fantastic tales about the landscapes and snakes of the Lone Star State.

Jim was a kindred spirit, passionate about ideas as well as about snakes. I've always enjoyed trying to understand *why* animals do such amazing things. It's fun to work out *how* they do things as well—but to me, the most fascinating issue is *why*. Translating the word "why?" into the jargon of biological research, we can try to identify the evolutionary advantages and disadvantages that fashion the diversity of sizes, shapes, behaviors, and ecologies out there in the natural world. The answer will involve the process of natural selection, as proposed by the great English biologist Charles Darwin more than 150 years ago. But that process plays out differently for different species in different places. For me, "the evolutionary play within the ecological theatre" is the most fascinating topic in biology.

Having Jim accompany me on snake-catching trips was a huge bonus: it gave me someone reliable to help with catching, as well as to discuss ideas as we searched for animals. But Jim had a more professional attitude to science than I had ever encountered in Australia, and he wanted our joint snake-collecting trips to yield real scientific outcomes, not just data and happy memories. I knew that scientists described the results of their work by publishing papers in learned journals (indeed, Hal Heatwole published very prolifically), but had never really thought about writing such a paper myself. The Armidale students I talked to were focused on writing their theses, not publishing papers. If they thought about turning a thesis chapter into a paper, it was a task for after the thesis was written, not before. But Jim had a different view. He had been taught that publications are critical for a young scientist's career—"publish or perish". If Jim was going to spend time catching snakes, he wanted to see a published result from the work.

That's a significant hurdle. To publish the hard-won information, it needed to be embedded in some wider context. Frankly, very few people are likely to get excited if you tell them that Blacksnakes have an average of 12 babies per litter, whereas Tigersnakes have an average of 23. But if you plug that information into mathematical models about population viability, a lot more people are interested. Your study clarifies the resilience of a species in the face of habitat change. So the challenge of research is not just to fill up a spreadsheet with numbers; it's to work out how those numbers give us new insights into the natural world.

Publishing the results of such an investigation in a scientific journal is far from easy. Non-scientists imagine that the process is a bit like writing an article for a magazine—you put together some words and pictures, send it to the editor, and they pay you for the opportunity to put the story into their magazine. Nothing could be further from the truth. You never receive a cent in payment. Instead, your job is to convince the journal that your paper is good enough to publish. Having a paper accepted for publication in one of the high-quality journals is a brutal process. After you submit the paper, the editor sends it to experts for evaluation. You don't know who these people are; and anonymity allows them to be open in their criticisms. Their job is to find errors, to rip holes in your logic, your analyses, your interpretations. A review can be many pages long, and it can be savage. Many aspiring scientists change careers after they receive that first set of reviews. You need a hide like a rhinoceros if the experts decide that your beloved study—the one that you just spent three years of your life crafting—is irretrievably flawed, useless, unpublishable. The top journals reject 95% of the manuscripts that are submitted to them, so your chances of having that prized paper accepted are about as high as winning the lottery.

Jim's hard-nosed professional perspective was a revelation. It forced me to think about the stories that might emerge from the data I was collecting—and so, to organize the way that I gathered information. Up until then I had gathered information without giving too much thought as to how to use it. The advantages of advance planning were obvious—with simple questions in mind beforehand, I could gather data to bear directly on those ideas. So, was there some straightforward questions that Jim and I could ask and answer about the snakes of Armidale, and so produce a published paper from our collaborative efforts? I hit on the idea of exploring a puzzling result I had stumbled across earlier—a bias in sex ratios. When I looked at my sample of Blacksnakes, there were twice as many males as females. Why was that? Surely, we would expect equal numbers of the sexes? Or more females than males, since that would help the species to build up its numbers faster (it only takes one male to fertilize many females)?

I cringe when I think of how simple-minded our thoughts were at that stage. In retrospect, it's clear that the relative numbers of adult male versus female Blacksnakes I encountered were the result of many factors—such as higher mortality rates in one sex than the other or (even more likely) behavioral differences. Sex-crazed males of most species ignore anything except females of their own species. That's a recipe for disaster if a truck turns up as the lovesick male is crossing the road. Male stupidity thus creates a huge bias in the sex ratio of snakes that are collected by a scientist or run over on the highway. The female snake sees the truck coming and leaves the road, whereas the male is still having erotic fantasies as the tires roll over him. And indeed, masculine myopia turned out to be the main reason why I was catching more male Blacksnakes than females. The females were outsmarting me and the males were not. But at the time, I thought it might be part of some interesting story about the evolution of snake reproductive tactics.

The topic that Jim and I chose for detailed study—biased sex ratios—was a perfect entrée into the newly emerging field of "evolutionary ecology". Mathematicians were beginning to build models of how biological traits evolve, and that task was a lot easier if the traits in question were simple numbers—like the number of sons versus daughters in a litter. And so, sex-ratio theory was the ideal way for two wet-behind-the-ears graduate students to learn about the complex interface between ecology and evolution. And we had another stroke of luck. At a conference in

Armidale, a fellow-student mentioned a book that discussed evolutionary forces on the sex ratio. Surprisingly, the small library at my university had a copy. "Adaptation and Natural Selection" by George C. Williams had been published in 1966, and so was likely to be out of date. But Jim borrowed it from the library anyway, and showed it to me before he took it home to read. We laughed at its simple language. In our arrogance, we doubted that the book would tell us anything we didn't already know.

Jim lived in a converted garage that he shared with his wife Becky, within walking distance of the university. He didn't come to the university the next day, nor the one after. Concerned that he and Becky might be ill, I dropped by to check on them. When I knocked at the door, Jim came out with a wild staring look in his eyes. He was at his taciturn best, sounding like an Old Testament prophet who had encountered the Word of God. As I started to ask him why he hadn't come into the department, he said "No. No. No. No talking. Read this, *then* we talk". "This" was the slim volume we had laughed about a few days before. He thrust the book into my hands, then walked back into his garage and shut the door.

Knowing Jim, I realized that argument would be futile. I took the book, went home, and read it. And my understanding of the natural world exploded. The book was superb. It was the work of a genius, writing in simple terms, asking fundamental questions in evolutionary biology. And then, answering those questions. People talk about "having an epiphany" or "blinkers being removed", but I didn't understand what those phrases meant until I read that book. It hit me with the force of a religious conversion. I realized that my thinking about biology had been woolly and convoluted, and I had misunderstood basic issues. Natural selection was much simpler than I had imagined, but also more complicated. It was elegant and beautiful. That little book described the processes that have sculpted life on Earth.

That sounds melodramatic, but I'm not alone in my homage to that book. In their autobiographies, many of the leading lights of evolutionary biology in the 20th century talk in starstruck tones about the impact of George Williams' little tome on their way of thinking. A few years later, after I finished my Ph.D. and moved to the United States for further studies, the first thing that Jim and I did was contact George Williams and plead for some time with him. He was on holiday at the time, and invited us to his small cabin in Ontario, beside the Skootamata River. I spent a mind-blowing week walking through the forest, catching Pike and Watersnakes, and talking to a shy, modest genius. George was tall and looked like Abraham Lincoln. He answered our naive queries with kindness and patience. He was puzzled by his superstar status among the world's evolutionary biologists. All he had done, he said, was to think about Darwin's theory of natural selection, in the light of more recent discoveries in science.

I learned a lot from George. Much of it was humbling: when George and Jim discussed complex ideas in mathematical terms, it was like listening to a conversation in a foreign language. Other parts were liberating: I learned that George was a slow thinker, like me. He would take an idea, reflect on it for a while, forget about it, think about it again, and so forth. And one day, the problem would be solved. Without conscious awareness, his brain had been exploring possibilities—and it had eventually produced a brilliant insight. If you pushed George for a quick answer, his response would be pedestrian. Many scientists are quick thinkers, able to frame a sophisticated response within microseconds. Terrific speakers. But those intellectual meteors don't change the way we think. Their ideas are superficial. George's deeper insights took longer to mature. I was emboldened. Perhaps an intellectual slowpoke like me really *could* do worthwhile science.

But that was years away. As graduate students, Jim and I were at the other side of the world from the centers of evolutionary biology in Europe and North America. Before the digital revolution, super-slow communications reinforced the isolation of Australia. If I wanted advice from someone in the United States—say, about how to surgically implant a radio transmitter into a snake—my letter would take weeks to reach them, and their response would take just as long on the way back. When the internet arrived decades later, it had far more impact on those of us in the "arse end of the world" (as an Australian Prime Minister once delicately described our geographic position), than it did for scientists in the already well-connected metropolises of the Northern Hemisphere.

Back to Armidale in 1972. George's book gave Jim and I a set of powerful intellectual weapons, built on a simple premise. George pointed out that characteristics evolve because they contribute to an individual's survival or its mating success, not because they make it more likely for the species to survive. George's perspective was the same one that Charles Darwin had espoused, but modern biologists had lost sight of that simplicity. Evolutionary theory had become tangled and complicated. If traits evolve because they benefit an entire species, then you can imagine "evolutionary advantages" to almost any trait (like the sex ratio among adult animals). And you can't test any of those ideas, because you can't measure whether or not a species benefits from having some characteristic or other. But if the only thing that matters is the consequence of a given trait to an individual animal, evolutionary biology is testable. Many years later, Richard Dawkins achieved international stardom with his masterful synthesis "The Selfish Gene" that championed the same argument. George's book was the catalyst that set many evolutionary biologists (including Dawkins) thinking along those lines.

These simple ideas ran counter to conventional wisdom. Jim and I had to jettison much of what we had been taught. Fortunately, we could use each other as sounding boards as we grappled with the revolutionary implications of gene selection. We talked for hours—in the lab, in the office, in the car, while walking through the bush—exploring how to think about evolution. The other graduate students dismissed our obsession as obnoxious intellectualism, and our friends were bored when the after-dinner conversation inevitably turned to natural selection—but Jim and I didn't give a damn. We were hooked on a dangerous new drug: individual-based evolutionary theory.

George's book gave birth to several new fields of research. For example, biologists had long investigated why animals exhibit different sizes and shapes, but not why animals have different behaviors, ecologies, and life histories. George argued that as long as a characteristic is heritable—that is, a child tends to resemble its parents—it can evolve. And so we can just as easily ask why Blacksnakes have a few large babies in each litter, whereas Tigersnakes have many small ones, as about why Blacksnakes are black and Tigersnakes are banded. For any question like that, you can frame dozens of possible answers—but George pointed out that the only sensible answers are simple ones. Animals do the things that helped their ancestors to survive or breed—not because an animal "wants" to do something, or because it helps the species. It's a simple brutal equation. Any gene that causes its bearers *not* to survive and breed soon disappears from the population. Today's living creatures are the descendants of "winners" across many generations, each subject to a powerful cull by natural selection. Today's Blacksnake (or Koala, or anything else) has inherited the characteristics that enabled its mother, father, grandmother, grandfather (and so on) to thrive. Thus, anything that an animal does is likely to have a payoff (in terms of survival and reproduction) for that individual. Our challenge as biologists is to identify those advantages.

That way of thinking means that many population-level characteristics—such as the relative numbers of adult males versus adult females—aren't targets of natural selection at all. There isn't a gene for "adult sex ratio"—instead, genes work by changing sex ratio at conception, or at birth. That might end up affecting adult sex ratio, but only indirectly. Ecological factors like differences in mortality rate between males and females can massively skew the adult sex ratio, even if it starts out as 50:50 at birth. George's book told us that my original observation—that we caught more male than female Blacksnakes—wasn't interesting from an evolutionary perspective. It was probably due to sex differences in catchability or mortality, not evolution. It would only be interesting if the litters inside pregnant female Blacksnakes contained more offspring of one sex than the other. And they didn't. When I dissected pregnant Blacksnakes, their litters contained equal numbers of sons and daughters. As a research project in evolutionary biology, it was a failure.

But we were lucky. I was also dissecting pregnant female snakes of other species, and one of them exhibited a remarkable sex bias among the offspring. In Tigersnakes, almost every litter contained more sons than daughters. That can't be due to a behavioral bias, because embryos inside a uterus don't have anywhere to hide. Genes control the sex ratio within each litter; and for some reason, pregnant Tigersnakes were investing more in male progeny than female progeny. Why?

Determining the sex of an unborn snake is easy. Each male snake has two penises, which reside in pockets in the base of his tail—and conveniently for sex-ratio researchers, those penises develop on the outside of the body until the embryo is full-term. When I dissected a pregnant snake, I could immediately score whether or not each embryo was displaying his manhood. All I had to do was count the boys versus girls. And because Jim was an expert in chromosome testing, he devised a genetic test as well. Even if the embryo was too small to have visible sex organs, or had died early in development, all that Jim needed was a single cell. His special stains showed up the sex chromosomes, providing a foolproof indication of the embryo's sex. So, our data were watertight. Pregnant Tigersnakes produce 1.5 sons for every daughter. But why would a mother who produces more sons than daughters end up with more grandchildren as a result? Why would evolution have favored a gene that says "produce more sons than daughters"? George's book told us that we could address that question using evolutionary theory. We hurled ourselves onto the challenge.

But although we made progress, we failed to solve the problem. Evolutionary ecology was a field in its infancy, and it would be several years before theoreticians devised an explanation for the kind of sex-ratio skewing exhibited by Tigersnakes. One of the giants of evolutionary biology, the British polymath W.D. Hamilton, broke the news to Jim and me when we had lunch with him after I had moved to the United States a few years later. Someone else had found the answer that we had searched for.

In the jargon of evolutionary biology, the answer was "local mate competition". Imagine that your daughters are home-bodies, settling down in the same area where they were born. In contrast, your sons are wanderers—they disperse far and wide. That fits with what we know of snake biology. Young males leave home, whereas females hang around. In this situation, your sons rarely compete with each other because they are scattered across the surrounding landscape. Producing an extra son doesn't make life any more difficult for your other sons. But your daughters stay close to home, so as they grow up they compete with each other for food and shelter. Every extra daughter you produce will reduce the survival chances of all of your other daughters. In other words, a mother's genetic payoff (in terms of grandchildren) is likely to be highest if she produces more of the sex that will wander off (sons) than of the sex that will stay close to home (daughters).

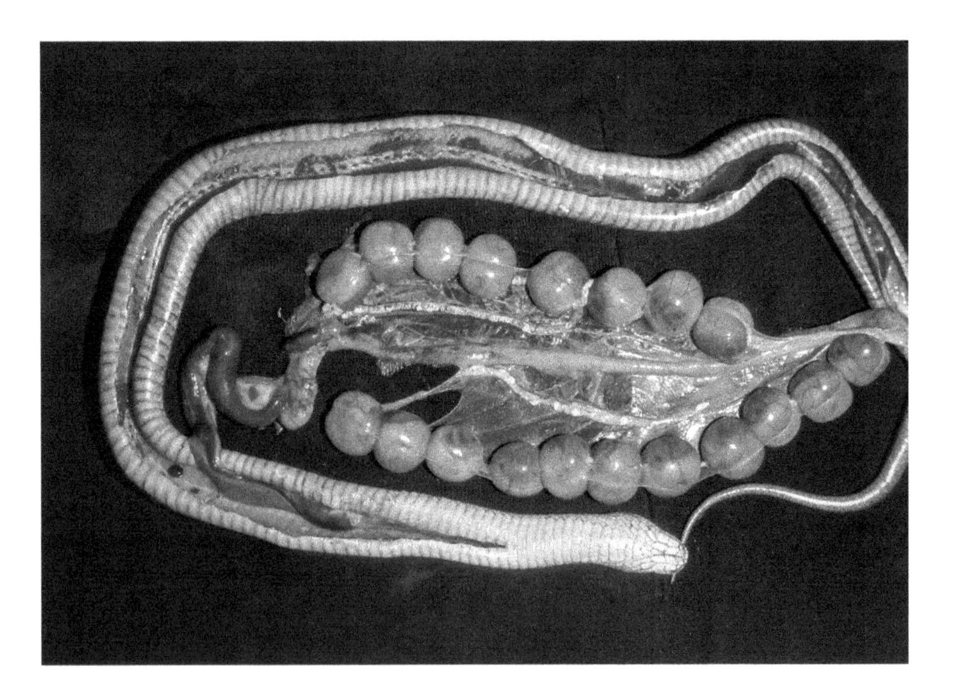

FIGURE 2.9 A dissected female Tigersnake, showing the embryos in each oviduct. Remarkably, Jim Bull and I found that these litters almost always contained more sons than daughters. Photograph by Rick Shine.

Although the explanation eluded us, Jim and I had identified a fascinating puzzle, so we were able to publish a paper on the topic. It was my first "real" scientific paper. I'd written a few stories for conservation magazines earlier, and I was a co-author with my supervisor Hal Heatwole on a small paper about mosquitoes biting frogs. Hal liked to have at least one joint paper with all of his students, and he was worried that I'd never publish anything else. With more than 1,000 papers to my credit since then, Hal's fears were unfounded—but I still appreciate his generosity of spirit.

****

Jim added a new intellectual dimension to my scientific world. Talking through new perspectives on evolution made a day in the bushland enormous fun even if we didn't find any snakes. And with my eyes opened to a powerful way of thinking, everything I saw around me was intriguing—what selective forces have shaped the color of the leaves in gumtrees? Why do kangaroos move around in groups? Why do some lizards lay eggs, whereas others give birth to live young?

Inevitably, though, snakes were the focus. There were plenty of adventures. Catching a deadly snake is risky business, no matter how carefully you prepare yourself for the task. There will be times when things go awry. One day, for example, I forgot to bring cloth bags with me when I was out collecting by myself—and as luck would have it, I caught four adult Blacksnakes within a few hours. How could I carry the snakes back to the car? I carefully peeled off a sock after I caught the first snake (hard to do with one hand!), popped the snake inside, and tied it up. Problem solved. And when I caught another snake, my second sock came off, and provided a secure home for that animal also. But when the third snake appeared, the situation began to get tricky. It's difficult to fit a snake into an already-occupied sock; the first occupant makes a bolt for freedom as soon as it sees a glimmer of light. And by the time I got my fourth snake, I knew it was time to head back to the car. I was pushing my luck. And my feet were hurting.

But the thrills didn't just come from live snakes; dissecting the animals afterward also provided revelations. Gradually, I began to understand how snakes spent their day-to-day lives. Finding fresh sperm inside the oviducts of females told me that Blacksnakes had a short mating season in spring, whereas Tigersnakes and Copperheads mated in autumn as well. By plotting out the body sizes of the smaller animals I caught against the dates that I caught them, I could follow the cohort of newborns as they grew larger over the next several months. That allowed me to estimate rates of growth and ages at sexual maturity. The male snakes matured a bit younger than the females—at 2 versus 3 years of age. Litter sizes were small in just-matured females but increased as the snakes grew larger because bigger females had more space for the developing litter. Pregnancy lasted three months, from late spring into late summer, and caused a massive shift in behavior. Grossly distended with their litters, pregnant females stopped feeding for a month or two before giving birth. They spent their time basking in the sunlight, remaining close to a safe shelter site in case a predator wandered by.

My growing datasets revealed some major differences between species. For example, all of the biggest Blacksnakes, Copperheads, and Brownsnakes that I caught were males. Females were much smaller. In contrast, a large Tigersnake was just as likely to be a female as a male. These patterns in "sexual size dimorphism" (in biological jargon) ended up becoming one of my major research interests, and I'll talk more about that topic in later chapters.

By the time I finished my Ph.D. in 1975, four years after arriving in Armidale, I had assembled the first detailed information on natural history of Australian snakes. And to explain the patterns that I saw, I had developed new ideas about the evolutionary forces that had molded those patterns. It had gone well, but what next? To continue a career in research, I needed a postdoctoral fellowship—and they were difficult to obtain. Although Jim had made me understand the importance of publishing my results in scientific papers, those manuscripts were still far from complete. So as far as the outside world was concerned, I had produced exactly zero evidence of any capacity for research. I applied for a postdoctoral fellowship in Australia, but failed to get it. Indeed, I doubt that I got past the first round.

My academic career was dead in the water. I couldn't get a postdoctoral fellowship on my own merits, and I didn't know any well-funded senior people interested in the kinds of reptile research that I wanted to pursue. It looked as though my decision to study snakes rather than more popular organisms had been a fatal mistake. My only option would be to look for a job in some "applied science" field, like wildlife management, and forget about a career in research.

At that point, a miracle occurred—in the unlikely form of my Grumpy Guardian Angel, John Legler. John returned to the University of Utah before I finished my Ph.D., taking Jim Bull with him. In general, John had a poor opinion of the students at Armidale, but he thought I had potential. I'm not sure why—John was a hard man to impress—but it mostly came down to one spoken presentation that I gave at a national scientific meeting. It was the first such meeting I had ever attended, so I was nervous. But I had a good story to tell, about the research that Jim and I were doing on sex ratios in Tigersnakes and Blacksnakes. John loved that talk, and decided that I was worth supporting. So before he left, John asked me if I would be interested in applying for a postdoctoral fellowship at Utah. I didn't need to be asked twice—it was an opportunity to travel overseas, see a different fauna, and continue working with Jim. And things fell into place very quickly. When John got back to Utah, he volunteered to chair the departmental committee that selected the next year's postdoctoral fellow, and arranged for one of his like-minded colleagues to join at the same time. There were only three people on that committee, so the decision was never in doubt. The postdoctoral fellowship was offered to a young Aussie based simply on John's enthusiastic evaluation of my potential.

But before I could embark on my overseas junket, I had to produce a Ph.D. thesis. The students of today prepare their theses on laptop computers, singlehanded, but the process was very different in 1975. First I had to conduct statistical analyses of my data, to check which results were "real" and which were just due to random variation. But in 1975, at a small rural university, facilities for data analysis were primitive. Lacking modern computers and user-friendly statistical software, I laboriously entered my data onto punch-cards and then deposited piles of cards (held together with rubber bands) into a pigeonhole in the mailbox in the Maths building that housed the university's sole computer. My cards were then run through the computer to perform the statistical tests that I needed. A few weeks later, I received reams of dot-matrix printout with the results.

Producing a legible thesis was an equally complicated process in the days before word processors. I wrote out drafts of my thesis chapters by hand, and then mailed them to my long-suffering mother. She resurrected her ancient typewriter out of a storage cupboard and typed my scribbles into neat scripts which she mailed back to me. Then I made corrections on those typed drafts before the departmental secretary (for some extra cash) typed the final copies for me during her lunch break. Chapter by chapter, those final versions accumulated on my desk.

At last, the thesis was complete. I was able to sit back and ask myself the bigger questions about my trajectory in science. I thought back to the arrogant Englishman who had challenged the very idea of research on snake ecology. He'd been wrong. It *could* be done. I'd done it. And my results showed the error in many of his dismissive comments. Although they all belonged to a single evolutionary lineage, the snakes around Armidale were not "all the same"—they lived in different habitats, ate different foods, and reproduced at different times and in different ways. And there were plenty of puzzling observations—like the sex ratios of Tigersnakes—that hinted at a greater subtlety and complexity than he had assumed. But on the other hand, I hadn't set the research world alight. I had been lucky—very lucky—to obtain that postdoc position in the United States. Without John Legler's support, my career would have been over before it truly began.

But all I could do now was ride my luck. In September 1975, I caught a taxi from the house I was sharing with fellow-students, dropped off four copies of my completed thesis at the university administration office, and then jumped back into the taxi for the trip to the airport, and the next phase of my career—swapping the sheep paddocks of the New England tableland for the high cold deserts of Utah.

3 Peering into the Love Lives of Blacksnakes

By definition, the outcome of a research program is uncertain. As Albert Einstein once remarked "If we knew what it was that we were doing, it would not be called research, *would it?" Research is a gamble. A tightrope that runs along the trade-off line between risk and reward. At one end of the scale is the pedestrian project—certain to work, but as boring as bat-poo. At the other extreme lies the project that is wildly ambitious—probably impossible. But if it works, you might win a Nobel Prize.*

To push back the frontiers, a young researcher has to take a few risks, and attempt at least a few projects that are exciting but may not be feasible. My first attempt to run a challenging project ended in disaster, and it hurt. I remember standing beside an irrigation canal in a semidesert landscape in 1981, watching a Blacksnake dying slowly. I had thrown everything—funding, time, and energy—into a research program that had gone belly-up, like the snake beside me. The gamble had failed.

My idea was simple: to use miniature radio transmitters to peer into the sex lives of snakes. All I needed to do was implant transmitters in all the snakes within a small population, and then monitor them to see which males fought with each other, who won those battles, who courted which females, who mated with who, and so forth. The emerging field of evolutionary ecology had spawned studies like this on a few birds and mammals, but reptiles had been ignored. If the study worked, I could conduct pioneering research. But the logistics were challenging. I would need to find a place where snakes were abundant, large enough to carry transmitters, and lived in open habitats where I could watch them. I knew the perfect place. The Macquarie Marshes are enormous swamps in central-western New South Wales, surrounded by arid plains. And the swamps are full of huge Blacksnakes. Some of the males were 2 meters (6 feet) long, and weighed more than 2 kg (4 lb) apiece. Those big bruisers could carry a transmitter without noticing it. And cattle kept the dry plains clear of grass, so the jet-black snakes could be seen from far away.

I obtained my first academic job—a lectureship at the University of Sydney—in 1980. Nervous about embarking on a risky research project, I decided on a practice run first. I scraped together enough funds to buy two transmitters and ran a "pilot study" during the mating season in spring. With research assistants Peter Harlow and Geoff Ross, I pushed small transmitters down the throats of two enormous male Blacksnakes and followed them around. I spray-painted each snake with a dash of color so that I could recognize it easily. The snake I dubbed "Silver" (based on his paintmark) was a shy creature, and hard to get close to. But the other male, "Yellow", was an extrovert, who flung himself into the mating season with great enthusiasm. I walked with Yellow through the open paddocks as he searched for females. Sometimes he'd crawl between my feet as I stood immobile. He wasn't

DOI: 10.1201/b22815-3

worried about me—a quick tongue-flick to my ankle, and he was off again to look for love. We saw male Blacksnakes fighting and courting, and my plans looked to be feasible. I recaught the snakes and gently palpated the transmitters back up out of their stomachs. As Yellow crawled away, I promised him I'd be back again next year.

The spring of 1981 saw me back in the Marshes, this time with a dozen transmitters and five helpers. I had begged, borrowed, and stolen all the equipment I needed, and used up every favor I could to enlist volunteer assistance. If it didn't work, my fledgling career would be in tatters. Tragically, the Marshes in 1981 looked nothing like the Marshes of 1980. Snakes were hard to find, and all of the big animals were gone. Dead. The surviving snakes were thin. And the reason was clear—a terrible drought. The sheep were walking skeletons, and the carcasses of stock littered the landscape. There was no grass, and no frogs for snakes to feed on. Like the sheep, the snakes were starving. None of the females had enough energy to breed, so the males were celibate as well. Snakes were lying out in the open, dying.

No matter how hard you try, you can't study snake reproduction if the snakes aren't reproducing. It was a hard lesson for a young academic. Australia is a tough country for ecological researchers, as it is for their study animals.

After my Ph.D., I spent three years at the University of Utah as a postdoctoral fellow. It was a steep learning curve—brushing shoulders with some truly brilliant people like my mentor Ric Charnov. Not much older than me, Ric is a mathematical biologist with an uncanny ability to visualize biological questions in terms of equations. Ric is an excitable soul, bubbling with passion for theoretical biology. Although his brain is wired differently to mine, we got along well. We shared an office for a while, and even an apartment.

Importantly, Utah showed me a new approach to research. The scientists in Armidale had been interested in evidence, not in theory. Our knowledge of Australian animals was abysmal, so Aussie biologists focused on gathering information rather than testing hypotheses. Theoretical speculations were viewed with skepticism, verging on hostility. The scientists I met asked me what species I worked on, not what questions I was asking. But American academia championed a different approach.

At the University of Utah, evidence was valued only to test ideas. And not all ideas were equal: the best-loved theories were those that could be expressed in mathematical equations. Evolutionary biology was flowering, in part due to the intellectual revolution spawned by George Williams' book. Zealous researchers believed that we were on the cusp of finding general "rules" of ecology. Once we identified those organizing principles, we could apply them to any ecosystem on Earth. Like chemistry and mathematics, ecology could be a "hard" science, where expert practitioners could accurately predict the detailed functioning of any specific system or species based on general principles.

Sadly, the natural world proved to be more chaotic than the true believers had hoped. Time and again, general theories failed to predict specific outcomes. Decades of disappointment eventually drove most of the enthusiasts—myself included—back into a more balanced view. We still hope to find "general" principles, but we accept that accidents of history make evolutionary outcomes unpredictable. As a result, we have to modify our ideas on a case-by-case basis to apply them to a local system. Of course, like the hippies in small communes stubbornly clinging to the social

FIGURE 3.1 During the mating season in springtime, male Red-Bellied Blacksnakes engage in spectacular combat bouts, with each male trying to push down the head of his opponent. Photograph by Geoff Boorman.

revolution of the 1970s, some biologists are still fighting the good fight, and looking for those elusive Universal Rules. God bless them.

I was entranced by this new approach—I liked the idea of answering a question rather than just gathering data. But as a naturalist at heart, I still spent a lot of time roaming around Utah looking at landscapes and animals. Some of the places I visited were magical. My favorite destination was a small valley in the desert of southern Utah, near the town of Saint George. The soft sand dunes held the impressions of the tracks of nocturnally active animals until mid-morning winds ruffled the surface of the desert. So if I rolled out of my sleeping bag at dawn, I could follow the overnight movements of Sidewinder Rattlesnakes, Desert Tortoises, and Gila Monsters. And if I was lucky, I would find the animals that had made those tracks. I met many of the wonderful American reptile species that Jim Bull had told me about when we were in Armidale.

I tried to conduct field research in Utah, just as I had in Australia—but I failed. Fieldwork requires a nuanced understanding of the local places, people, and animals. As a blundering foreigner, my attempts at American fieldwork ended in frustration. In my first summer in Utah, for example, I launched a research project on sex ratios in the litters of American snakes, to compare to our Tigersnake studies in Australia. I couldn't find enough snakes around Salt Lake City, but a friend told me that Addicks Dam near Houston, in eastern Texas, was a hot spot for Watersnakes. It was 2,400 km (1,500 miles) away, but I could drive my second-hand VW station-wagon from Utah to Texas. I convinced one of Ric Charnov's graduate students (Jim McKinley) to come with me, but my plans were thrown into chaos when, just before we were due to leave, Jim acquired a golden retriever puppy named Gus. Jim couldn't leave his new canine friend behind. No problem, I said. Gus can come with us.

That trip taught me more about canine bowel malfunction and Texan culture than about snakes. The first night after we arrived in Texas, Jim and I left a sleeping Gus in our hotel room and drove my venerable station-wagon out on country roads near Houston, spotlighting for snakes. I had met some grumpy landowners around Armidale, but the Lone Star State took it to a whole new level. I had firearms pointed at me three times that night—once by a wildlife ranger (who thought I was illegally hunting deer), once by prison-guards when I accidentally drove too close to a state prison, and finally by a policeman. The police officer was the most frightening. He saw my car parked by the side of the road, then spotted me lurking nearby. In the darkness, he mistook my snake-tongs for a rifle and was understandably nervous. His hand-gun trembled as he aimed it at me, and his voice wavered as he screamed out at me to come toward him with my hands above my head. My attempts to explain that I was a harmless snake-catcher were futile—perhaps my foreign accent was a problem—until I had the bright idea of emptying my hoop-bag to show him a huge Cottonmouth Moccasin that I had just caught. A large Cottonmouth is a fearsome creature, and this was the biggest specimen I had ever seen. More than a meter (3 feet) long, and as thick as my arm. When I dumped the snake onto the road, it reared back with its mouth wide open, revealing huge fangs and the white lining of the mouth that gives the species its name. The police officer took a step backward, returned his revolver to its holster, and uttered the memorable line (in a thick Texan drawl) "Boy, you got more hair on your balls than I got". Then he climbed back into his patrol car and drove away. I thanked the snake as I returned it to its swamp.

Although I survived my American fieldwork without bullet wounds, my projects failed. I struggled to find snakes or to understand the local ecology. But those fieldwork catastrophes were a blessing in disguise. In Utah, I was surrounded by clever people who enjoyed discussing evolution: an opportunity for me to try my hand at hypothesis-driven biology. I had access to superb library facilities (far better than at Armidale), and I decided to test ideas by using already-published information rather than gathering my own data.

What hypotheses should I try to test? I started with an idea I had in Armidale, to explain a puzzling situation. Why do males grow larger than females in some species of snakes (like Blacksnakes), whereas females are the larger sex in others (like Death Adders)? The explanation that I devised was simple: the relative sizes of the sexes depend on whether or not males fight with each other for access to females. If males engage in physical combat, larger body size pays off for them—it helps a male to defeat his rivals. A successful fighter has more opportunities for mating, so he sires more offspring. The best size for a male to be is larger than for a female. As a result, natural selection favors genes that switch expression depending on whether they find themselves inside the body of a male or a female. If they end up inside a male, the genes say "keep on growing after you mature"—and so, such a male reaches a large size (if he survives for long enough). But the gene sends out different instructions if it finds itself inside a female; there, it says "stop growing fairly early in life, and put your energy into producing offspring instead of growing larger". The end result of that evolutionary process is that body size averages larger in males than in females. The genes that cause that result leave behind more copies of themselves than genes that code for equal size in the two sexes, or for bigger females than males. Hey presto, males grow larger than females.

But what happens if males don't fight with each other? If bigger males don't produce more offspring, there is no evolutionary advantage to larger body size in males. Better for a male to stay small, mature early, and not waste energy on growth. For females, the situation is simpler—and it doesn't depend on whether or not the males are beating each other up. For a female, bigger is always better. A larger female can carry more offspring in her body, because she has more room to do so. As a result, bigger females will have higher evolutionary fitness. The genetic message that works best is to grow large if you are a female, but not if you are a male. So, we can make a simple, testable prediction. *Female snakes will grow larger than males of their species, unless males fight with each other during the breeding season. In species with male–male combat, males will grow bigger than females.*

Now I had a prediction, how could I test it? I needed information on the body sizes of males and females in as many snake species as possible, to compare to their reproductive behavior (specifically, whether or not males engage in combat bouts). This was before the internet, so assembling information was an arcane art. My breakthrough came in realizing that scientists who wrote papers about the classification of snakes (describing new species and so forth) often provided information about sexes and body sizes of all the specimens that they had examined. By going through those papers, I found data about sizes of males and females in hundreds of snake species. And by reading the wider literature, I could assemble records of male combat. To my delight, the pattern I had expected turned up in every group. In snake species where males fight with each other, the males tend to grow bigger than females. But in species where males don't fight with each other, the males are smaller than their partners. The prediction worked so well that I followed it up with similar analyses of turtles, frogs, and salamanders. And it worked every time. I had discovered a general

FIGURE 3.2 Combat between rival males has evolved in several evolutionary lineages of snakes, including elapids like these Lesser Black Whipsnakes. And if a species exhibits male–male combat, males usually grow larger than females. Photograph by David Nelson.

principle about how a species' mating system influenced the relative body sizes of males and females. It was the kind of "general rule" that my Utah colleagues valued, an exciting outcome for a young Aussie trying to make his mark in international science. I wrote papers describing my results, and those papers were published in major journals.

I was lucky to have Jim as a co-author on some of those first few manuscripts. Having already published several papers himself, he was well aware of the confronting nature of the review process. Most of the reviews were positive, and the papers were accepted, but some of my efforts were absolutely trashed and rejected. Frustratingly, whether or not my papers were accepted seemed to depend as much on luck (who was chosen as a reviewer) rather than the caliber of the work. I began to realize that reviewers are fallible—sometimes overly kind, sometimes overly critical; sometimes putting very little effort into their reviews, and other times putting in far too much effort. On the one hand, it meant that a set of highly positive reviews didn't necessarily mean that my paper was wonderful; on the other hand, a series of harsh reviews might say more about grumpy experts than about the paper itself. Nobody likes having their efforts attacked and rejected, but I became philosophical about outcomes of the review process, and confident about the value of my outputs.

Ultimately, relying on other people to evaluate the quality of my research would only lead to perpetual soul-searching and deep frustration. That ability to distance myself from harsh criticism saved me heartache when my papers were rejected; and also it helped me keep a sense of perspective later in my career, after I began to win awards and high praise. Much of that praise was as undeserved as were the scathing criticisms of aggressive reviewers.

When I had completed my analyses on sex differences in body size, I turned to another topic: one that later became central to my research career, although I never anticipated it at the time. The subject was a classic one: the evolutionary transition from egg-laying to live-bearing reproduction. That transition has happened in many lineages of animals, from flies through to mammals, but reptiles offer the best opportunities for examining the process. In most species of lizards and snakes, the female lays eggs that incubate for weeks or months in the nest before they hatch. But in about a quarter of lizard and snake species, the female retains those eggs inside her oviducts until they are fully developed, and she gives birth to fully formed offspring. Many biologists had speculated about how and why that shift in reproduction has evolved.

Jim and I had joined that throng when we were still in Armidale. Using the ways of thinking espoused by George Williams ("what is the advantage to an individual animal that exhibits this new trait?"), we came up with some new ideas about the evolutionary pressures that drove this switch from oviparity (egg-laying) to viviparity (live-bearing). Ideas are great, but tests of those ideas would be even better. I decided to test our predictions by sieving through the huge literature on this topic. In particular, I wanted to combine information about reproduction (was the species an egg-layer or live-bearer?) with information about evolutionary relationships between species (who was related to whom?). That way, I could work out when one reproductive mode had evolved into the other. Had live-bearing only evolved a few times? If so, today's live-bearers (although comprising hundreds of species) are the result of just a few evolutionary origins of the trait. Or alternatively, did live-bearing evolve many times, whenever an egg-laying species encountered some particular environmental challenge?

Today, most evolutionary biologists use this "comparative" approach to test their ideas. They embed the information about their trait of interest onto an evolutionary tree, so they can see how and where that trait evolves. But back in the late 1970s, that method was in its infancy. Again, I was a pioneer without intending to be.

Months of library work revealed vastly more transitions in reproductive mode among reptiles than anyone had suspected. I concluded that egg-laying has given rise to live-bearing more than 100 times within lizards and snakes. That's an amazing number. Mammals only made the switch once (between the egg-laying Platypus and Echidna and the live-bearing Everything Else), and birds never made it at all (they are all egg-layers). But lizards and snakes have jumped from egg-laying to live-bearing so often that it beggars the imagination.

Why has live-bearing evolved so often? My analyses suggested that climate is the key: live-bearers thrive in colder areas. Egg-laying doesn't work if the soil is too cold for an embryo to develop properly, whereas live-bearing protects the babies from that frigid nest. A mother who retains her babies inside her body (instead of laying them

FIGURE 3.3 Viviparity (live-bearing) has evolved many times within reptiles, with the result that we sometimes find species that lay eggs (like the Collett's Snake, upper picture) that are closely related to species that bear live young (like the Red-Bellied Blacksnake, lower picture). Photographs by Peter Harlow.

in eggs) can keep them warm by basking in the sunlight. She can produce healthy babies even in an environment where the soil is too cold for any embryo to develop.

Unfortunately, Jim and I couldn't claim any credit for the idea that cool climates favor the evolution of live-bearing. That hypothesis had already been around for almost 50 years. In the 1930s, three workers—a Russian, a German, and an Australian—all suggested (independently) that live-bearing was an evolutionary advantage in cold climates. The Australian, Claire Weekes, was a pioneer in terms of both reptile reproduction research and as a woman in a man's world. A staff member of the University of Sydney, she shocked the male hierarchy by conducting field-trips without a chaperone. Based on the lizards she collected during those trips, Weekes wrote a series of landmark papers that remained the gold standard for research in her field for decades to come. The German, Rudolf Mell, was a remarkable figure as well: a schoolmaster in China, with a prodigious handlebar moustache. Like Weekes, he delighted not only in fieldwork and careful analysis, but also in developing ideas to explain the patterns that he found. Those ideas remain influential. But perhaps the most extraordinary of all was the Russian, Alexei Sergeev. While he was still a student, Sergeev wrote papers that were decades ahead of their time, foreshadowing topics that would later take a central place in the scientific study of reptiles. For example, he demonstrated that reptiles can regulate their own body temperatures by basking in sunlight. Unfortunately, Sergeev's career was cut short by the Second World War and, eventually, he committed suicide in a prison-camp. Years later, when I finally had Alexei Sergeev's massive treatise on reptile viviparity translated from Russian into English, I was astonished at the insights of this tortured genius. I felt like a charlatan. Many of the ideas that Jim and I suggested in the 1970s, thinking they were new, were laid out clearly in Sergeev's monograph of 1940.

****

But my time in Utah wasn't all about research. I joined a soccer team in Salt Lake City, both for the exercise and to deal with the only downside of my life in Utah—social isolation. After my active social life in Armidale, Utah was a stern challenge. Especially, for an atheist. Salt Lake City is the center of the Mormon Church, and no true-blue Mormon girl would get too friendly with a non-Mormon—at least in those days. So, my social life was quiet. When Jim and I wrote a scientific paper about animal species with low frequencies of reproduction in 1977, I added a sentence to the Acknowledgments section saying "The junior author thanks the women of Salt Lake City for teaching him the true meaning of the phrase 'low frequencies of reproduction' ". An irate reviewer insisted that I remove the sentence.

But, things were to change. The high point of my time in Salt Lake City came later that same year, when I met a young woman named Terri Griffith. Smart, elegant, energetic, funny, and drop-dead gorgeous. And not a Mormon. Before long I was in *the* relationship of my life. Being with Terri in Utah was great fun, and I considered staying in the United States. But I was homesick, and so after three years in Salt Lake City, it was time to head back to Oz. Terri and I were married in 1978, and then caught the plane to Sydney a few weeks after the wedding. More than 40 years later, she is still enriching my life every day.

What kind of woman marries a snake-obsessed field biologist? People often assume that Terri shares my passion for reptiles, but she doesn't. Terri is a social worker, not a snake-wrangler. When we first met, she was frightened of snakes—and in truth, she still doesn't like handling them. But after decades as my unpaid field assistant, she's long ago conquered any squeamishness. I've even seen an affectionate glimmer in her eyes from time to time when she is close to a harmless snake. Or maybe she's just smiling at her own foolishness in spending her life with a man who is besotted with these peculiar creatures.

When it was time to leave Utah, I faced a tough job market back home. Academic vacancies were few and far between, and it looked as though my decision to focus my research on snakes would bring my career to a premature end. Other fields of science had better job prospects. For example, my brother John was a pioneer in the exciting new field of molecular genetics, and he was fending off job offers. But for me, the situation was different. I was publishing lots of scientific papers, but they were on non-trendy topics (natural history) and on unpopular organisms (reptiles). I would have to go wherever I could find an opportunity.

I applied for every postdoctoral position that was advertised, and finally received a letter offering me a job. It was in New Zealand not Australia, and so not on snakes (which don't occur in New Zealand). But there are many fascinating lizards there, as well as the iconic Tuatara. And although it's cold, New Zealand is stunningly beautiful, with a vibrant culture. I accepted the job, and I resigned from my position in Utah. But then, international politics intervened. The United Kingdom terminated a financial agreement that had been keeping the New Zealand economy afloat for decades, and the economy of the small island nation went into free-fall. My prospective employers warned me that I'd have a salary, but no money for fieldwork. Look elsewhere, they said—no point coming here. In desperation, I applied for a postdoctoral fellowship that had just been advertised at the University of Sydney, Australia's oldest university. My hopes of getting that job were close to zero, so Terri and I bought our own plane tickets back to Oz, and I resigned myself to unemployment. A few days before we were due to fly home, though, a letter arrived. It offered me the fellowship at Sydney.

I had been lucky. After long arguments about who should get that fellowship, the top rank had gone to a desert ecologist from Melbourne. I came in at #2, and that should have been the end of the story. In most years, the Biology Department would only be allocated one fellowship. I was at the head of the reserve list, with no real prospect of an offer. But the scheme was university-wide, so I still had a faint chance—one of the successful applicants from another department might turn down the offer if they obtained a job elsewhere. If so, I would be offered the vacancy. And that's what happened. When I arrived in Sydney, people in the department were surprised; they didn't even know that I had been awarded the postdoc. But they were very supportive, and the desert ecologist became (and remains) a close friend.

My official supervisor at the University of Sydney was Professor Charles Birch (1918–2009), an outstanding biologist with a formidable international reputation. Despite those accolades, Charles looked and acted like a typical scruffy academic. His hair was unfashionably long, because he forgot to get it cut, and his beloved sports car was full of scrapes and dents because he was more interested in biology and ethics than in driving safely ("I had another car accident this morning on the way to work. My fault, again"). Charles refused to learn the names of any students, on the grounds that his brainspace was limited, and the neurons required to recall that name would no longer be available to remember the name of a biological species. Nonetheless, that brainspace was impressive. Early in his career, Charles played a pioneering role in the development of ecological theory. In 1954, when I was just 4 years old, Charles and his mentor Herbert Andrewartha wrote a seminal book entitled *The Distribution and Abundance of Animals*. It was a slap in the face to conventional (that is, European and American) approaches to ecology.

The Northern Hemisphere scientists saw natural ecosystems as stable, with animal populations in equilibrium with their food supply. The balance of nature was maintained by feedback loops; any increase in animal numbers reduced the food supply per animal, bringing the system back into stability by decreasing per-capita survival and reproduction. Elegant, static, and un-Australian. When Andrewartha and Birch looked out the window in Adelaide, they saw a different and more chaotic world. In Australia, shit happens. Instead of being in serene equilibrium, animal populations lurch from one catastrophe (droughts, floods, or just bad luck) to the next. The numbers of animals change from year to year, victims of environmental variation. The Great Tits of Oxford may live in a consistent and predictable world, where ecological interactions generate stability. But the flies buzzing around Australian fruit trees, and the snakes that slither beneath those trees, eke out a more dynamic and brutal existence.

If Charles had still been pursuing those issues in 1978, I would have happily sat at his feet and gloried in his wisdom. But by the time I arrived in Sydney, Charles had moved into philosophy, trying to find middle ground between evolution and theology. Religious scholars loved his writings, whereas biologists thought they were nonsense. Science and religion see the world in very different ways. As a hard-nosed scientist, I view religion as the shrinking remnant of phenomena that we do not yet understand. Charles Darwin's great defender Thomas Huxley once said "*extinguished theologians lay around the cradle of every new science like snakes around the body of Hercules*".

Ideas about the nature of science were not the only principles under challenge at the University of Sydney in 1978. The academic world was in transition. The conservative, hierarchical, male-dominated ivory tower was grappling with social inclusion and equality of opportunity. Society was changing, as age-old traditions were challenged by the Woodstock generation. In days gone by, elderly white male professors decided the career prospects of younger workers, over a glass of sherry and behind closed doors. Advancement depended upon gender and social class as much as ability. Those days were coming to an end, killed by societal shifts. Add in some ambitious young scientists keen to build their own careers, and you had the ingredients for an academic system in fierce conflict.

Although I wasn't interested in Charles' pseudo-spiritual discussions, we became friends. Charles inspired fear in my colleagues, but he enjoyed my passion for evolutionary theory, my irreverent approach to authority, and my willingness to explain aspects of popular culture that he (as a true academic nerd) found puzzling. Once a week, Charles held court at "ecology morning tea". Attendance was strictly by personal invitation from The Great Man. Charles would sit like a headmaster at one end of the long table, posing abstruse quasi-ecological questions ("How many tons of dogshit are deposited every year in New York City?"). Being nominated as the person to respond to such a question was frightening, but it provided an opportunity to demonstrate either knowledge or wit (or ideally, both). Charles was naive, and he struggled to understand jokes. Once, he pompously complained "Even when you ask a graduate student a simple question, he always answers a different question. Isn't that true?" An urbane young student immediately answered "Half past twelve". As Charles shouted "What? What? That doesn't make sense!", the room dissolved in laughter. The professors didn't realize it yet, but they were no longer gods.

One of the lessons I learned from my early days at Sydney University, as different worldviews collided and the academic system was reshaped, was that science doesn't operate in a social vacuum. It's tempting to see research as an objective search for truth, with impartial intellectuals scrupulously following "the scientific method" to obtain the hard evidence needed to determine which ideas are true and which are false. The reality is different. Research is conducted by fallible human beings; putting on a lab coat doesn't alter personal belief systems. And the 1970s were chaotic; long-cherished ideas about the innate superiority of the upper class over the working class, of men over women, and of whites over other races were coming under strong attack. I read Germaine Greer's feminist diatribe "The Female Eunuch" and was ashamed to recognize my own chauvinism. It was deeply humbling.

How does any of this societal mumbo-jumbo relate to science? The fundamental theme of 1970s ideology can be summarized in a single phrase: *"different" does not mean "inferior"*. For hundreds of years, mainstream society had advocated that one way of doing things was better than another; that one kind of person deserved more respect than another; and so forth. Although the focus of the new social debate was on gender, race, and sexual preference, the broader message—that we should question the innate superiority of some things (or people) over others—affected the way that scientists thought. And those winds of change blew away two myths that had dominated mainstream thinking for centuries and had been an unseen part of my own attitudes. These were the ideas that "cold-blooded" (ectothermic) animals like reptiles and amphibians were inferior to "warm-blooded" birds and mammals (endotherms), and that anything Australian was inferior to anything European or American. My attitudes didn't change overnight, but the scene was set. As I embarked on my research career, I was beginning to discard the crippling assumption that I was studying "primitive" creatures on a second-rate continent. If anyone had challenged me about the usefulness of studying Australian snakes, as the English postdoc had done during my undergrad days, I would have mounted a spirited defense.

****

What kind of research should I do? I needed to develop my own directions in research—but that didn't daunt me, because I had selected my own projects ever since my Honors year. My supervisors had been supportive without trying to push me in particular directions. Dick Barwick in Canberra, Hal Heatwole in Armidale, Ric Charnov in Salt Lake City, Charles Birch in Sydney—they all let me frame my own projects. I'm grateful. They gave me the space I needed to develop my own ways of doing science. Many of my ideas didn't work out, but by 1978 I had substantial experience with conducting independent research.

But I only had two years of postdoctoral salary before I had to find a "real" job, and there was no room for complacency. The job market was tough. The only reasons I still had a scientific career were two accidents. The first was John Legler liking me enough to arrange a fellowship at the University of Utah. I had won that position because of John's opinion, not any evidence of my own ability. And my second postdoctoral fellowship, at Sydney, was an accident as well. I hadn't been good enough to win a fellowship in the first round, and I was only funded because somebody else dropped out.

That kind of luck was unlikely to keep me afloat a third time. To have any chance of further employment, I needed to be *very* productive during my postdoc. In other words, "publish or perish": I needed my papers to be accepted for publication in international journals. So, regretfully, snakes were not an option. Too difficult to find and to work with. Lizards are easier; and lizard research was more respectable than snake research. It was increasingly clear to me, from conferences and discussions, that most professional scientists didn't take snake research seriously. Some people were supportive, but others were condescending. Remembering the same attitudes from years before, I could hear the dismissive comments of the English postdoc ringing in my ears. Major universities in the United States sometimes employed academics who studied lizards, but the only snake biologists I knew were unemployed, or they were tucked away in small institutions in the backwoods. So, I decided to focus my research on lizards—until I could obtain a permanent job. Hard times call for hard choices.

While in Utah, documenting the link between cold climates and the shift from egg-laying to live-bearing, I also thought about why that evolutionary transition has occurred. Although some of the ideas had been around for years, nobody had tried to test them. For example, just how cold did the ground need to be for egg-laying to become unsuccessful? How much warmer could a pregnant female keep her embryos by basking in the sunlight? How much did that warmer incubation modify the offspring's rate of development? I could measure all of those things if I could find a field site that contained both egg-laying and live-bearing lizard species, living side-by-side. Preferably, a site that was almost-but-not-quite too cold for egg-layers to persist. And I remembered a field trip from my undergraduate days, when the University had transported a bus-load of students from Canberra to a nearby mountain range. Steep mountains gave access to a wide range of climatic conditions, and lizards were abundant.

Terri and I bought a small tent and headed off to the magnificent Brindabella Range, so that I could explore the lives of alpine lizards. The Range encompasses a huge area of forest, dense in the gullies but sparse on many ridges, with a bewildering variety of eucalyptus tree species. But most of my work was along the edges of the forest, in long open corridors beneath power lines. On the lower slopes, these "hydrocuts" were regularly cleared by workcrews to stop trees from fouling the powerlines. And on the higher peaks, ski resorts cleared trees to provide downhill runs. Regardless of the motivation, the removal of trees created ideal forest-edge habitat for sun-loving lizards. And the eucalypts that had been cut down lay about in their thousands, slowly rotting away, and providing perfect hiding places for reptiles and amphibians.

FIGURE 3.4 My wife Terri turning over logs in the Brindabella Range, looking for lizards and their eggs. Photograph by Rick Shine.

I turned over many thousands of logs during those two years of research, to find lizards and their eggs. It was a magnificent natural laboratory. Eggs were easy: if I found them, I could do whatever I liked with them. But capturing adult lizards was a challenging task. Hot lizards are fast lizards, and my capture rate was woefully low if I turned logs in the middle of the day. My best strategy was to begin work at dawn, as soon as there was enough light to see my way, while the lizards were still chilled by the cold Brindabella night. I'm not an early riser, so it took all my self-discipline to roll out of the sleeping bag when the alarm went off, struggle into warm clothes, and emerge from the tent into a dark and frozen wasteland to start the day. On some mornings, I had to break the ice in my billy before I could boil the water for a cup of coffee.

But the pain of those cold mornings was rewarded by encounters with wildlife in the half-light. Stocky Wombats returning to their burrows after nocturnal grazing; kangaroos and dingoes wandering past on their way to daytime retreats. A chorus of birdcalls as the light broke over the horizon. And fascinating creatures aplenty under those logs that I turned. The most spectacular were Corroboree Frogs, so named because their extravagant black and yellow stripes and blotches resemble the body-paint used by Indigenous Australians. The frogs were abundant at one of my sites, Coree Flats; I often found several under a single log. Tragically, the species is now on the edge of extinction, wiped out by an invasive fungus brought to Australia. The memory of those living jewels amongst the sphagnum moss is bittersweet. I feel privileged to have seen them, and I am devastated that they have gone.

But frogs were not my quarry. I was looking for reptiles. I worked as fast as I could, rolling logs before the sunshine turned torpid lizards into speedsters. By mid-morning, my back was aching and the lizards were beginning to outsmart and outrun me, so I returned to the tent for another cup of coffee, a chat with Terri, and a changeover in collecting gear. Chasing active lizards through the forest was an exercise in futility, but I could catch them with a fishing rod and a small noose made of dental floss. The method required extreme patience, but with luck I could slip the noose over a basking animal's head and pull it tight, then grab the wriggling creature before the noose slipped off. And Terri was right beside me with our prized electronic thermometer, ready to record the lizard's body temperature before it could cool down after capture. We used the same thermometer to record temperatures in the natural nests that we found: small clusters of pure-white eggs under logs and rocks.

The research program worked smoothly. By measuring temperatures inside natural nests, and in pregnant lizards, I could estimate the impact of live-bearing on the embryo's thermal environment. And by incubating eggs and embryos at known temperatures in the laboratory, I could work out how Mum's Heating Service affected when a baby lizard hatched, and what kind of a baby it was. We could measure the consequences of that evolutionary shift from oviparity to viviparity. And sure enough, I was able to show that if the ground was cold, retaining developing embryos inside your uterus was the recipe for successful motherhood.

But if live-bearing was so beneficial, why hadn't all of the local lizard species abandoned egg-laying? It didn't take much thought to identify a "cost" of live-bearing that might outweigh its advantages. As any pregnant woman will tell you, carrying developing offspring around inside your body is a costly business. The litter of a pregnant lizard can comprise half of her total body weight, and surely constrains her athletic activity. That burden might render her vulnerable to predators, or incapable of feeding. So how can a female live-bearer minimize the costs of pregnancy? During my doctoral research in Armidale, I'd noticed that female snakes stopped eating when they were pregnant—probably because it was too risky to search for food when they were heavily burdened. Instead, they concentrated on basking, and stayed close to shelter. In scientific jargon, they were swapping one type of "cost" (the risk of being eaten by a predator) for another (a reduced food intake). I realized that "costs of reproduction" were important, but nobody had ever measured them in the field. I expanded my study on the Brindabella lizards to encompass these ideas as well.

I built a wooden raceway from scraps of lumber, and I purchased a cheap video camera so I could film how quickly female lizards could run. By comparing running speeds during pregnancy versus afterward, and studying non-pregnant lizards at the same time, I could measure the impact of pregnancy on a female's mobility. To see how reproduction changed a female lizard's basking and feeding behavior, I caught pregnant and non-pregnant lizards and brought them back to Sydney. Then I set up a child's plastic-sided swimming pool in my backyard, filled it with grass and logs to mimic natural habitats, and released the lizards into it. After the lizards had settled in to their new home, I sat beside the enclosure in a lounge chair recording their activity patterns and habitat use. Simple science, but I was replacing vague speculation with real numbers. And I could plug those numbers into mathematical models, that I could publish in scientific journals.

As my two-year fellowship came to an end in 1980, I was thrown back into the maelstrom of worry about obtaining a "real job"—that is, an academic position. The job market was still awful, and although I spent hours scanning journals

FIGURE 3.5 This Tussock Grass Skink is so heavily pregnant that she can barely waddle, and it's easy to see why pregnant females change their behavior to cope with the burden. Photograph by Mark Hutchinson.

and newspapers where such jobs were advertised, I rarely found anything I could apply for. I enjoyed Sydney, and would have loved to stay there, but the chances were slim—universities were under increasing financial pressure, and were rarely appointing new people even if existing staff moved away or retired. But in the final year of my postdoc, two of my senior colleagues left to take up jobs elsewhere. That left a hole in the teaching program, so the department had to hire someone to replace the folks they had lost. An academic vacancy was advertised—not only in my own department, but in the field in which I specialized (biology of vertebrates).

This time, I didn't need to rely on the old-boy network or on serendipity. I had finally gotten enough experience, and produced enough research, that I was a plausible contender. My burgeoning publications helped but more importantly, some guest lectures I had given to undergraduate students had impressed people on the selection committee. I was offered the job. A real academic position! I had never imagined being employed by Australia's oldest university. And even better, a senior colleague was there to mentor my first faltering steps into professional life. Gordon Grigg had conducted research on many kinds of animals, but reptiles (especially crocodiles) had a special place in his heart. Gordon smoothed my way through the challenges that confront an early-career academic.

Again and again, throughout my career, well-meaning colleagues had advised me to switch from studying snakes to "better" types of animals—the kinds of beasts that many other people already studied. And I had resisted those calls, and I had somehow got a permanent academic job anyway. A reward for perseverance, and for taking on a challenging task. But although it would be fun to bask in self-congratulation, I was beginning to realize that my success was due to other factors as well. As a heterosexual white male, I had received favored treatment all the way along—the dice had been loaded in my favor. Equally talented, hardworking people had been denied professional advancement because of discrimination—either intentional or unintentional—against their gender, race, or lifestyle. I was delighted to have the lectureship at Sydney, but I couldn't pretend that I had won that job on a fair playing field. And the same was true for my earlier jobs, especially at Utah. The process that had yielded me that fellowship was more about politics than scientific merit.

At the time I was appointed to the lectureship at Sydney University, I was keeping a Blacksnake in a large glass tank in my office so that I could feast my eyes on my favorite species as I worked. Good for my mental health, but that pet snake jeopardized my career. A week after Charles Birch told me that I was the successful candidate, I came into the office one morning to discover that my pet had given birth, and there were suspiciously few babies in the cage with her. Had any of them escaped, through small ventilation holes in the side of the cage? Sure enough, Charles came into my office a few hours later with a wastepaper basket containing an indignant baby Blacksnake. "My secretary found this snake in the corridor", he said. "She thought it was a Blacksnake, but I told her you wouldn't be stupid enough to let a venomous snake loose in the Department, so I knew it must be harmless. I picked it up and popped it into this bin". Speechless, I just nodded my thanks.

Life as a young academic was a steep learning curve. Giving lectures to hundreds of first-year students was stressful at first, but I enjoyed it, and writing lectures on diverse subjects filled in some major gaps in my own knowledge. I learnt by doing, because in the 1980s nobody taught academics how to teach, or how to conduct research. Nowadays, a young academic must attend an avalanche of "how to" courses before they can do anything. And the nature of research has changed as well. In the bad old days, preparing a research paper was a collaboration between the researcher who wrote the paper, the departmental secretary who typed the paper, the departmental artist who drew the figures, the university librarian who helped to locate relevant publications, and so forth. Today, I can do it all on my laptop.

The biggest difference from modern academia, though, was in the more leisurely pace of life. Morning tea was a sacred obligation, when every academic in the department looked forward to animated debates in the tearoom. By the time I left Sydney University in 2018, though, the tearoom was deserted; my colleagues were teleconferencing their international collaborators instead. When the coronavirus epidemic forced everyone to work from home in 2020, it had little effect on the academics at Sydney University. They were already living in their own isolated bubbles.

****

In 1980, a week before my 30th birthday and with a real job at last, I began my first major snake project since my Ph.D. work had ended five years earlier. I recalled those giant Blacksnakes of the Macquarie Marshes, and decided to embark on a radio transmitter–based enquiry into the snakes' private lives. It might tell me whether or not I could build a career as a snake ecologist.

FIGURE 3.6 Abundant along the streams and swamps of eastern Australia, the fearsome warning colors of Red-Bellied Blacksnakes belie their gentle disposition. Photograph by Rob Valentic.

The bonus in choosing the Macquarie Marshes as a study site was its allure. Being there was a thrill. The swamps cover a huge area—almost 5,000 km^2 (2,000 square miles) during a major flood—west of Quambone and north of Dubbo, in the geographic center of New South Wales. It's far from typical habitat for Red-Bellied Blacksnakes, which prefer cooler, well-watered coastal regions. The parched landscape surrounding the Marshes is home to semidesert species, like the Blacksnake's relative the King Brown Snake and a host of smaller species such as the DeVis' Snake (which looks like a pocket-sized Death Adder), and the elegant Grey Snake (called "Quambone Snake" by the locals). My old friends the Eastern Brownsnakes were also abundant, and remarkably variable in color. More than once after I seized a jet-black tail disappearing into the reeds, an explosion of lithe athleticism told me that I had mistakenly grabbed a dark-colored Brownsnake rather than the sedate Blacksnakes that were my intended quarry.

It was the abundance of Brownsnakes that had first drawn me to the Marshes, during my Ph.D. The area around Armidale was too cool for these sun-loving speedsters, so in order to build up sample sizes I traveled further afield. The landscape was stunning. I had grown up in a world of dense coastal forests and mist-shrouded mountains, and sheep paddocks clinging to the sides of hills. Suddenly I was in the midst of a much harsher and more exotic world: the western plains. Flat, open, with huge river gums lining the watercourses. The light was harsh, and the glare off the sun-baked ground was painful. Hot and dusty most of the time, and shade was a scarce luxury. Sheep, cattle, and kangaroos gathered in those few shady places. But occasionally—

FIGURE 3.7 The flat, hot, dry landscape around the Macquarie Marshes is atypical habitat for this water-loving species. Photograph by Rick Shine.

once every few years—the Marshes were transformed into a huge swamp. After heavy rains hundreds of kilometers to the north, the water slowly pushed its way through once-dry channels and spread out across the low-lying pastures.

By luck, my first trip to the Marshes coincided with a major flood. Rising waters had pushed the wildlife up to higher drier ground, concentrating animals on the small patches not yet underwater. After a 7-hour drive from Armidale, it was nighttime before I arrived and began to set up my tent. As I moved branches on the ground to clear a space, my flashlight beam revealed snakes of three species. And when I crawled out of my sleeping bag to boil the billy in the morning, I could see hordes of kangaroos, emus, and feral pigs, as well as snakes. It was like watching a TV documentary.

Never, either before or since, have I encountered such large Blacksnakes, nor so many. Gleaming monsters, basking in the morning sunlight. Subduing such huge muscular animals was a physical battle, but the challenge paled into insignificance compared to the Brownsnakes. More vigilant and faster-moving than the heavy-bodied Blacksnakes, Brownsnakes rarely went into the bag without giving me an adrenalin jolt. Once, I found two adult male Brownsnakes tightly intertwined in ritual combat—the first time I had seen the behavior myself, even though I had already published several scientific papers about male–male combat.

How could the Marshes support so many snakes? You can't sustain predators without prey, and this abundance of snakes was built on an unimaginable plethora of amphibians. Literally, tons of them. As dusk fell, every soil crack was lined with thousands of small froggy faces waiting to clamber out and begin feeding on bugs. In the light from my head-lamp, the reflections from frog eyes looked like the Milky Way. For any predator who liked to feast on amphibians, the Macquarie Marshes was an "all you can eat" restaurant. When I dissected road-killed snakes, their stomachs were full of frogs.

The property I worked on—"Sandy Camp"—was owned by the Robertson family. By the standards of western New South Wales in 1980, the Robertsons were environmentally enlightened. Most farmers killed every snake they saw, but the Robertsons took a kinder view. They executed Brownsnakes, but left Blacksnakes unmolested. That tolerance was driven partly by the Blacksnake's laid-back attitude and low venom toxicity, and partly by the belief that Blacksnakes keep Brownsnake numbers in check by eating them. Like most stories about snakes, there's not much evidence to support it. Blacksnakes do eat Brownsnakes, but the reverse is true also. Regardless, it's a useful myth—it stops people from persecuting my favorite animal. And once you decide to coexist with snakes rather than massacre them, you soon learn that they are not the aggressive demons of folklore. Sue Robertson kept a broom on her verandah to sweep Blacksnakes out into the garden in the morning.

The Macquarie Marshes were perfect for radio-tracking. The snakes were abundant, supersized, easy to see in this flat open country, and mellow. They didn't flee when I approached them, perhaps because emus are common. To a snake, a person probably looks like an emu. And researchers were welcome on Sandy Camp, so we could stay in shearing sheds rather than tents. Conversations with the landowners gave us invaluable information. For example, Sue told us about a bypass channel that ran through the heart of the Marshes, and was skirted by a dirt road for many

FIGURE 3.8 Catching a Red-bellied Blacksnake in the Macquarie Marshes. Photograph by Peter Harlow.

kilometers. The banks of the canal were festooned with snakes: a perfect study area. To find and catch those snakes, we developed a simple technique. Geoff Ross drove the field vehicle, while Peter Harlow and I stood on the vehicle's rear bumper-bar, looking down into the canal from our high perches. Whenever we sighted a snake, we'd tap on the vehicle's roof, and Geoff would stop the car (unless he was listening to the radio, in which case we'd have to keep thumping until he heard us). Then we'd jump down, catch the animal, measure it, give it a distinctive paint-mark, and release it. And repeat the process. Our elevated perches provided a breathtaking view of kangaroos, emus, wild pigs, black swans, and all the other wildlife of the Marshes. The only downside to this idyllic activity was our precarious perches; a bump in the road had us flying off into the nearest patch of thistles. It was a small price to pay.

Our capture sessions were restricted to the early mornings and late afternoons, because it was far too hot for snakes to be active at midday. In the stifling summer heat, the bypass canal was the only place we could cool off. The channel was about 2 meters (6 feet) wide and 50 cm (20 inches) deep, large enough to sit in. Because cows routinely used the channel as a latrine, the fluid passing along it was more like mud than water; Geoff renamed the "bypass canal" the "alimentary canal". And if you sat still in the water, a crayfish was likely to pinch you where you least appreciated it. On one trip, we brought a small inflatable dinghy to row along the canal. I have vivid memories of a large Blacksnake climbing into the boat with us, while Terri took refuge by clambering up onto my head. It's difficult to steer a boat with a snake coming in over the side and a woman standing on your shoulders, especially when you can't stop laughing.

To estimate the numbers and sizes of snakes in this population, we captured as many as we could, measured them, and gave them individual paint-marks by spray-painting each snake with bands to form a unique color-code. When we saw the snake again, we could identify it from its color-code without needing to recatch it. That told us how far the animals were moving, and how many different individuals lived in the area. But it puzzled the local farmers. Once as I was spray-painting a large Blacksnake beside the channel, a stockman rode up. He sighed with relief. He'd been seeing rainbow-colored Blacksnakes, and was wondering if it was time to cut back on his alcohol consumption. He was pleased to discover the real reason for the sudden proliferation of vivid serpents.

Not all of the landowners were so supportive. When Sandy Camp was sold to new owners, profits were given higher priority than biodiversity conservation. A CSIRO research team was working on Sandy Camp at the same time as we were, on birds rather than snakes. On one trip, the new property owner asked the CSIRO team which parts of the property were most important for bird conservation. They told him that the "zoo" paddock, so called because of its abundant wildlife, was the jewel in the crown. A major breeding site for several species. The property owner immediately cut down all of the trees in that paddock, so that the government wouldn't try to acquire it if they decided to expand the national park.

Redneck developers weren't the only threats facing the Macquarie Marshes. Feral animals and weeds were common. The swamps teemed with wild pigs:

classical "wild boars", jet-black, with large heads and formidable tusks. And they were thriving—many sows were followed by a conga-line of piglets as they trotted across the open fields. The adult pigs were too fast and formidable for us to tackle, but the piglets were slow enough for us to run down. Lacking refrigeration, fresh meat was hard to come by—and the piglets solved that problem for us. We had pork for dinner every night.

Our catching-and-painting study was a sidelight to the main project, which was to radio-track snakes so I could tease apart their social lives. The basic methods of a radio-tracking study are straightforward. Catch your animal, fit a transmitter to it, and release it. Then at regular intervals, go out with your receiver tuned to the animal's unique transmitter frequency. Every time you locate it, you can see what it's doing (and who it's doing it with), how far it's traveled, what habitats it's using, and so forth. But how do you attach that transmitter? If you're studying Lions or Foxes, you fasten the unit to a collar around the animal's neck. But a collar around a snake's neck wouldn't work. If you're tracking Eagles or Herons, you can glue your transmitter to the animal's back. But you can't attach a transmitter to the outside of a snake; that bulge will get stuck as soon as the snake crawls into dense scrub, or into a narrow crevice. As a result, your best option is to put the transmitter *inside* the snake.

How can you do that? The usual method is to implant the unit surgically. Our transmitters were small (the size and shape of a bullet), so we could anesthetize the snake, make an incision in the belly, pop the sterilized transmitter inside, sew up the wound, and let the animal wake up. That's an excellent protocol for a long-term study, but not for short-term work. A snake that has been disturbed like this will hide for a few weeks, and perhaps shed its skin (to aid the healing process) before re-embarking on its usual activities. By the time the snake had recovered, I would be back in Sydney teaching next semester's students. To study snake behavior over the short term, I needed to get transmitters inside undisturbed animals.

I had no idea how to overcome that challenge, but my research assistant Peter Harlow devised a fiendishly clever solution: insert a transmitter inside a freshly killed mouse, and dangle that mouse in front of a hungry snake. Blacksnakes are gluttons; like Labradors and Cocker Spaniels, they never stop thinking about food. So, Pete's idea just might work. With luck, the transmitter would lodge in the animal's stomach for long enough for us to get our data.

It worked. The first snake we found in that trip was a large male Blacksnake, cruising through the grass looking for a mate or a morning snack. He was flicking out his tongue as he traveled, a sure sign that he was looking for either sex or food. We hastily dispatched a mouse, removed its innards, put the transmitter inside, and sewed the rodent back up. Then I tied fishing line to the mouse's tail, with the other end of the line attached to a 3-meter (9-foot) fishing rod. I dangled the mouse in front of the snake, and the big Blacksnake didn't hesitate. It had never seen a white mouse before, but that package in front of it smelled like food. That was all that the snake cared about. It threw itself onto the "lure" with the enthusiasm of a politician who locates a kissable baby during an election rally, and it began gobbling down the electronically enhanced rodent.

FIGURE 3.9 Dangling a dead white mouse in front of a Red-Bellied Blacksnake usually resulted in an immediate strike by the snake. Photograph by Peter Harlow.

At this point, I realized that I had a problem. I had tied the mouse's tail to the fishing line, so the snake was going to end up firmly connected to the rod. By this time the mouse was almost gone, with just a leg poking out from the Blacksnake's mouth. I seized a pair of small scissors and crawled on my belly toward the snake, fearing that it would either take off in a hurry (in which case it would discover that its delicious meal was connected to an immovable object) or launch an attack (in which case I might have trouble evading it). Fortunately, neither of those things happened. The snake concentrated on swallowing its lunch, and I was able to cut the fishing line a few millimeters from the mouse's tail. After that, I attached the mouse loosely to the fishing line, so that it could be separated by a quick jerk of the rod once the snake had the rodent in its jaws.

As I've explained earlier, the mating-behavior study was a failure. By the time I had everything in place, the drought was in full swing and the snakes were dying. But we discovered a great deal about Blacksnake biology, so it wasn't a waste of time. Transmitters are a real eye-opener. Even if you know a species intimately, the ability to relocate animals every day provides novel insights into their private lives. You can find a telemetered snake wherever it goes, even if it is well-hidden. The telemetry signal often led me to places that I had never imagined a snake would use. Because I had always found Blacksnakes close to water, I thought that they were streamside specialists. But the transmitters told me that these snakes often wander off into the bush—sometimes for kilometers. The animals are harder to see in dense scrub, but they're out there.

Another lesson from telemetry is a snake's ability to fit itself into a tiny hiding place. Often, I zeroed in on a transmitter signal coming from an open field, with nowhere for a large snake to hide. Has the snake been killed and eaten by a predator, leaving just a transmitter lying on the ground? But no, close inspection shows a tiny patch of scales—the snake is coiled around the base of a small plant, hidden by the leaves. Or a tiny burrow—surely too small for the snake to fit into? But no, that's where the signal is coming from. And tomorrow morning, the snake crawls out from that hole to bask in the sunshine.

The third lesson from telemetry was that snakes have individual personalities, even within a single population. Some are bold, some are shy. Some are homebodies, some are travelers. Some start the day early, and some sleep in. Some prefer the open plains, some like the swamps, and some seek out tangled heaps of fallen timber. Even two animals of the same sex and size, studied at the same time in the same place using the same methods, can be as different as chalk and cheese. Some of that variation is due to genetics, and some to an animal's experiences early in life—but whatever its cause, this variation in personality adds a lot of interest to studying snakes. And a few challenges, such as when you encounter a psychopath in a species that is usually well-mannered.

The behavior of radio-tracked snakes also depends on weather. Blacksnakes are inactive during winter, sleeping in shallow burrows or under logs and tree-roots. In spring, the males emerge first, bask outside their winter homes, and their hormones go into overdrive to produce sperm in readiness for the mating season. By late spring, the males are fighting. Near a receptive female, rival males intertwine their bodies and rear their heads up as high as they can reach, each trying to force his opponent's head to the ground. The battle is a test of physical strength, and biting is rare. Bouts can go on for hours if the two combatants are equally sized, leaving the edges of their scales frayed and bleeding. Like two testosterone-soaked farmhands in the country pub on Friday night, fighting male snakes ignore everything else around them. As a result, it's easy to get a front-row seat at these epic battles. Sometimes, the snakes even continue to fight after you haul them back out of the scrub onto open ground to take a photograph.

People who witness these spectacular combat bouts often misinterpret the behavior as courtship—but foreplay is different, and far more subtle. It typically involves a male crawling beside a female, tongue-flicking her back and trying to position himself for mating. But the activity can be much more vigorous if a female Blacksnake is not in a romantic mood. To escape her suitor, she takes off at full speed and he hurtles after her, trying to pin her against a log so that he can press his attentions. Mating itself is a more relaxed affair. As soon as the male manages to insert one of his twin penises into the female's vent, he concentrates on transferring sperm. If the female crawls away, the male is dragged along backward by his intimate organ. It looks uncomfortable.

But the sex and violence are short-lived. By early summer, mating is over. The female Blacksnakes are pregnant, and the males have reverted to a quiet bachelor lifestyle. Food re-emerges as their main focus. As long as the weather is cool and frogs are available, both sexes emerge from their overnight retreats to bask for an hour or so in the morning before wandering off to look for tasty amphibians. Tongue-flicking every crevice they encounter, they search all day. But as midsummer approaches, foraging becomes less intense. Hotter drier weather drives frogs underground, so they are harder to find. And a jet-black snake warms up quickly in the sunshine, so overheating becomes a real danger. Once while I was radio-tracking a Blacksnake in the Marshes on a very hot day, the snake suddenly raced forward to lie in the thin shadow created by a slender upright branch. I almost walked on the animal before I saw it—my foot was right beside its head—but the snake didn't move an inch. If it left that patch of shade, it would cook in the sun-drenched open plains around us.

And as summer progresses, Blacksnakes become indolent. They don't need to bask; it will be warm enough by mid-morning for them to be active if they want to. Basking is a dangerous business—especially when you're cold and slow, and thus are vulnerable to predators—and so they skip that risk if they can. The other reason why snakes stop moving around is that females are increasingly distended by their developing offspring, so they stop feeding. And at this point, Blacksnakes do something surprising. Snakes are usually solitary creatures, but pregnant female Blacksnakes come together in groups to share a single overnight burrow and spend the days sprawled around the entrance to that retreat.

Up to six females can be found in a single crèche. Their grouping behavior is motivated by temperature. A tangled mass of snakes retains heat, so the group cools down at night more slowly than a single female. Similarly, a large pile of snakes heats up more quickly in the morning. A group also provides more eyes to detect any incoming predators, and more fangs to discourage them. The usual disadvantage of gathering together—competition for food—doesn't apply, because the pregnant females have stopped feeding. And so female Blacksnakes from a wide area hang around together until the offspring are born.

The warm summers of the Macquarie Marshes reduce the thermal advantages of grouping, so some females go through pregnancy without looking for company. To find a really big pile of pregnant Blacksnakes in their maternity ward, you need to look in colder areas. In the Canberra region, the New England tablelands, and the New South Wales south coast, encountering a large crèche can be a real shock for a bushwalker. The snakes are heavy-bodied because of their pregnancy, so they look huge; and it's disconcerting to encounter a pile of venomous snakes rather than the usual one-at-a-time. One dramatic example happened when I was radio-tracking a female Blacksnake at a site south of Sydney. The telemetry signal came from a Bandicoot's daytime nest, or dray—a football-sized lump of dried grass and twigs. I gave the dray a casual poke to encourage the female to exit, and she did so immediately—together with four friends. Suddenly, there were snakes thrashing about all around my feet.

****

Although people often assume that snakes love hot weather, it's not true. Overheating is a major risk for them. As a result, it's a lot less interesting to radio-track snakes in midsummer than in spring. In summertime, my radio-tracked snakes stayed deep within their shelters while I spent the day in the nearest patch of shade, wishing that my animals would do something. Anything. Or that I'd remembered to bring a cold drink with me, and insect repellant. Fortunately, other wildlife provides entertainment. During one summer trip to the Marshes, a huge mob of Grey Kangaroos—hundreds of them—filed past me for almost an hour as I sat quietly under a large River Redgum. The 'roos didn't see me, and it was marvelous to be in the midst of so many of these iconic marsupials.

Smaller creatures provided entertainment while the snakes were sleeping. The dominant vegetation around low-lying swampy sites in the Marshes is a reed called "cumbungi" (or bullrush, or cat's tail). The snakes often sheltered deep within those reedbeds where it was damp, cool, and safe. Checking out those reedbeds while our radio-tracked snakes were snoozing, I discovered small green frogs sitting on these reeds, fully exposed to the sun's rays. It was uncomfortably hot for me even in the shade—how could a frog survive without boiling? The frogs were reluctant to move, so we could measure their body temperatures by pushing a thermometer gently against their thighs without causing them to flee. There wasn't much else for us to do—just record telemetry signals from our sleeping snakes every hour—so we had plenty of spare time to gather data on frogs. And sure enough, these young Green Tree Frogs had body temperatures around 36°C. They were warmer than our Blacksnakes. I had assumed that frogs would be killed by direct sunlight on a hot day, but I was wrong. A waterproof layer on the frog's skin prevents it from drying out.

Although radio-tracking snakes in the Macquarie Marshes was great fun, the drought continued. The study I had planned was impossible. So, reluctant to surrender, I shifted the study to a site near Goulburn where the drought had been less severe. Bubalahlah Creek is as lovely as its name suggests—a clear stream running through a grazing property, with a conveniently located cottage for accommodation. Magnificent forest, with Koalas and Crimson Rosellas in the trees. But again, the project foundered. The local snakes were too small for my transmitters. The next year I tried the Jamison Valley, in the Blue Mountains, but again the snakes weren't large enough. Where could I find giant snakes, like those in the Marshes? A friend told me about Coomonderry Swamp, near Shoalhaven Heads on the coast of Sydney, where the snakes were huge. Only a few kilometers from the ocean, the swamp is the largest freshwater coastal wetland in southern New South Wales. There are few tracks within the area, so as soon as you plunge into the dense thickets around the lagoon, you are isolated from the outside world.

I was dubious about the stories of huge Blacksnakes, because the vegetation seemed too dense to support these sun-loving reptiles. But there were small open patches among the thickets, and a silent approach through the bushes and reeds would often reveal a giant snake basking in the sunlight. Perfect, I thought, and gave it one more attempt. The thick swamp vegetation made it difficult to observe snakes but I felt that finally, my luck had changed. In fact, what changed was the weather. In 1984, the drought finally broke. Torrential rain encouraged an explosion of plant growth, and Coomonderry became a soggy thicket. I could barely see a giant Blacksnake even if I was standing on it. And that was the end of my ill-fated attempt to conduct a study on the love lives of Red-Bellied Blacksnakes.

Those field studies on Blacksnake sex taught me a lot. I never obtained the data that I wanted, but I learned how to organize a field project in remote areas. And importantly, my failure demonstrated the catastrophic impact of weather variation on the lives of snakes. The pioneering ecologists Andrewartha and Birch had been correct, all those years ago. In Australian environments, unpredictable variation in rainfall drives massive changes in animal ecology. To really understand my snakes, it was futile for me to focus on measuring "average" values for traits like mating systems and life history. The data I obtained would depend on local conditions; if I repeated the same study two years in a row (as I had in the Marshes), I would get different results.

That wake-up call changed my career. It "Australianized" my approach to research. The harsh ups and downs of the Aussie bush don't render short-term studies useless, because knowing something is better than knowing nothing. But I now understood that I couldn't ignore variation. My results would depend not just on what species I was working on, but where and when I studied it. To understand these dynamic systems, I would need to study them over long periods, or in several sites. A short fieldtrip to a single place could provide a nice dataset, but I needed more. I had to find new ways to study snakes—ways that would overcome the problems of huge variation in time and space.

One enduring memory from the Macquarie Marshes came late in that fateful trip in 1981, when the drought hit hard and the Blacksnakes stopped breeding. My wife Terri was experiencing outback Australia for the first time. Sleeping in a small tent under giant River Redgums and spending the day uncomfortably close to large venomous snakes. A challenge for a young woman from urban America, but she loved it. After Terri returned to Sydney to work, we maintained contact via phone calls. Mobile (cell) phones were yet to be invented, so the nearest telephone was a public phone booth in the town of Quambone, an hour's drive away. A week after Terri left, I drove into town in the evening, put a few coins into the public callbox, and put in the call. Terri was there, and she had exciting news. She was pregnant! In honor of the magnificent backdrop for his conception, we named our first son James Macquarie Shine. Although the Blacksnakes weren't producing offspring, the Shine family was about to.

4 Long-Dead Snakes at the Museum

I became a biologist because snakes strike a deep chord in my psyche. They are so *elegant and attractive. The sight of a Red-Bellied Blacksnake shimmering in the morning sunshine beside a mountain stream makes me stop and stare, and gaze in reverence. Put that same snake into a cage, and the magic is lost. It's the combination of snakes and their habitats that excites me. At the core of my passion is a mixture of intellectual curiosity and aesthetics. How do snakes function within their natural context? How can I better understand the snake as a complex, stylish player within the awesome intricacies of a natural ecosystem?*

So, you might think, I would find looking at a dead *snake as interesting as talking with a sedated accountant, or watching a TV show about the indiscretions of a drunken footballer. And yet, here I am deep in the bowels of the Australian Museum, sitting at a lab bench with bottles of alcohol-preserved snakes in front of me. And methodically, with an efficiency born of long experience, I extract yet another corpse from its bottle, record its tag number, and measure its body length and head size. And then an incision with my scissors, down the snake's belly, and I am looking at the stomach contents and reproductive organs of a snake that was collected more than a century ago, and that has remained coiled in its alcoholic mausoleum ever since. Untouched, until I came along. Those extensive collections of dead snakes are a goldmine for an ecologist. I can learn more about snake biology in an hour at the museum bench than I could in a year in the field.*

And today is a red-letter day. I've been looking at small burrowing snakes from Western Australia, posted to me by the Curator of Reptiles at the Western Australian Museum. I expected to find small lizards inside the stomachs of these boldly banded little serpents, because most small Australian snakes eat nothing else. Skinks, skinks, skinks. But although some Simoselaps *species are indeed skink-eaters, the stomachs of a few species contain only the eggs of lizards, not the lizards themselves. Could these snakes be specialist egg-eaters? Prowling around at night looking for freshly laid eggs, and ignoring the adult lizards that had laid those eggs? That kind of ecological specialization had been reported in snakes from other continents, but never from Australia. But with every snake I slice open, the pattern is stronger.*

If I had to rely on catching these snakes myself, it would take decades to reach any conclusions. My friend Ric How has captured more of these little banded burrowing snakes than anyone else ever has, during a mammoth pitfall-trapping study in bushland remnants near the city of Perth. He caught more than 500 of these little banded snakes in small buckets dug into the sandy soil, but it took him 11 years to do so. For three of the five species, he averaged less than one capture per 1,000 days (three years) that a trap was open! These snakes are either incredibly rare, or

DOI: 10.1201/b22815-4

incredibly hard to catch. Or both. You can't do "normal" ecological research with capture rates like that. But the alcohol-filled jars in the museum contain hundreds of these snakes, collected over many decades. Well before lunchtime on that first day, I am certain that the Australian snake fauna contains three species of specialist egg-eaters.

In the early 1980s, two forces drove me away from fieldwork for my next set of studies. One of those forces was scientific, and the other was personal.

The personal one was the need to balance my research with the other parts of my life. I finally had a permanent job, so the ever-present fear of a terminal career collapse had disappeared. I had the challenge of lecturing to hundreds of students; and most of all, the kaleidoscope of emotions associated with being a husband and a father.

Even before my sons were born, though, my life was changed by another event. Terri and I decided to get a dog. I'd never had one as a child, and now that we planned to live in the same house for a while (for the first time in my life), here was my chance to become a dog-owner.

Terri and I spent an hour in a secondhand bookstore peering through books that provided advice about "which breed of dog is best for you?", comparing the advice offered by those volumes. We finally settled on a breed that sounded ideal. Cute, good with children, small enough for a suburban yard, not yappy. The breed we chose was the West Highland White Terrier—the small white version of the Scottish Terrier. The only statement that puzzled me at the time (but now makes perfect sense) was "Westies are possessed of no small measure of self-importance". A puppy named Cooper entered our lives a few months later, and promptly took over. Thirty years and three generations of West Highland White Terriers later, our lives are still supervised by these confident little dictators. As a friend remarked to me a few years ago, "a home without a dog is just a house".

Cooper was a mellow dog, but we worried as the time approached for our first child to be born. How would a pampered terrier behave toward the new arrival? To our relief, the dog never showed any resentment. One afternoon as Cooper lay sleeping on his back on the living-room floor, our baby son Mac crawled over, seized Cooper by his two most precious assets, and dragged him down the corridor. The dog just yelped and waited to be released. I figured that if Cooper wasn't going to react to that kind of treatment, Mac was safe.

Many years later, my infatuation with Cooper stopped me from being elected as a Fellow of the Australian Academy of Science. It was my own fault. During the Marshes fieldwork, I tried to make the work more enjoyable for my assistants Pete and Geoff by involving them in the entire research process, including writing papers. A chance for them to be authors themselves. I remembered our small dataset on frogs sitting out in the sunlight. A simple demonstration of animals doing something interesting and unexpected: the ideal basis for a first publication. I even analyzed the data and drew up the graphs—but no matter how much encouragement I offered, neither Geoff nor Pete wanted to do any work on the manuscript. I finally wrote the paper myself in a few hours—but then was stumped as to who should be given credit as an author. My assistants had helped gather the data, so it wasn't fair to leave them off completely. I complained that my dog Cooper had contributed as much to that manuscript as the research assistants had—and so on a whim, I made Cooper the leading author. He was later to author a few other papers as well, in line with an old tradition in evolutionary biology. In an age before political correctness, young scientists invented fictitious authors to poke fun at their stodgier colleagues. The great British evolutionary biologist John Maynard Smith published several papers with a door-knob ("Isidore Knobby"), and my French colleague Xavier Bonnet has a well-published camera-bag named Rex Cambag.

Two decades later, that juvenile tomfoolery came back to bite me in the bum. When I was being considered for election to the Australian Academy of Science (a great honor for any scientist) in 2002, a spiteful colleague black-balled me by claiming that I was guilty of scientific fraud. My crime was those papers that had been co-authored by my dogs (and yes, I admit, a subsequent one by my pet chicken Nancy as well). Everything in those papers was factual, but he viewed the canine co-authorship as "fraud". It was a humbling lesson—not least, about the dangers of humor in the workplace. I contacted the editors of the journals involved, admitted my crimes, and asked for their forgiveness. I offered to publish admissions of guilt, if they so desired. All of the editors reacted the same way. They told me I was an idiot, then laughed and recounted stories of similar jokes that had enlivened their own careers. Nobody saw it as a hanging offence. And when my name was put forward again to the Academy the following year, I was elected—and handed an apology by the colleague who had "spat the dummy" about my dog-authored papers the year before. He finally saw the joke, but I restricted the authorship of my papers to human beings ever since.

Cooper was child-friendly but with small children around the house, venomous snakes were a bad idea. I had been keeping two Blacksnakes in a cage on the balcony, but the potential for catastrophe was clear. Regretfully, I took the snakes down to the bushland site where I had caught them, and I let them go. I was surprised at how much I missed those snakes. People who share my passion for snakes will understand the strength of those (admittedly, one-sided) bonds between a snake enthusiast and his/her pet serpents. The feelings are beautifully captured in the autobiography of Japan's most famous snake biologist, Hajime Fukada (born in 1913, died in 2002). Fukada's classic book *Snake Life History in Kyoto* describes his long-term efforts to capture, mark, and recapture snakes in rice-fields close to the city of Kyoto in southern Japan. Urban expansion brought his studies to an end, as his rice-fields were transformed into suburbs. The most poignant line in his book, though, relates to a time near the end of his life, when he was recovering from major surgery. Hajime wrote wistfully that "*my health does not allow me to keep snakes any longer. The life without live snakes is a prosaic one*". I know exactly what he meant.

Mac was born in 1982, and Ben in 1984. With delight, I watched my sons grow from babies into toddlers. Terri and I had a fantastic time with our two little men, but it wasn't easy. We were perpetually exhausted. When I look at scientific projects I conducted in the early 1980s, I can't remember much about gathering the data or writing the papers. I was chronically tired from child-rearing. Terri and I worked together well as parents, under Cooper's supervision, but it wasn't easy, especially for Terri. So increasingly, I modified my research to make it more compatible with family life. No reading scientific papers or working at the computer in the dreaded early-evening hours when the boys had to be fed and bathed. But the big problem was fieldwork. It seemed like every time I took off on an extended field trip, one of the boys would fall ill, or a major appliance (refrigerator, washing machine, etc.) would break down.

Like almost all fieldworkers, I was torn between my commitment to family and to research. Sydney is a huge city, with over five million people, and the traffic is horrendous. All of the good field sites were far from the city, so it was simply not possible for me to set up a field project within an hour or two's travel. Disappearing into the field for weeks at a time wasn't an option. I needed a snake-ecology project that could be done in bouts of an hour or two, so that I could gather data but still be home in time to pop the kids into the bathtub.

And I realized that a perfect opportunity existed, just down the road. I could dissect preserved specimens that were held in museums rather than collecting the snakes myself. I could make new discoveries about Australian snakes, and still be back for dinner. Without needing to kill any snakes, I could obtain data on vast numbers of animals in a short period of time. That time-efficiency was attractive to an ambitious young academic who was determined to publish rather than perish. The biggest museum of them all, the Australian Museum, was 10 minutes by bus from the University of Sydney. So in between giving lectures and supervising practical classes, I could zip down to the museum and examine preserved snakes.

The idea of using museum specimens for ecological data is so obvious that it raises a question: why hadn't other researchers already exploited this opportunity, either in Australia or elsewhere, or with other types of animals? I think it's because the people who choose ecology as a career are the ones who love being out-of-doors. Live animals are more attractive than formalin-soaked corpses, and the bush is a nicer workplace than a fume-filled laboratory. And if your interest lies in mammals and birds, museum collections won't tell you much about the animal's lifestyle. Their corpses don't contain much ecological information because the collector has skinned the specimen and discarded most of the innards, like the stomach and gonads. All that the museum keeps is the animal's outside covering. But frogs, lizards, and snakes are so small that it's easy to retain the entire animal. Preserved in formalin or alcohol, it sits in its bottle in the museum (with a tag, to identify where and when it was caught) until a researcher asks to examine it.

But of course, looking inside a preserved reptile isn't the same as looking inside a freshly killed reptile. Many museum specimens are in far-from-pristine condition. Some snakes were collected by scientists, humanely killed, and professionally preserved; but others were hacked to pieces with an axe, or run over by a truck (or two, or three). A couple of days putrefying on the highway might not destroy the animal's bones and skin, but its insides will look like soup. Fortunately, though, my doctoral studies in Armidale had given me an intimate familiarity with the internal organs of snakes. I knew how to navigate my way around inside a road-killed snake's abdomen, determining its sex and reproductive condition from small bits drifting inside the shattered carcass. Like a serpentine CSI, I could decipher an animal's sex and reproductive state from tiny grey blobs floating around its body cavity.

And there was also a compelling scientific rationale for looking at the museum specimens: a breadth of information. Three years in Armidale had yielded data on half-a-dozen species, but that approach wouldn't let me extend my studies to the other 200 or so snake species found in Australia. And even worse, my studies on Blacksnakes had convinced me that a quick "snapshot" wouldn't be enough. Any given population in any given year might do something atypical. I dreamed of assembling information on the ecology of all of the snakes of Australia—something that had never been achieved for the serpent fauna of any continent.

To achieve my aim, I'd need to examine snakes from a vast range of sites and years—and to get those animals I'd need an army of collectors, spread out across all of Australia, working for decades. And even if that was possible, I couldn't justify killing thousands of snakes just to find out about their ecology. But a volunteer snake-collecting army had already been killing snakes and preserving their bodies for two centuries, and perhaps I would be able to exploit their efforts.

As pioneering pastoralists spread out across the Australian continent in the 1800s, snakebite was an ever-present danger. To minimize that risk, the colonists needed to find out what types of snakes occurred around their properties. That way, they would know whether or not a bite needed to be treated; and could evaluate the risks involved. Motivated by that fear, explorers and pastoralists routinely killed and pickled any snakes that they encountered, and shipped them off to the nearest museum for identification. Over the decades, the collections accumulated. Researchers used them to describe new species and to work out geographic distributions of each type of snake, but nobody had recognized the potential of those collections for ecological information. If I was allowed to examine the preserved animals—to look inside their stomachs and reproductive organs, in particular—it would be the Holy Grail. I could explore samples of every species, collected over 150 years and all over the continent. By averaging out local variations, data from those collections could give an overview of the diets and life histories of Australian snakes.

FIGURE 4.1 Snakes often eat very large prey items, and—conveniently for researchers who want to identify dietary habits—almost always consume their meal in one piece rather than tearing it to pieces. For this White-Lipped Snake, a Metallic Skink was its last supper. Photograph by Mark Hutchinson.

But one major obstacle remained: permission. Natural history museums are conservative establishments, and the curators see themselves as guardians of a priceless legacy. They welcome researchers who want to measure the museum's irreplaceable treasures, or to count their scales. But being torn apart by a brute with a pair of scissors was a very different prospect. How would the curator at the Louvre react if you asked to dissect the Mona Lisa?

I expected the curators to show me to the door as soon as I broached the subject, but it was worth a try. And the obvious place to start was the Australian Museum, just down the road from Sydney University. The man in charge—the reptile curator—was Hal Cogger. I had met Hal during my Ph.D., when he came to Armidale to catalogue specimens there—including the snakes I had collected. They ended up in the Australian Museum, and Hal was grateful. So, I decided to ask Hal if I could look inside his preserved snakes.

On the plus side, Hal Cogger is a kind, approachable, modest person. He is unfailingly polite, so I expected that even if he refused my request, he wouldn't yell at me. On the minus side, Hal is a daunting figure: a giant in the field. He has written several influential books, including one—*Reptiles and Amphibians of Australia*—that is revered by herpetologists (reptile and amphibian enthusiasts) all over the world. Widely called "the bible", Hal's book was the first to pull together information on every species of frog, lizard, snake, turtle, and crocodile in Australia. This was no easy task. Scientific papers about these animals were contradictory and had critical gaps in information. Hal had to decide which animals deserved to be called separate species; what names to give them; where they occurred; and how to identify them. To his eternal embarrassment, successive generations of Aussie reptile biologists idolize Hal. As researchers discovered more and more species of Australian frogs and reptiles over the course of Hal's career, the number of species in his bible almost doubled (from 664 to over 1200)—but throughout it all, the book has remained the solid-gold standard.

I had a long speech prepared to justify the proposal, but it wasn't needed. Hal's response to my nervous query was immediate and positive. As soon as I explained the basics, Hal said "yes, of course, great idea". Fearing rejection, I had carefully chosen a specific group of snakes to examine: the Golden-Crowned Snakes. These graceful little snakes are mildly venomous but reluctant to bite. They are common in the northern suburbs of Sydney, often drowned in swimming pools or brought into the living room by the pet cat. So, the Australian Museum had many specimens (hopefully, encouraging Hal to view them as less-than-priceless). And there was an interesting question to ask about them as well. Scientific publications (including by Hal himself) claimed that these snakes reproduce by live-bearing, but I was dubious. All of the species that are closely related to the Golden-Crowns are egg-layers, and the stories about live-bearing Golden-Crowns were based on vague reports. Looking inside preserved animals in the museum could resolve the issue. Did the oviducts of adult female snakes contain hard-shelled eggs, or developing embryos without an eggshell? A few days later I came into the museum to examine the snakes. I felt like a barbarian in the cathedral—dissecting the specimens felt like sacrilege—but the other reptile biologists at the museum fully supported Hal's decision. And by the time I left to catch the bus home that afternoon, the issue was resolved—Golden-Crowned

Snakes are egg-layers. More importantly, a whole new research vista had opened up. The curators of the largest reptile collection in Australia encouraged me to continue my work. I had the keys to Fort Knox. A way to conduct research on snakes, and still spend time with my family.

What characteristics should I score when I was looking inside a preserved snake? Reproductive mode (egg-laying versus live-bearing) was just the start. Other obvious opportunities were to record the body sizes of males and females, to see which sex grew larger; to determine the number of offspring inside reproductive females (by counting the eggs, babies or large follicles in the ovary); to identify the body sizes at which males and females matured; to compare the relative sizes and shapes of heads and tails between the sexes; to enumerate the types and sizes of prey eaten by snakes of each size and sex; to clarify the seasonal timing of reproductive cycles; and so forth. I could extrapolate my Armidale studies to an entire continent. There were lots of things that I *couldn't* learn from a dead snake, but it was one hell of a start.

Within two years I had dissected most of the snakes in the Australian Museum collection. Would the curators of other museums allow me to expand my studies to their collections also? Yes they would, because Hal Cogger was prepared to vouch for me. His support not only got me into the Queensland Museum, but even encouraged the curator Jeanette Covacevich to invite Terri and I, and our son Mac, to stay at her apartment while I was examining specimens. And with recommendations from Jeanette as well as Hal, the curators at other museums agreed to my plans also. I had

FIGURE 4.2 Golden-Crowned Snakes were the first species that I examined for ecological data in the Australian Museum. Photograph by Stephen Mahony.

access to tens of thousands of snakes, of every Australian species. I could assemble comprehensive ecological information on the snake fauna of Australia.

There were hiccups, of course. The curator of the Northern Territory Museum never replied to my letter asking for permission to dissect King Brown Snakes. Instead, Graeme Gow decided that he'd do it himself. No need to let that damn southerner into *his* collection. But Graeme discovered that it's difficult to tell the difference between the ovaries and the kidneys in a snake that's been massaged by several trucks and has decomposed under the tropical sun before being preserved. He abandoned the attempt, and soon afterward he left the museum. By the time I was ready to come to Darwin and look inside snakes, I had unfettered access there as well.

Museums attract unusual people. In the golden days of wildlife taxonomy, every field trip to a remote area yielded at least one or two species that were new to science. These days the obviously "new" species are harder to find; it takes detailed analysis (usually of genetic data) to discover that "species X" actually consists of two types of animals that haven't interbred with each other for a million years. In earlier days, it was common to find animals that, even at first glance, were obviously undescribed. I once found specimens of two "new to science" frog species, the Wallum Sedge Frog and the Cooloola Tree Frog, in a 5-minute period, on Fraser Island, in 1971. The first frog was sitting on a path at night as I walked toward the outdoor toilet, and the second was sitting on the toilet seat. The payoff to surveying uncharted territory encouraged museum collectors to embark on bold adventures. Generations of curators had remarkable careers, rivaling the Indiana Jones stereotype—months of dangerous fieldtrips to unexplored areas, followed by years of painstaking analysis to examine the specimens from each place. Spending every day alone, surrounded by specimens in jars of alcohol, was too much of a temptation for some of the early workers. One renowned reptile curator at the Australian Museum in the 1880s, Douglas Ogilby, was sacked when he drank the alcohol from the collection jars. As a result, the specimens were ruined. But Ogilby soon got an equivalent job at the Queensland Museum, a safer workplace for him because the Queenslanders preserved their specimens in formalin not alcohol.

During my working lifetime, museums have changed beyond recognition. Nowadays, researchers examine DNA sequences not scale counts. Budgets have been slashed, and staff numbers have fallen. Few young scientists are attracted to the "old-fashioned" science of identifying and classifying organisms. The experts at identifying particular groups of animals and plants have retired, and nobody is replacing them. The general public sees the museum's role as education and entertainment, not research.

But during the 1980s, things hadn't yet fallen into such dismal disrepair. Most institutions had full-time reptile curators. They often talked with each other, and so the message spread—Rick is careful, and minimizes damage to the specimens he dissects. So, in the end, I was given permission to look inside thousands of snakes. For some species, I've looked inside almost every specimen known to science.

There was plenty of excitement, and lots of tedium. I developed an allergy to formalin, from too much exposure to fumes. For example, in those olden times, the Northern Territory Museum kept their largest specimens—crocodiles, pythons, and so forth—in tubs of concentrated formalin, inside a small metal shed that was exposed to full sun, with no air-conditioning. The air inside was around 50°C (120°F), with as much formalin as oxygen. To access the animals, I had to take a deep breath while standing outside the door, run into the shed, haul a specimen or two out of the tubs with my snake-tongs, then sprint back into the open air. A single breath inside that shed would have been fatal. Preservative killed the skin cells on my fingertips, eliminating my sense of touch for weeks. My sense of smell was gone as well, and every meal tasted like preserved snake. But in return, extraordinary amounts of information poured in.

Among that vast number of animals I examined, a few stand out in my memory. For one species of very rare snake, a single collector in the 1800s had found, caught, and preserved most of the known specimens. Since then, the species has declined to near-extinction because of agricultural development and feral pests. But because of that individual collector's efforts, I could paint a picture of the life of those animals—so that over a century later, we can develop evidence-based plans to conserve them. I felt a real affinity with that collector. He died long before I was born, but our "collaboration" contributed to conserving an endangered species. Occasionally, though, there was a darker dimension. For example, one day after I had dissected a juvenile Brownsnake—a perfectly unexceptional animal—I noticed a tag attached to the animal's body, stating that the snake's bite had killed a young woman. It was chilling to reflect on her death, and on the hundreds of Brownsnakes I had captured, and the bites I had so narrowly avoided.

Other findings were simply weird. For example, one Brown Tree Snake had a ceramic egg in its stomach. In the early 1900s when suburban chicken-yards were in vogue, such eggs were used to discourage hens from eating their own eggs, and to keep hens brooding—the hen was deceived into thinking it still had eggs to incubate. Clearly, the snake was fooled as well, and came to an untimely end when the chicken's owner went to check why the backyard poultry were making such a fuss. And that wasn't the only mistake I encountered. One Curl Snake had a huge burr (a "three-cornered jack") in its stomach, eaten accidentally with a prey item. The stomach was badly lacerated. If the snake hadn't been collected, it would have died of starvation or septicemia.

Perhaps the strangest thing I ever found inside a snake's stomach was a small glass vial inside a Snouted Cobra in a museum in Pretoria, South Africa. I had been working away for three weeks solid with collaborator Bill Branch, research assistant Peter Harlow, and student Jonno Webb. It was tiring, repetitive work, enlivened by humorous chitchat as we sat side by side at the dissecting tables. One morning, the jokes revolved around the lack of hair on my head. And when we came back to work after a quick lunch break, I found a glass vial inside the stomach of the next Cobra that I began to examine. Inside the vial was a newspaper clipping, with an advertisement for treating baldness. Pete had slipped it inside the animal as we left the lab 30 minutes before.

The museum work went smoothly—and unlike a field study, could be done all year regardless of weather. Even after I initiated other field projects, I stopped by the Australian Museum to dissect snakes posted to them by other institutions. As I traveled around the country for conferences and fieldwork, I added a few days onto every trip so I could sit at the local museum from dawn to dusk, measuring, cutting, and counting. I stopped trimming my fingernails a few weeks before each trip, so I could

FIGURE 4.3 Brown Tree Snakes are common in the suburbs of Sydney, and often are found inside houses. Because the corpses of many of those trespassers are sent to museums for identification, and later preserved, I had access to large sample sizes when I dissected these animals for ecological data. Photograph by Matt Greenlees.

pry apart preserved tissues without needing forceps. Well-preserved small-bodied snakes were easy—I could do scores in a day—but it was a painstaking process with older, brittle specimens, and a frustrating one with the contorted corpses of big species. Wherever I could, I supplemented the museum data with information on field-collected animals. I kept pregnant females so that I could incubate their eggs and measure incubation periods, offspring sizes, and so forth. Friends with captive snakes helped as well, recording and donating information from their pets. But the museum data were the centerpiece. Within ten years, I had information on the ecological characteristics of every species of snake that occurs in Australia.

****

By the late 1980s, I was able to piece together a comprehensive picture of snake ecology across an entire continent. At the time, we didn't have this even for well-studied regions like North America or Europe. What did I learn from poring over 22,000 preserved snakes? First, the information on each snake species was useful in its own right. For most of these species, there were no published data on any ecological topics until my work. Or if there *were* published reports, they were wrong. The writers of "popular" books had guessed at what was going on, and later books copied the earlier ones, conferring respectability on the "pseudo-fact". Some statements, although repeated in many books, were based on a single early guess. One of my favorite examples concerns an unlikely dietary record. The founder of the Australian Reptile Park, Eric Worrell, wrote in his 1963 book *Reptiles of Australia* that Water Pythons eat Freshwater Crocodiles. Now, that's pretty damn unlikely. I have studied Water Pythons for decades, and my collaborators and I have captured thousands of them, and we've never found a croc-shaped bulge inside a python. So where did the story come from? Worrell's biographers Kevin Markwell and Nancy Cushing tracked it down. Eric once sent a box by post from Darwin to New South Wales, and inside the box were a live Water Python and a hatchling crocodile. When the box was opened a few weeks later, the crocodile was missing, either eaten or stolen. That single event, involving a snake jammed into a box with a crocodile, was the basis for Worrell's remark; but even today, books about Australian snakes say that Water Pythons eat crocodiles.

The most widespread error in the earlier books was the claim that insects are important prey for small Australian snakes. That's a reasonable guess, but it's wrong. The only specialist insect-eaters in Australia are the wormlike Blindsnakes, who consume the eggs and larvae of ants and termites. The elapids (venomous snakes), colubrids (harmless snakes), and pythons leave insects alone, except for occasional lapses when a hungry snake encounters a large grasshopper. In other parts of the world, insects are important prey even for large snakes. I had expected that Australian snakes would be the same, and I was resigned to having to identify half-digested beetles or spiders inside snake stomachs. But it never happened; fortunately for me, the small Aussie snakes eat mostly lizards.

FIGURE 4.4 Although they look like insect-eating burrowing snakes from the American deserts, Australian burrowers like this Bandy-Bandy eat reptiles rather than insects. Photograph by Matt Greenlees.

The lack of insect-eating and spider-eating snakes in Australia raises some broad evolutionary issues. One of them is about "ecologically equivalent species"—the idea that even if two continents are colonized by different types of animals, evolution will produce similar results. That is, we will end up with near-identical species in each place, despite their separate evolutionary origins. Textbooks of evolution often feature paired photographs of animals that have evolved to look alike. Thylacines (marsupial "Tasmanian Tigers") and wolves are a classic example. The simple body plan of snakes makes convergence even more likely. For example, there are remarkable similarities between some of the boas in Brazil, pythons in New Guinea, and vipers in Sri Lanka. All are bright green and lie in tight coils on tree branches as they await their next meal. In every continent, you can also find short fat brown snakes with large triangular heads that are beautifully camouflaged in their terrestrial ambush sites—but the species involved is an elapid in Australia (the Death Adder), a boa in Papua New Guinea, a viper in Japan, and a pit-viper in North America. And it's not just ambush foragers—the slender fast-moving diurnal "whipsnake" that zips past your feet is a colubrid in South America and an elapid in Australia.

Those resemblances are impressive—but if we look in enough detail, there are always some differences as well. For example, although some of the small nocturnal snakes of Australia look incredibly like their North American "counterparts", the Aussies eat lizards and the Americans eat insects. Convergence has limits. The trajectory of evolution is strongly affected by historical accidents, creating different solutions to the same ecological opportunity. That complex mix of similarities and differences gives natural history much of its fascination. Whenever I travel to a new country and see the wildlife, it is like meeting the extended family of my friends back home. I am struck by both the similarities and the differences.

The lack of insect-eating elapids raises another thorny issue as well: the idea of "empty niches". Have the Australian snakes simply failed to "discover" this niche over the last few million years? If so, an invasive species with that kind of diet might flourish in Australia—without any competition from the native snakes. Or is there some reason why the "insect-eating snake" niche doesn't work in Australia? Ecologists disagree about the answers to those questions, but I believe in the reality of empty niches. Many invading species thrive in their new homes by using resources that the locals ignore. For example, invasive Cane Toads and House Geckos live in the Australian suburbs, where few native species occur. Suburban pests thrive when introduced to new homes in other parts of the world, because one city is very like another.

As you'd expect, my museum dissections revealed errors about snake reproduction as well as feeding. For example, the older books were wrong about which species laid eggs and which ones gave birth to live young. That confusion was due partly to mistakes in species identification. Most people can't tell the difference between various species of snakes. For example, most purported "Blacksnakes" aren't the "real" Red-Bellied Blacksnake; they are dark-colored individuals of other species. If you're scared of snakes, you don't approach closely enough to notice if that scaly intruder in your backyard has a loreal scale next to its nostril, or divided versus single scales beneath its tail. All you get is a quick glimpse of a long dark body.

The most misunderstood snake species were the small ones. The authors of those early books didn't care much about small drab snakes; they spent their time talking about Taipans instead. More exciting material. But the scarcity of research on small drab snakes was an opportunity as well as a frustration. One of my first research students was Jonno Webb, a good-natured reptile enthusiast who did his undergraduate degree at Sydney University. Jonno grew up on the New South Wales coast, surfing, fishing, and snake-catching. In 1990, Jonno approached me about doing an Honors project on snake ecology. Most serpent-obsessed students are quivering with desire to work with Taipans, or Tigersnakes, or anything else large and deadly. My response is to shake my head sadly. If they really want to commit suicide via snakebite, I'd rather they enrol in a project on deadly snakes with someone else as their supervisor. As a young man, I had the same obsession with deadly snakes—as evidenced by my doctoral studies in Armidale. The species I studied were among the most dangerous snakes in the world. But working day after day with deadly snakes had taught me a valuable lesson: for a scientific research project, venom is a disadvantage not a selling point. Everything I did with venomous species—catching them, keeping them, measuring them—took longer than it would with a harmless species. To minimize that snake's opportunity to kill me, I had to work slowly and carefully. That resulted in less data per unit effort. And if things go wrong with a venomous snake, you are looking at some uncomfortable time at the hospital. Or worse.

So, I was delighted when Jonno told me that he wanted to study a harmless species. Like all Blindsnakes (family Typhlopidae), the ones around Sydney are strange little serpents. Pink or gray in color, they look like earthworms. No obvious head or tail, no visible eyes (just small dark spots beneath scales), and their long cylindrical body is covered by tiny thick scales. If you look closely, though, you can find a mouth well behind and below the tip of the snout, and a very snake-like forked tongue flicking in and out of that mouth. Blindsnakes spend their lives underground, and to this day they remain an almost complete mystery to science. Our museum dissections told us that the Blindsnakes around Sydney (like most Blindsnakes worldwide, we later discovered) feed on the eggs and larvae of ants, including the ferocious "Bulldog Ant". The only place to find ant eggs and larvae is deep inside an ant nest. How could a sightless snake find an ant nest? And having found that nest, why wasn't the snake immediately torn to pieces by fearsome worker-ants? To resolve those questions, we needed live animals not preserved specimens. We were lucky, because it was an unusually wet year. Blindsnakes are easiest to find after heavy rain, when the flooded soil drives them up out of burrows to shelter under rocks. Splashing around in gumboots, Jonno and I soon caught all the Blindsnakes we needed.

The first question we asked was: how do Blindsnakes find ant nests? Lacking functional eyes, the snakes surely must rely on smell. Ants lay down scent trails as they travel away from the nest, to help their nest-mates find food sources. Can Blindsnakes recognize and follow those invisible scent trails? Jonno laid out sheets of paper in the lab, and allowed ants to wander across the paper to deposit scent trails. Then at night, he released Blindsnakes and recorded their movements. Sure enough, they immediately located the ant trail (even if it was a week old) and followed it closely. Jonno's snakes ignored trails left by earthworms, termites, or other small invertebrates. They were after ants.

FIGURE 4.5 Looking more like earthworms than snakes, species of the family Typhlopidae spend most of their lives underground. Photograph by Sylvain Dubey.

The next question that Jonno investigated was: how can a small snake enter an ant nest without being ripped apart? Soldier ants are formidable guardians, with painful stings and aggressive attitudes. How could a Blindsnake survive against them? When Jonno put Blindsnakes into containers with ants, the answer was immediately obvious. The snake's small, overlapping, smooth scales were impregnable. Ants couldn't get a firm grip with their pincers on the highly polished scales of the Blindsnake, nor insert their stingers between those tightly connected scales. The only place an ant could get a firm grip would be the snake's eyes, and that's why evolution has reduced the Blindsnakes' eyes to small dots. Inside a Blindsnake egg, the embryo has large eyes—as in any other snake. But by the time the egg hatches, those eyes have almost disappeared. For a Blindsnake, eyes are a vulnerability.

Decades later, Jonno's Honors work remains the only detailed analysis of Blindsnake behavior—testifying to everyone else's obsession with large scary species. But it provides a terrific example of combining museum dissections with behavioral work, and it shone a bright light on a secretive little animal that is virtually impossible to study in the field.

****

By the beginning of the 1990s, my scientific career was taking off. A stream of papers about Australian snakes had built my professional reputation. When I traveled back to the United States to attend a conference, at least some of the delegates recognized my name. And because email had abolished Australia's academic isolation,

I could exchange ideas and information with colleagues in Cambridge and Berkeley as easily as with my Aussie friends. But unlike my international colleagues, I had the snake fauna of an entire continent to play with, all by myself. Almost too good to be true. On a whim, I added a sign-off line to my email messages: "So many snakes, so little time".

The huge database from my museum dissections provided a rich vein of insights into snake biology, ideal for contemplating during free time between meetings and lectures. Because I was interested in the evolution of viviparity, for example, I paid particular attention to which species were live-bearers and which were egg-layers. And a fascinating story emerged, embedded within the history of snake colonization of the island continent. The ancestors of Australia's elapid (venomous) snakes swam across from Asia more than ten million years ago. Slowly moving northward, the Australian landmass was about to bump into the Indonesian region. The first scaly colonists that crossed the ocean from Asia to Australia were similar to modern-day Sea Kraits. It's easy to imagine a few Sea Kraits being blown off course and ending up on Australian shores.

The Australian continent was a supermarket full of edible lizards, and those ancestral snakes didn't pass up the dietary opportunity. Sea Kraits are elapids, and so their descendants—the snakes that spread out to colonize the Wide Brown Land—also are elapids. That accident of history gives Australia a singular claim to fame: it is the only large landmass in which most of the snakes are venomous.

The early elapids were egg-layers, like their amphibious ancestors, but eight million years ago there was a substantial cooling event—an Ice Age. The colder conditions made it difficult for snakes to find warm places to lay their eggs, especially in the southern mountains. So, female snakes from those areas evolved a new way of reproducing: instead of laying their developing eggs in a nest, they retained them inside the uterus (where they could keep them warm by basking in the sunlight). Viviparity (live-bearing) evolved twice. One of those origins gave rise to a diverse range of species, including Tigersnakes, Copperheads, and Death Adders. The other origin had more modest consequences. It occurred within the genus *Pseudechis* (the Blacksnakes), resulting in a single viviparous species—the Red-Bellied Blacksnake—in cooler regions, and a few species of egg-layers (like King Brown Snakes and Spotted Blacksnakes) in warmer regions.

And the females weren't the only ones to try out novel tricks in reproductive biology. The males experimented with lifestyle diversity as well. As I mentioned earlier, one of the first scientific papers that I wrote had been about differences in average body size between males and females within the same species. Based on the idea that bigger size only pays off for males if they have to defeat their rivals in combat, I had predicted that male snakes will grow larger than females only in species where the males fight with each other. That had worked surprisingly well in my survey of already-published information about snakes from various parts of the world, but I hadn't been able to include Australian species in that analysis (except for the ones I had studied myself during my Ph.D.). Would I find the same pattern in the Aussie snakes that I was measuring in museums?

To answer that question, I needed information not only about body sizes (which I could get from my measurements of long-dead snakes), but also about which species exhibited male–male combat and which did not. The dead snakes couldn't help me in that quest, so instead I asked naturalists about their observations in the field, and pet-keepers about the behavior of their captive snakes. I accumulated stories and photographs about snake battles, giving me strong hints about which species engage in serpentine fisticuffs and which are pacifists. Sure enough, the species where males fight either have males bigger than females (as in Taipans, Blacksnakes, and Scrub Pythons) or the same size as females (as in Tigersnakes and Swamp Snakes), whereas in species where males don't fight (like Death Adders and Bandy-Bandies), the females grow much larger than the males.

Because I had looked inside so many snakes, I could make some broad generalizations. Most Australian elapids are small, brown, and slender, and they wander around actively at night looking for sleeping skinks. They find the lizards inside their burrows, kill them with venom, swallow them headfirst, and then go looking for the next morsel. Females grow almost as big as males, or a bit bigger, and they mate through the cooler months (including hanky-panky inside their overwinter retreat sites). The females ovulate in spring, then either lay eggs in early summer or give birth in autumn. They don't take care of their offspring, but the progeny grow rapidly, maturing in two years. Males mature a bit younger and smaller than females, even in species where the males eventually grow to be larger. Litter size increases as a female grows larger (because she has more room for babies); and females of small species have a few relatively large babies, whereas females of large species have many little ones. Small species keep eating lizards all their life, but individuals of larger species switch across from skinks to larger prey as they grow bigger. The only Aussie snakes that get their fangs into furry meals are the large agile ones (like Taipans and Brownsnakes) that are quick enough to catch a rat; and the sneaky Death Adders that lie camouflaged in ambush for an unwary rodent.

Death Adders are the most unusual elapids. For example, they are short and stout, not long and thin, and have large triangular heads. Their tail-tips have peculiar little appendages that resemble insects, and the snake waves its tail-tip in front of its nose as it lies hidden in the leaf-litter. Woe betide any lizard, frog, or mouse that thinks it has stumbled onto a free meal. Those characteristics I've just mentioned are very un-elapidlike. All across the world, elapids (such as Mambas, Cobras, and Coral Snakes) are long thin animals that search actively for prey rather than lying in wait. The key to the Death Adder's divergent lifestyle is its way of catching prey. If you opt for ambush, the evolutionary pressures change dramatically. A short fat body provides muscle for a fast strike, and a large head and stocky body enable you to swallow large prey. And you better be able to handle any prey item that comes along, even if it's large—you could be waiting a long time before the next edible item strolls past. And of course, a lure on your tail isn't any use if you are moving around looking for sleeping lizards—it only works for a snake that lies in ambush. The way you catch your food even affects how you reproduce. Death Adders don't rely on speed to catch their food or escape from predators; indeed, they don't move around much at all. As a result, carrying a large litter of developing offspring during a long pregnancy isn't a significant "cost" to the female. Death Adders can afford to be viviparous not oviparous, and to carry large litters. An American colleague has even suggested that ambush predation affects your bowel habits. Many ambush forager snakes retain their feces in the hindgut for weeks rather than voiding them, and that extra weight anchors the back part of the body, providing a stable base from which to launch a strike. So, if you want to strike fast enough to catch a fast-moving rat, avoid defecating for as long as you can. The idea of "adaptive constipation" appeals to my smutty schoolboy mind.

At the other extreme from Death Adders, some of the Aussie elapids are slender streaks of lightning. The aptly named "whipsnakes" of the genus *Demansia* differ from Death Adders in every way. They rely on large eyes and acute vision to spot a prey item, and they use their stunning speed and agility to chase down fast-moving lizards. They are active by day not by night, and lay eggs instead of giving birth to live young. A whipsnake's life is built around speed. While the Death Adder lies in ambush, the whipsnake spends its days in hyperactive wandering. Genetic studies suggest that the ancestor of Australia's whipsnakes made the trip across from Asia to Australia separately from the other elapids; the Aussie whipsnakes' closest relatives are from Papua New Guinea. Given the land bridges that formed between Australia and its northern neighbors whenever an Ice Age dropped sea levels, animals often moved backward and forward. Evidence for these migrations is accumulating for many groups, not just snakes.

One whipsnake species in particular has been a focus of great joy and immense frustration to me. In 2007, my colleague Glenn Shea revised the species within a confusing group of desert-dwelling whipsnakes, describing some new species in the process. Glenn is a legendary figure in Australian herpetology, a tall man whose

FIGURE 4.6 Unlike most elapid snakes, Death Adders like this Floodplain Death Adder are stocky in build, and they lie in wait for their prey rather than searching actively for food. Photograph by Rob Valentic.

face is largely hidden by long hair and a luxuriant beard. He looks more like an Old Testament scholar than a reptile biologist, but his knowledge of snakes and lizards is incomparable. And he enjoys difficult challenges, like sorting out the correct species definitions and names for confusing arrays of organisms that everyone else has put into the "too hard" basket. The arid-zone whipsnakes fell into that category. After years of painstaking study, counting scales, and examining diaries from the 1800s expeditions that collected the first specimens, Glenn worked out what was going on. He realized that there are several separate species of snakes that until then had all been lumped together into a larger catch-all species name. That's not unusual—the number of "official" species of Australian reptiles has doubled during my professional career, making it difficult to keep up with all the new names. If someone discovers a new species of mammal or bird anywhere in the world, it's front-page news. But if a biologist looks in detail at any group of Australian reptiles or frogs, they often find an undescribed species or two. Some of them are spectacular animals—for example, a new species of Taipan was found in 2007, the same year that Glenn was dividing up the whipsnakes.

One of the snakes that Glenn recognized as new to science was an elegant little animal from desert areas of the Northern Territory and Western Australia. It's slender, brown to red above, and has a distinctive yellow patch on its neck. To my everlasting delight, Glenn named the species in my honor: *Demansia shinei*, or "Shine's Whipsnake". My ticket to immortality! But I am yet to see a live specimen of the snake that bears my name. Few *Demansia shinei* have ever been collected despite its large geographic distribution. I've never conducted fieldwork in that part of the world, but whenever a colleague (or even a friend of a friend) is heading off to work in that area, I embarrass myself by asking them to keep an eye out for this superb serpent. So far, to no avail. I've heard stories of snakes zipping across in front of cars and disappearing into the scrub; and I was sent one tragic photograph of a snake that didn't move quickly enough to escape the car in question. But so far, not a single live *Demansia shinei*. I've promised myself that one day I will make a pilgrimage to the desert myself to track down this elusive animal.

Inevitably, my museum data didn't solve every question I wanted to ask. For example, one enduring puzzle is why Australian snakes are so inconveniently deadly. Lists of the "Top Ten Deadliest Snakes Worldwide" in TV shows and internet chat rooms are dominated by Australian elapids. There are many problems with such comparisons—for example, most of the information comes from studies on lab mice. That may not tell us much about the venom's ability to kill natural prey (or humans). But still, the toxicity is mind-blowing: a bite from an Inland Taipan contains enough venom to kill 200,000 mice. It's massive overkill. Weapons of Mouse Destruction on an awesome scale. What evolutionary pressures could explain that incredible firepower? Frustratingly, I still can't answer that question, although a more powerful venom may help a snake to kill its meal very quickly. Put yourself in a Taipan's place. You don't have any hands, and the large rat you've just bitten will turn your face into hamburger unless you can immobilize the rodent instantly.

Peering into the innards of dead snakes, like a fortune-teller exploring the entrails of chickens, gave me a broader perspective on the ecology of my favorite animals. From time to time, I still visit museums to look inside dead snakes.

FIGURE 4.7 Although Shine's Whipsnake occurs over a large area of arid habitat in tropical Australia, I have yet to see a live specimen of the snake that bears my name. Photograph by Steve Wilson.

There aren't many unopened snakes left for me to look at in Australian museums, but there are still quite a few in other countries. The papers that I published from that work are in run-of-the-mill scientific journals, not high-impact publications. They are old-fashioned natural history, but I'm as proud of them as I am of more "important" conceptual work that I have done. Those early papers are still useful. If people need to know something about the biology of Australian snakes to develop conservation plans, or to test some new theory, they turn to my natural history reports from the 1980s.

Scientific research has its fads and fashions, like every human endeavor. Much of my own professional success—awards, promotions, funding—has been built on new ideas, and papers in the top international journals. But a decade after you publish that blockbuster paper, the concepts in it already seem dated, and the field has moved on. My museum-based papers remain useful long after my theory-testing papers have faded into obscurity, as scientific fashions continue to change. And that enduring value is a testimony to the importance of natural history, and of natural history museums. Those once-mighty institutions have fared badly in recent decades. To a casual tourist, a room crammed with jars of preserved animals looks (and smells) awful. Why did so many animals need to die for science? It's a valid question. But those jars are not trophy rooms full of animals that were slaughtered by cruel scientists in search of glory. Instead, the collections are an irreplaceable record of information about the natural world, and how that world has changed within the last two centuries.

It took me years to work my way through the preserved snakes in museum collections, and by then my sons were old enough to travel with us. Increasingly, Terri was able to get involved again in the research, as my unpaid assistant and photographer, and the boys soon became part of the research team as well. By the time our sons were teenagers, the Shine Family Snake Team could gather information from snake carcasses quicker and more accurately than any other group I ever worked with. I would do the dissections while Ben opened the bottles, Mac measured the animals, and Terri wrote down the data. So overseas research trips could be family trips at the same time, with a week visiting castles in France followed by a week dissecting vipers. A week on the beach in New Caledonia followed by a fortnight chasing sea snakes. Every family has to find its own solution to work-life balance. The one that the Shine Family evolved was unusual, but it worked.

5 A Plethora of Pythons

Although they lack venom, giant pythons are powerful creatures, with jaws full of sharp recurved teeth that would make a Great White Shark proud. And although many pythons are amiable, some are not. The Carpet Snakes of coastal Queensland are especially grumpy. On one field trip, a large Carpet Snake claimed ownership of a forest path that was the only way to get to my camp. Every time I used the path, the snake hissed and lunged at me in a theatrical but impressive display of belligerence.

Fortunately, most pythons are mellow. Around Sydney, for example, the local version of the Carpet Snake—the Diamond Python—is a calm and friendly snake. I never appreciated that tolerance more than one Friday afternoon in the late 1980s. I was driving across the Sydney Harbour Bridge in rush-hour traffic, amidst the chaos of drivers hurtling homeward. It was a bad time for a 2-meter (6-foot) Diamond Python to escape from its bag on the back seat, crawl up over my shoulder, and turn to look me in the eyes from a few inches away.

It was my own fault. I'd put the snake into the bag at the lab and I had tied it up myself. As the snake flicked my face with its tongue, climbing up onto my head to look around, all I could do was remain nonchalant, as if I often drive in rush-hour traffic with a python around my neck. Stopping on the Harbour Bridge would be far more hazardous than ignoring the snake. If anyone from a passing car had noticed my serpentine headwear, it may have precipitated a traffic accident. But luck was with me, and 10 minutes later I was able to pull over on a quiet street, gently untangle the snake from my body, and pop it back into the bag.

My hands were trembling a little as I retied the bag, took a deep breath, and continued on my way to the university's field station at Pearl Beach. I had brought the snake along with me to photograph in a bushland setting. A group of biology students were spending the weekend there, and they had offered to help with snake-wrangling during the photographic session.

It was evening when I arrived at the field station, so I left the snake inside its bag in the back seat of the car. Soon after sunrise the next morning, I came back to retrieve the snake, and take my photographs in the early-morning light. Unfortunately, the bag was empty. The snake had repeated its disappearing act, forcing its way to freedom through the poorly tied knots that held the bag closed.

It's surprisingly difficult to find a large snake inside a small car. A snake can squeeze itself into a crevice that looks too small for a worm. During my Ph.D. studies, several snakes escaped from bags inside my car, and they proved difficult to recapture. One large Copperhead lived in my VW for two weeks. Several times a day I'd sneak up and throw open the doors, only to see a glossy brown tail disappearing

DOI: 10.1201/b22815-5

under the back seat. Finally, he was slow to react when I threw open the door, and I was able to grab his tail before he could escape.

The most memorable "Snake in a Car" experience during my Ph.D. occurred as I was driving back from a day's collecting, when I felt something crawl over my ankle. A large Tigersnake had escaped from its bag, slithered over my feet, and curled up in the space beneath the brake, clutch, and accelerator pedals. I was on a quiet country road with no other traffic, a far better situation than Sydney Harbour Bridge. But if I depressed the brake, I would squash the snake, a provocation that might induce deadly retaliation. I eased my foot off the accelerator and waited for the car to slow down on the next hill. Then with judicious use of the handbrake, we coasted to a stop and I slowly extracted my other foot from beneath the snake's coils, hopped out of the car, and returned the snake to its bag.

My Pearl Beach Python Problem wasn't such a death-defying scenario. Although it was large, the snake was not venomous. I finally located it up behind the dashboard, wedged deep inside a twisted mass of wires and connecting cables. I got hold of its tail, and soon freed up the last meter of its body. But it's impossible to pull a python out of a tight crevice by its tail. These muscular snakes can cling to any tiny irregularity. I had to dismantle the wiring to remove leverage points, loosening the snake's body inch by inch. There was only room for one person up under the dashboard, so the students just sat around eating breakfast and laughing at me. That snake had a hundred opportunities to bite, but it didn't. When I finally managed to get the snake out, it just hissed a little in frustration.

The turn indicators on my car never worked again. But you can be sure that when I retied that bag, I did so with passion and care. I even took an extra length of strong cord and put several turns around the bag's top. Houdini himself couldn't have escaped, and the snake was still inside its bag when I returned the escape-artist to its cage on Monday morning. It took me 5 minutes to untie the bag, though.

Responsibility to my growing family precluded long-term fieldwork in the 1980s, but I had enough time to supervise someone else to do the work. Now that I was a university lecturer, I had access to an unpaid army of research assistants—students. They conduct most of the scientific research that is done in universities, with the supervisor just providing guidance. If I could find the right student and the right species of snake, I could dabble in fieldwork without deserting my family. What kind of project might be feasible within the suburbs of Sydney? It would have to be based on telemetry: a dozen animals would be enough if I could relocate them frequently, and so build up detailed information on each one of them. And what kind of snake should I study? I needed a species that grows big enough to carry a powerful (and thus large) transmitter, easily found even from a distance.

I remembered that huge Diamond Python I had captured in suburban bushland during my childhood. Even in Sydney, the largest city in Australia, there were pythons to be found. Somehow or other, there must be a way to study those magnificent snakes.

For most people, the word "python" conjures up the image of a giant snake in a tropical rainforest. But in reality, pythons come in a wide range of shapes and sizes, and they occur in a diverse array of habitats. Some live in the deserts, and some extend into cool climates. Diamond Pythons are found all the way south to the border

FIGURE 5.1 Surely one of the world's most magnificent snakes, Diamond Pythons are large, colorful, and generally amiable. Photograph by Sylvain Dubey.

between New South Wales and Victoria. These huge serpents are truly spectacular—jet-black, with yellow rosettes. Some individuals glow so brightly that they look like they have been splattered with yellow fluorescent paint. Add in their genial disposition and their occurrence on the fringes of Australia's largest city, and you have all the ingredients to make the species a sought-after domestic pet. Snake-keeping became popular in Australia in the 1980s, as people discovered the joys of having a snake instead of a budgerigar in the living room. In some parts of Sydney, there was a snake-keeper in every suburban block. Nobody knew how to breed snakes in captivity, so all of these pets were captured from the wild. If a splendid black-and-yellow python turned up in Aunt Mary's chicken-pen, the snake was likely to end up in a fish-tank near the television set.

The stage was set for an almighty battle between snake-keepers and government authorities. In New South Wales, the National Parks and Wildlife Service (NPWS) is responsible for managing protected fauna, including snakes. The authorities had no idea of how many Diamond Pythons lived around Sydney, but such supersized snakes surely must be uncommon. Otherwise, pet cats would be disappearing, and backyards would be full of pythons with a mid-body lump the size and shape of the missing Garfield. Zealous NPWS staff feared that pet-keepers were threatening the existence of these giant snakes. If Diamond Pythons were rare, and hundreds of them were being hauled out of the bush and put into cages, something had to be done.

The NPWS authorities decided that prohibition was the best response. Pet pythons should be outlawed! The snake-keepers protested that this was an overreaction; they believed that Diamond Pythons were common. The two sides swapped arguments in heated meetings, and I was asked to mediate. Was pet-keeping a threat to python populations?

But there wasn't much that I could say. The only real information about the biology of Diamond Pythons came from my dissection of museum specimens, and that didn't tell us anything about the species' abundance. I had seen quite a few Diamond Pythons in suburban bushland (mostly males wandering around in springtime, searching for females), so I suspected that truth was on the side of the pet-keepers than the government authorities. But I had no evidence about what the snakes were doing, or how many were out there.

As I was pondering the problem one day in my office, a student dropped by to talk to me. In his early 20s, Dave Slip was a large, powerfully built man with thick glasses, long blonde hair, and a profuse white beard. I had lectured to his class the week before, and the lecture had piqued his interest. Dave was looking for a research project, and I suggested a radio-tracking study on Diamond Pythons in the suburbs. Dave would be able to do the project part-time, in between working as a landscape gardener. No need for long field-trips. If we implanted radio transmitters into a few snakes, Dave could obtain sufficient data for a thesis. And it would be cheap. In my office drawer, I had the transmitters left over from my ill-fated project on Blacksnakes. Plus I could scrape together enough funding to buy some supersized units that would produce stronger signals—much easier to radio-locate.

Dave liked the idea. Finding snakes was our first challenge, but I put the word out among my snake-loving friends. We need pythons! Gerry Swan phoned one day to say that he'd found a large female Diamond Python crossing a busy road in the Sydney suburb of Belrose. A respectable insurance executive by day, Gerry was also a passionate reptile enthusiast, and his evenings and weekends were devoted to scaly rather than commercial targets. He was in his "executive" persona, driving to work, when the python made the dangerous decision to cross a highway during rush-hour traffic. Fortunately, Gerry noticed her before she ended her life under the tires of a fast-moving vehicle. He stopped in the middle of the road, jumped out of his car, and

FIGURE 5.2 Dave Slip was my first research student. He is handling a Water Python here (rather carelessly; it bit him a few seconds after I took the photograph) but most of his snake research was on Diamond Pythons near Sydney. Photograph by Rick Shine.

caught the snake—to the frustration of the other commuters, who had to wait while he wrestled her under control. When he got to work, he phoned me to say that we were welcome to her. So, the project was up and running. Our surgical procedures to implant the transmitter were primitive but effective, and within a week we released the snake near where Gerry had found her. The road she had been crossing was a major highway for urban commuters, but it was surrounded by dense forest and steep sandstone cliffs. Suburban development had intruded, with houses along the ridgetops—but the backyards of those houses projected into bushland. A mosaic habitat for a snake, where even a giant python could remain undetected as she moved among the bush and the backyards.

It was early spring, and the big snake was living a quiet life. On cool days she stayed inside a rocky crevice, emerging for an hour or two at midday to bask. For some students, that would have been boring. But Dave is a patient man, content to sit back and watch the world roll by, and he would come back bursting with enthusiasm after spending a full day watching a sleepy snake doing very little. Perhaps those special cigarettes that he smoked helped him chill out.

That first big female python taught us a great deal, beginning with the fact that a Diamond Python is virtually invisible among the undergrowth even when you know (from the transmitter signal) that she is there. The second lesson was that Diamond Pythons prefer backyards to bushland. Our big female often moved through the bush, but only in transit to a backyard or roofspace. And when we looked at what Diamond Pythons eat (from scats, and dissections of museum specimens), their fondness for backyards was easy to understand. The suburbs are a python smorgasbord, because humans subsidize the ecosystem. We add water and nutrients to make our garden plants flourish, creating snacks for Brushtail Possums—the favorite prey of Diamond Pythons. A python can find more possums in a pergola than in a gum-tree. Cats, dogs, chickens, guinea pigs, and parakeets don't taste bad either.

Unlike the Blacksnakes I had studied previously, Dave's big Diamond Python generally didn't move from one day to the next. Instead, she coiled up at the edge of the shrubbery, facing out toward a clearing, and just sat there day after day. Like Death Adders, these pythons rely on ambushing prey rather than chasing it down. A juvenile Diamond Python ambushes lizards by day, until she grows large enough to ambush mammals by night. Larger body size lets a python retain heat in its body for longer—especially when it is tightly coiled—so that when a possum wanders past at dusk, the adult python is still warm enough to detect it and launch a strike. Pythons are devastatingly efficient foraging machines, especially for warm-blooded prey. The snake can detect a temperature difference of less than a thousandth of a degree centigrade, using heat-sensing pits below its jaws. Inside the python's brain, the messages from nerves leading to those facial pits merge with signals from its eyes: the snake "sees" temperature as well as shape and color. And its forked tongue samples the air and ground for scent molecules that are then transferred to an organ in the roof of the mouth to be analyzed. If a hot and smelly mammal passes by, a python knows all about it.

Pythons are choosy about where they lie in wait for their prey. The ambush site must offer security from predators but enable the python to detect the approach of edible items. Snakes often choose a spot at the edge of dense vegetation, with the body hidden but the head peering out into the open area in front of it. Using its tongue to detect chemicals left by animals that have passed by earlier, a python can select a high-traffic area such as the intersection of two mammal-trails. Snakes seem to prefer spots where a passing possum has to jump over an obstacle, and so must focus on the trail not the hidden killer off to one side.

The muscular body of a python lets it cling to almost any structure, so the posture of an ambushing snake depends on the local landscape, and what kind of prey it is after. Hanging head-downward from a bush, tree, or rock is a popular tactic. A rodent or marsupial traveling along the ground isn't aware of any danger until an open mouth full of needlelike teeth thunders down from above. In a second or so, it takes the unlucky mammal to begin thinking about retaliation, its options disappear. The snake wraps its coils around the prey's body, immobilizing teeth and claws.

These days, pet-keepers rarely feed live rodents to their snakes. There's no reason to subject a mouse to that ordeal; and pythons happily eat dead prey. But in the bad old days, the conventional wisdom was that snakes would only eat live prey. So as a teenage snake-keeper, I put live mice into the cages of my pet snakes, and then watched the outcome. At the time, I wondered why the prey didn't retaliate sooner. That wasted second after the initial strike is critical—if the rodent immediately bit the python, the outcome of their battle might be different. Why didn't rats respond faster? A young Water Python answered that question for me many years later, on the wall of Fogg Dam in the Northern Territory. For our research on these pythons, a colleague and I drove slowly down the dam wall at night, jumping out of the vehicle to pick up snakes on the road. Water Pythons are feisty, but they aren't difficult to handle. If you pick one up behind the head, it immediately coils around your hand and arm, but it can't bite you. And if you plunge your arm into a cloth bag, and then let go of the snake's head, it will drop off into the bag. I've done it hundreds of times. But that night, when I jumped out of the car, I encountered three small snakes close to each other. So, I grabbed all three. One of the snakes went into its bag with no problem, but while I was manipulating the second one to do the same, I took my eyes off the third snake. It pulled its head loose, turned around, and bit into my hand—right across the middle of my palm.

My initial reaction was "ouch, but no big deal". I've been bitten by plenty of pythons. It leaves small puncture-wounds, but it doesn't hurt too much. I focused on getting the other snake into the bag, and let the biter chew away on my hand in the meantime. And a second or two later, I realized my mistake. At first, those small teeth only penetrated a few millimeters into my palm. But in the extra time I gave it, the young snake threw a body coil around my hand. That created a sandwich with my hand as the filling, the snake's muscles as the outside layer, and the snake's head in between. When those powerful body coils contracted, they pushed down on the animal's head and drove the snake's teeth deep into my hand—deeper than I would have thought possible. The pain was incredible. And two thoughts immediately came to mind: (1) Arrgh! Somebody help me! (Nobody did. By then, my colleague was convulsing in hysterical laughter.) (2) So *that's* why a rodent bitten by a python doesn't retaliate instantly. It's too preoccupied with the agony inflicted as those muscular coils drive the snake's teeth deeper and deeper.

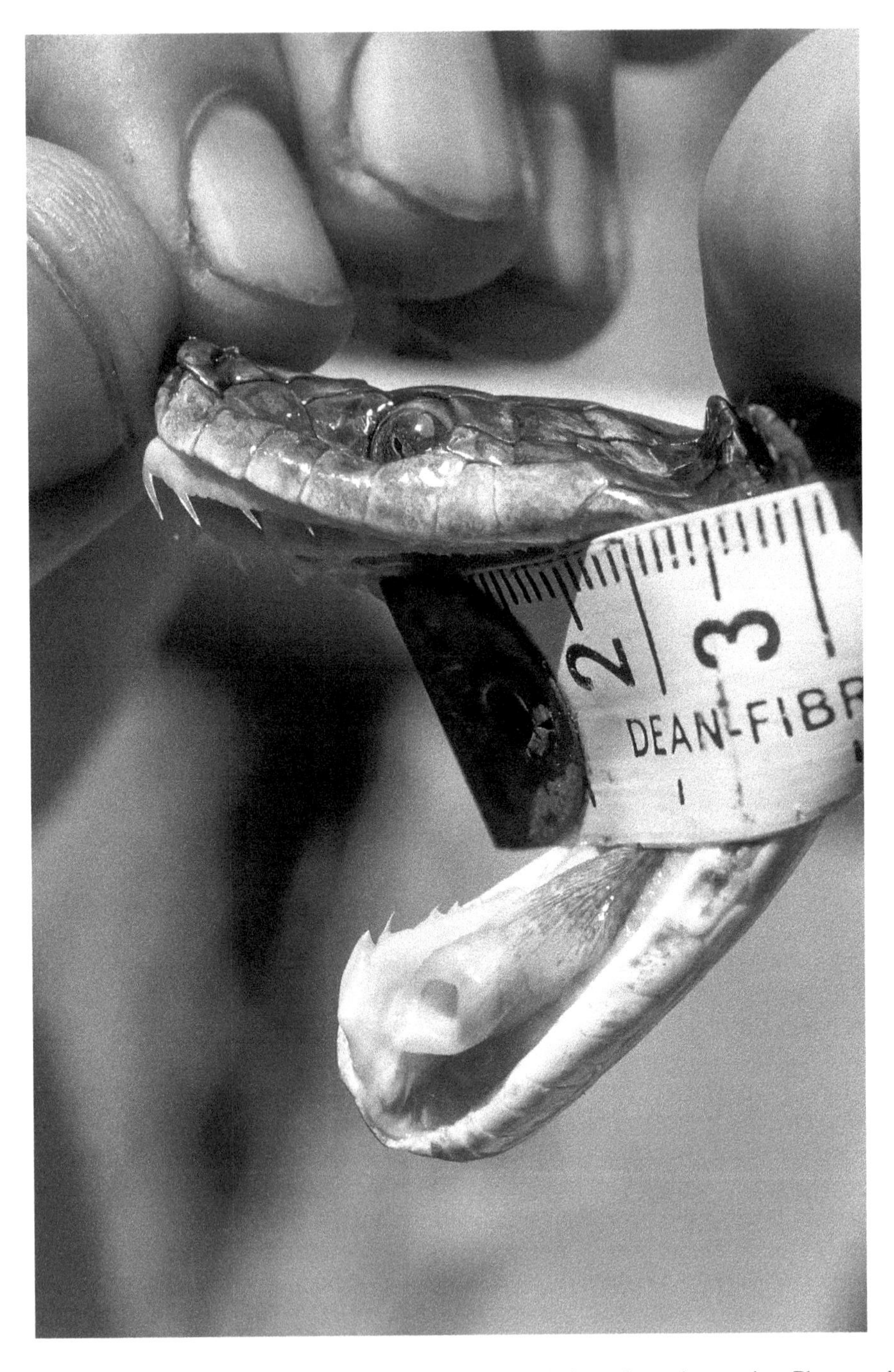

FIGURE 5.3 Water Pythons are not venomous, but their teeth are impressive. Photograph by Rick Shine.

But, back to Diamond Pythons in the backyards of Belrose. In that suburban setting, Dave faced some cultural problems I had never encountered when I was studying Blacksnakes in the Macquarie Marshes. What if you track your snake through the forest and find it sitting in an apple tree in someone's backyard? If you don't tell the owners of the property, they'll get a terrible shock when they find the snake. Not knowing that it's harmless, they may kill it. Even though snakes are officially protected, there's a loophole: you can kill a snake if you believe that it's a threat to you or your family. Given the widespread paranoia about snakes, many people believe that they are in danger if they encounter a snake the size of a shoelace. A giant python arouses far more panic.

So, Dave's policy was to knock on the front door, introduce himself, and let people know they had a new (and harmless) reptilian friend in the backyard. But it didn't always work as planned. In those days, Dave was a seriously strange-looking man, tall and powerfully built with a luxuriant white beard and long pale hair hanging down to almost obscure his face. Once his knock on the front door was answered by a well-known television comedian, Paul Hogan (later of "Crocodile Dundee" fame). Hogan was convinced that this was a joke created by a rival TV network. He looked high and low for hidden cameramen, and was only convinced when Dave showed him the snake beside the swimming pool.

The next lesson from that first big female Diamond Python was that this species is into communal sex. To Dave's delight, his radio-tracked snake attracted a friend in late spring—a male who, although adult, was much smaller than the female. They were curled up together when Dave radio-located the female one day. Dave brought the male into the lab, surgically inserted a transmitter, and then let Romeo go again. He wandered away, deciding that this particular Juliet was dangerous company. But before long another male joined her, then another. We equipped them all with transmitters, and before long we were watching a very well-mannered slow-motion snake romance. The female mated with each of the males in turn, with the other males ignoring the sexual shenanigans beside them.

Although immune to jealousy, the males were attentive lovers. One morning the female python wandered away across the ridge to a distant gully, and we doubted that the males could ever find her again. But one after another, the males emerged from their overnight retreats, tongue-flicked the ground where the female had been lying, picked up her scent, then took off along exactly the same path that she had followed. Clearly, they were detecting the faint chemical traces she had left behind on the ground. With each male following close behind the one ahead of him, several meters of python were spread out along that line. By nightfall, the group was reunited.

Even better, some of those males led us to other females—who in turn attracted other males, and so on. So, our sample size of radio-tracked snakes grew rapidly. And we now knew that our first observation of a mating aggregation wasn't a fluke—Diamond Pythons have an unusual mating system, different from anything that had been reported in other snakes.

Those extra animals enabled us to look at other issues as well, including the hazards of suburban life. Even a large snake is at risk from ignorant homeowners, dogs, cars, and construction crews. My saddest memory of Dave's project is of the day when we went out to relocate our original female python and found her mangled body amid a pile of logs that had been pushed together by a bulldozer. Simple mathematics tells us that in order for populations of Diamond Pythons to thrive, they must produce enough offspring to balance those high mortality rates. This leads me to the next big insight from Dave's work—nesting behavior.

Once again, Diamond Pythons did the unexpected. When a female is ready to lay her eggs in early summer, she constructs a nest. By crawling in under the leaf-litter and then coiling, a female python creates a small mound—the size and shape of a football—with a single entrance hole. The nest is difficult for a predator to find, and the leaf-litter provides good insulation. The female can bask in the sunlight, bringing her temperature up to over 30°C (86°F), then crawl into her nest and coil up tightly. She can stay warm for hours within her insulated igloo—and so, keep her eggs warm as well.

Staying warm confers a huge benefit for a snake embryo. Even a tiny increase in incubation temperature accelerates development, so the offspring hatch earlier, giving them time to catch a lizard or two before their first winter. But also a warm nest creates a "better"—bigger, quicker, smarter—baby reptile. We don't know exactly how that happens, but it's true for many kinds of reptiles. If I turn up the temperature of the laboratory incubator a couple of degrees, the baby lizards or snakes hatch sooner, run faster, and learn quicker. The ideal temperature varies among species, and some lizards from alpine habitats develop best at lower temperatures. But for pythons, the optimum temperature is around 30°C, day and night.

FIGURE 5.4 After she lays her eggs, a female Carpet Python wraps around them to protect them from predators, and to keep them warm. Photograph by Steve Wilson.

How can a mother python keep her eggs at the ideal temperature? In most reptiles, the mother's job begins and ends with finding a nest-site where the eggs will stay warm. It's all about "location, location, location". When it's time to lay her eggs, a female reptile may travel long distances to find a hot spot. In the Galapagos Islands, Land Iguanas migrate to the slopes of volcanoes, where subterranean molten rocks keep the soil warm. Lacking convenient volcanoes, most lizards and snakes simply search for a sun-drenched clearing in the forest, perhaps beneath a rock that will heat up rapidly by day and cool down only slowly at night. The monitor lizards ("goannas") lay their eggs inside termite mounds where the insect homeowners keep everything warm year-round. Reptile embryos themselves have a few tricks as well, moving within their egg to find the warmest spot. But in pythons—and pythons alone—mum has an extra talent. While she is caring for her offspring, she becomes "warm-blooded".

For most of her life, a female python regulates her body temperature the same way as every other snake. She heats up by basking in the morning, then moves to the shade to cool down. Although we often call snakes "cold-blooded", a Brownsnake sprinting across the road at midday is hotter than a human being. Snakes control their body temperatures precisely, by shuttling between warm spots and cool spots. The source of heat is the outside world, so reptiles are called "ectotherms" ("ecto" means "outside, and "therm" refers to temperature). In contrast, mammals and birds are "endotherms" ("endo" = from within), because we generate our own body heat. Relative to body weight, we eat about ten times as much as a snake does. We burn up that extra 90% just to keep ourselves warm.

In the 1970s when I was a student, textbooks asserted that ectothermy was "primitive" and endothermy was "advanced". Reptiles are ectotherms, we were told, because they are too simple and unsophisticated to be endotherms. But snakes like the Diamond Python have overturned that comfortable assumption of endothermic superiority. Remarkably, the female Diamond Python turns herself into an endotherm—a warm-blooded animal—while she is taking care of her eggs. She shivers to warm herself up, like birds and mammals do when they get really cold. And by putting small electronic thermometers into python nests, Dave showed that these paragons of maternal devotion keep their eggs warm throughout the long incubation period.

Why don't snakes just keep themselves warm 24/7, as mammals do? Because it takes a lot of energy, and usually isn't worthwhile. Female Diamond Pythons expend so much energy in brooding their eggs that they need years to regain condition before they can breed again. Female pythons are emaciated and celibate for a year or two while they recover. Even with access to a steady supply of plump possums, a female python needs years to build up enough energy reserves to produce a large clutch of eggs.

In terms of our concerns about python conservation—the initial stimulus for Dave's work—the news was good. Diamond Pythons were more abundant than anyone had guessed. First, we found many snakes within a small area. Second, the snakes were almost undetectable; they were difficult to see even when we had the transmitter signal to guide us. Diamond Pythons live secretive lives, unseen by the people who walk past them every day to hang out the washing. The managers at the National Parks and Wildlife Service looked at our results, and they agreed that Diamond Pythons were not at risk of extinction. With so many snakes out there in the suburbs, collecting a few pythons as domestic pets isn't a big issue—especially, compared to processes like forest-clearing for new urban developments.

I don't blame the National Parks staff for taking action. It was a legitimate concern: over-collecting might indeed have been imperiling populations of Diamond Pythons. Who would have guessed that giant pythons can be successful city-dwellers? Common sense suggests that these spectacular snakes are too big, slow, and easy-to-see to persist on the fringes of our urban landscape. Giant pythons belong in pristine forests, not backyards. But in this case, "common sense" is wrong. Many major cities in mainland Australia are home to at least one species of the genus *Morelia*—Diamond Pythons in Sydney, Carpet Pythons in Perth and Darwin, and huge Scrub Pythons in Townsville and Cairns.

How can large pythons flourish in degraded suburban ecosystems? Their success is a tribute to their pragmatism. A python doesn't mind if the shelter it calls home is the roofspace of a house rather than a tree-hollow, or if the warm furry menu item that wanders past is wearing a collar engraved "Fluffy". A roofspace may be safer than a hollow log, and edible mammals are more common in the suburbs than in the forest. But the combination of large snakes and city-dwellers creates interesting situations. The Diamond Python's unusual mating system, where a group of males spend springtime with the female of their joint affections, results in a lot of reptile crammed into a small space. A terrifying spectacle for the homeowner who pokes his or her head through the manhole in the ceiling to find out what's making all the noise. On the plus side of the equation, the irritating possum that ate your roses is gone for good.

Many of my scientific colleagues don't share my fascination with the intersection between "civilization" and wild predators. I was told that studying suburban pythons was "artificial", and biologists should work only in wilderness untouched by humans. According to those ideologues, scientists need to document "natural" phenomena and to focus on what animals do when humans aren't around. I disagree. Every environment on the planet has been affected by human activities, and we need to incorporate those impacts into our studies. A python is no less a python if it lives in a roofspace, not a tree-hollow. If the snake population is being wiped out because the animals are being hit by cars, we should study that threat (and find ways to minimize it) in the same way we would study any "natural" threat. Encouragingly, research on urban ecology has increased over recent decades. Ecologists are becoming a bit less prissy.

The 1980s—the first decade of my employment at Sydney University—worked out well in terms of my scientific research as well as my home life. Before 1980, the only detailed ecological information on Australian snakes came from my fieldwork in the New England area, on half-a-dozen species of elapids. But now, a decade later, we had data on almost every snake species in Australia, based on the museum dissections. And Dave's Diamond Python study, plus my Blacksnake telemetry, had told us about the day-to-day lives of pythons and elapids in other parts of Australia. Jonno Webb had clarified the feeding biology of the most poorly known snakes of all—the wormlike Blindsnakes. The picture was far from complete, but the 1980s had transformed our knowledge base.

FIGURE 5.5 One of the best things about living in Sydney is that every once in a while, a large Diamond Python wanders across your backyard. Photograph by Rick Shine.

Unfortunately, those new results were inaccessible to the people most interested in them. Very few people read scientific journals, because the jargon is off-putting. I wanted to bring this information to the general public. A better understanding of snake ecology might help pet-keepers to improve the lives of their captives and might encourage the broader community to appreciate these maligned creatures. I decided to write a "popular" book on the ecology and behavior of Australia snakes, with photographs of snakes foraging, eating, courting, and mating. The only snake books available at the time were identification guides, illustrated by portrait shots that showed off each species' distinctive features, but didn't tell us anything about snakes as animals. So, in 1990, I wrote a book about snakes, and then spent several months accumulating photographs to illustrate it. I spread the word: "I want pictures of snakes doing things, but I can't pay for the photographs". The response was extraordinary. People I'd never met—friends of friends—mailed me their precious pictures.

Reed New Holland offered to publish the book, and *Australian Snakes: A Natural History* appeared on the shelves in 1991. It's now gone through a couple of editions, and multiple reprintings, and has been translated into German. I still get a warm glow when I see a copy of that book. The only speed-bump came in 2009, when the publisher reprinted the book with a new cover photograph—but of a legless lizard not a snake! They had used one of the photographs in my book, without checking with me and without reading the caption. After my anguished complaints, the publisher recalled the volume and replaced the cover—but not before book-collectors snapped up every copy they could find. It's the only book about snakes that has a lizard on the cover.

In addition to providing information for the snake-loving public, writing that book taught me a great deal. Explaining something makes you think about what you are saying—and in the process, you notice things that otherwise escape your attention. As an example, I realized that the common names given to snakes in Australia are, frankly, boring: "Brownsnake", "Blacksnake", and so on—simple descriptive terms. Nothing to compare with "Shark of the Forest" for an elapid in the Solomon Islands, or "Bushmaster" for the giant viper of Central America (a snake also known as "the Pineapple Bitch" because of its large rough scales). Many of the spectacular and iconic names for snakes—like Cobra, Mamba, and Boomslang—came from languages other than English. So, why don't we use more Aboriginal names for Australian snakes? Fortunately, a few early scientists went to the trouble of finding out what Indigenous people called each type of snake. For example, we have anthropologist Donald Thomson to thank for "Taipan", a striking name taken from the language of the Wik-Mungkan people of Cape York Peninsula.

Museum curator Ludwig Glauert gave us "Dugite" and "Gwardar" for the deadly Brownsnakes of Western Australia. "Gwardar" supposedly translates into "go the long way around" (that is, stay away from this snake!). Of course, transcribing Aboriginal names into English is famously difficult. The "dugite" is called "Torn-ock or Tookyte" by Edward John Eyre, in his delightfully named 1841 treatise *Journals of Expeditions of Discovery into Central Australia, and Overland from Adelaide to King George's Sound, in the years 1840–1; Sent by the Colonists of South Australia with the Sanction and Support of the Government including an Account of the Manners and Customs of the Aborigines and the State of their Relations with Europeans, Volume 1.* The same document also provides Aboriginal names for other serpents, including "Bardick", still used for this miniature doppelganger for the Death Adder. Others such as "Norne" for the Tigersnake are less often heard; and yet others have disappeared completely. I've never heard anyone use "Werr" for a Whipsnake, "Wakel or Wa-a-kel" for a western Carpet Python, or "Kerry-gura" for a Burton's Legless Lizard (although one of my Indigenous friends from the Kimberley calls these lizards "pencil snakes", which is a fine name). The boring names of our wonderful snakes are a downside of Australian pragmatism.

******

My next foray into python biology took place in northeastern New South Wales, in the lush green hills around Mullumbimby Creek. Only a day's drive north of Sydney, but it's a very different landscape—warmer, wetter, subtropical. Abundant rain and fertile soil encourage vegetation growth. That wall of green outside the car window as you drive along is beautiful to look at, but so dense that it is difficult to walk through. When I think of fieldwork around Mullumbimby, my first mental image is of scrambling up muddy gullies, tripping over thorny vines. And besides being a farming community, Mullumbimby was the epicenter of hippie life in Australia. You could see a diverse assortment of people at the neighborhood store. A matronly lady wearing jeans, khaki shirt, and riding boots is chatting with a younger woman wearing a tie-died dress and facial piercings. Her barefoot child is wearing a Che Guevara t-shirt. Marijuana smoke drifts through the air. And beside a laundromat, there's an organic-produce store painted in rainbow hues and plastered with New Age slogans.

The area around Mullumbimby has a special place in my heart. In 1973, the small town of Nimbin hosted the Aquarius Arts Festival, Australia's equivalent of the much-celebrated Woodstock festival in the United States four years earlier. As a young Ph.D. student, I spent a week at the Nimbin Festival. We didn't have Joan Baez or Bob Dylan, but there were lots of peace, love, and public nudity. It was enormous fun, and a real eye-opener. I didn't realize how much one human body can differ from another until I saw a thousand nude people on a riverbank.

My friend Mark Fitzgerald lives in a wooden house beside Mullumbimby Creek near Nimbin. Mark is a longtime resident of hippie heartland. He moved there to farm bananas and raise a family—a lifestyle choice—after pursuing a university degree in English literature. And that training still shows, because Mark has a far larger vocabulary than any other snake-catcher I ever met. In the early 1980s, I met Mark through a mutual friend, Greg Mengden. Greg is an affable Texan who came to Australia for his Ph.D. (on snakes, of course). Mark is an outstanding naturalist and snake-catcher, and he and Greg made epic journeys around Australia in search of serpents. Eventually tiring of the banana-farmer's life, Mark embarked on a university science degree, and then a Ph.D. (with me as supervisor, even though we are about the same age) on Stephens' Banded Snakes. More of that later in the book. But in between Mark's undergraduate science degree and his Ph.D. studies, I employed him to solve a mystery in python biology. The snakes at the center of the puzzle were abundant around Mullumbimby (including in the ceiling space of his own house), so Mark was ideally placed to do the work.

Dave Slip's pioneering study had shown that groups of male Diamond Pythons gather around the female, take turns mating, and behave in tolerant, non-violent ways. The males are smaller than the females. That's what we expect from evolutionary theory: if you don't need to beat up other males, there isn't any benefit to being the biggest bloke. You should put your efforts into courting females rather than brawling with testosterone-soaked rivals. Be a lover not a fighter. The hippies of Mullumbimby would nod their heads approvingly at this lifestyle choice, and smile at this example of tolerance in the Animal Kingdom. But ironically, the pythons that were slithering around the marijuana plants in the gardens of Mullumbimby were treating each other in far less kindly ways.

The yellow-and-black peaceniks we call "Diamond Pythons" are the southeastern race of a very widespread species. All of the rest are called "Carpet Pythons" or "Carpet Snakes", because their richly patterned brown blotches make them look as if they are wrapped in a Persian carpet. Despite that color difference, the two forms are closely related. Both can grow to massive sizes, sometimes more than 3 meters (10 feet) long, and those muscular bodies can exert a constricting power that has to be experienced to be believed. Diamonds and Carpets interbreed readily in captivity. Indeed, all of them are brown and splotchy when they emerge from the egg; a Diamond Python doesn't develop its distinctive color until it is several months old. The geographic transition from one type of python to another is abrupt, though. If

FIGURE 5.6 Over most of Australia, Carpet Pythons are splotched in shades of brown, not bright black-and-yellow like the Sydney subspecies. This specimen is a Murray-Darling Carpet Snake. Photograph by Rob Valentic.

you drive north from Sydney for a few hours, the giant snake crossing the road in front of you will be brown-on-brown, not yellow-on-black. They both belong to the species *Morelia spilota*, but they are very different animals.

What's going on here? A few million years ago, the southern group was separated off from the others, and started to evolve into a new species as it adapted to its local environment. But the process didn't proceed to complete separation; so when the two races met again, they could still interbreed. Diamond Pythons ended up as a local variant of a wide-ranging species, instead of becoming a separate species in their own right. Coastal Carpet Snakes—like the ones around Mullumbimby—grow bigger than their southern relatives, and are more heavyset, with larger heads. And their personalities are different. Unlike their laid-back Sydney relatives, Coastal Carpets are belligerent. If they have a poster hanging on their bedroom wall, it's a portrait of a heavy metal band rather than of a pair of dolphins sharing a wave on a sunlit evening. These are snakes with attitude. I'd love to know what happens at that intergrade zone, when the polite black-and-yellow southerners bump into the rowdy brown-splotched northerners. I suspect it's like a terrible family reunion that brings people together from far-flung cultures with different customs. The gentle Diamond Pythons will be pushed to the back of the room, emitting feeble protests as their nasty northern cousins grab all the food, get drunk, fight each other, then trash the place.

It's hard to imagine an intimidating bruiser from Mullumbimby composing romantic poetry while the bloke from next door makes love to his sweetheart. And indeed, our museum dissections hinted at different sex lives in Carpet Snakes than in Diamond Pythons. The biggest Carpet Snakes in the museum collections were males—the reverse of the situation for Diamond Pythons. I predicted that male Carpet Snakes must fight with each other—a dramatic difference from their pacifist relatives a few hundred kilometers away.

To test my prediction about fisticuffs between male Carpet Snakes, Mark caught 19 pythons from two areas near his home. One site was along Mullumbimby Creek itself, in a mosaic landscape containing banana plantations, shrublands, regenerating forests, and human settlements. Mark caught some of the snakes at his own house, when he found them inside his roof. Very convenient fieldwork, although the proximity of large pythons and domestic pets brings some anxious moments. Once Mark heard loud squeals from the verandah and arrived in time to rescue his new puppy Daisy from the coils of a hungry snake.

Mark's other study site was near Bangalow, in an orchard with regularly spaced peaches, nectarines, and macadamia trees in a wide expanse of short grass. The manicured lawns are penetrated by gullies of thick vegetation, ideal habitat for feral rats: unlimited food in the orchards by night, with dense cover to use as a hideout by day. Good news for a rodent-eating python: a rat that feeds on peaches and macadamia nuts surely must taste delicious.

Like their Sydney relatives, the Coastal Carpets are ambush foragers and mostly stay within dense cover. The snakes are invisible in the tangled undergrowth of the gullies, but obvious if they leave their home thickets. Every once in a while, a python decides to go wandering through the macadamia orchards. Despite their size (up to 7 kg [15 lb]), it's risky business. Grass-slashing tractors killed a few of our snakes, and predators took others. One large snake was killed by a Lace Monitor, and the same lizard species drove another female python off her nest and ate most of her eggs. But the main predators were foxes and wild dogs that killed and ate snakes as they were crossing the open orchards. If you'd asked me beforehand, I would have bet on the python winning that contest. Perhaps one or two did, and wandered off afterward with a brushy tail hanging out of their mouth. But clearly, some of those battles ended in victory for the mammal not the reptile.

In the thick vegetation around Mullumbimby Creek, pythons are less vulnerable to predators. Their main risk comes when they cross roads, and hence are in danger of being run over by rainbow-painted kombivans with hippies at the wheel. Fortunately, the giant size and distinctive colors of the Coastal Carpets make them easy to distinguish from deadly Brownsnakes, even to someone in an altered state of consciousness. Most drivers let the snakes slither unmolested across the road.

But that culture also raised difficulties when Mark needed to relocate his radio-transmittered pythons along Mullumbimby Creek. Many of the snakes spent their time in and around small houses scattered among the rainforest. For the residents of those houses, the sight of a stranger waving a telemetry aerial in the air fueled fears of government surveillance. Perhaps a new technology for drug detection? To succeed in his radio-tracking, Mark needed social skills as well as local contacts. Fortunately, Mark is a born diplomat. It's a pleasure to hear him talk, even if it's only about the weather. He reads a thesaurus for entertainment and as a result, his vocabulary includes words that have me scratching my head in confusion.

As I'd predicted, social interactions among Carpet Snakes were more brutal than among Diamond Pythons. None of that namby-pamby southern tolerance. If a male Carpet Snake finds a reproductive female, he immediately begins courting her. If a rival male turns up, attracted by the female's alluring scent, fierce battles ensue. Combat bouts in venomous snakes (like Blacksnakes, Tigersnakes, and Copperheads) look more like a dance than a fight—but when two male Carpet Pythons take each other on, it's no-holds-barred. If snakes had fists, a Carpet Snake Battle would be a bare-knuckle slugfest. The main tactic is to twine around the other guy, and squeeze as hard as possible—and nobody can squeeze like a python.

Unlike venomous snakes, fighting male Carpet Pythons also use their teeth as weapons. And very impressive teeth they are! You don't catch bandicoots and possums with feeble dental equipment and weak jaw muscles. The teeth of male pythons create gaping cuts as they pull across their enemy's body. The result is parallel open wounds, ripping right through the skin and into the underlying muscle, until one of the snakes decides that it can't take any more punishment. These deep cuts leave permanent scars, which male Carpet Pythons appear to wear with the same pride an outlaw biker feels toward his tattoos: Stripes of Honor. Larger body size is a critical advantage in these epic battles, with the result that genes for large body size in males have been favored by natural selection. That's why male Coastal Carpet Snakes grow larger than females.

So far this all sounds neat and tidy. I framed a hypothesis (male–male combat favors larger size in males), made a prediction from it (male Carpets fight each other, even though male Diamonds don't), and then Mark went out and gathered data that confirmed my prediction. Males are the larger sex in the combative subspecies and the smaller sex in the more peaceful subspecies. But as is usually the case in biology,

FIGURE 5.7 Carpet Pythons are common over a huge area of eastern Australia, even in the suburbs of major cities. Photograph by Shannon Kaiser.

there are some inconvenient exceptions to this general pattern. Not everybody is playing by the same rules.

First, Mark found two cases where a female Carpet Python was being courted by more than one male. On one occasion, two males were intertwined with a girl on a tree-branch. In the other, three males were simultaneously courting a female on the ground. That sounds more like tolerant Diamond Pythons than macho Carpet Snakes. Are there some sensitive New Age males out there among the barbarians? Maybe—but those group-sex male Carpets had battle-scars, so they must have been involved in previous combat bouts. And years later, I came across another confusing case: an adult male Diamond Python in bushland near Sydney, who was crisscrossed with combat scars. He was irascible (unusual for a Diamond Python), and I wondered about his background. People have been bringing Coastal Carpet Snakes to Sydney as pets, and for rat-control in barns, for many years. Maybe that snake was fathered by an escaped pet with genes for aggressive machismo? Or perhaps he was a gentle soul who had recently been beaten up by an escaped Carpet Snake? That might make anyone irascible.

The story took another turn more than 20 years later. Mark was still living in the same house, and Carpet Snakes were still his neighbors. That summer of 2017, as in many others, a female Carpet Snake took up residence inside the roofspace, and a male joined her. But remarkably, that male hung around, battling and vanquishing a succession of other hopeful suitors over the next two months. Instead of wham-bang-thank-you-ma'am, the resident male spent his time curled up with the female, and on the lookout for rivals.

FIGURE 5.8 In southwestern Australia, female Carpet Pythons grow impressively large. Peter Harlow is demonstrating their generally benign disposition. Photograph by Rick Shine.

Those "exceptions to the rule" hint that the sex lives of these giant snakes are more complex than we currently understand. Most male Carpet Pythons in north-eastern New South Wales opt for aggression, but there may be a few pacifists. And an occasional male who guards his mate instead of heading off to find another female. And although most male Diamond Pythons are sweethearts, the sort of snake you'd bring home to meet your mother, there are a few renegades. A male python may adjust his tactics depending on local conditions. We haven't studied enough animals, for long enough, to pick up subtle differences. And we have very little idea about why that difference in mating systems evolved in the first place.

To get a broader perspective, I needed to look at pythons in other parts of Australia. "Carpet Pythons" of the species *Morelia spilota* occur across most of the continent, and broader comparisons might shed more light. Frustratingly, though, information was sketchy—especially about the most distinctive subspecies, the Western Australian form *Morelia spilota imbricata*. These are spectacular animals, with reports of snakes well over 3 meters (10 feet) long, heavy-bodied like all of the Carpets, and with delicate yellow-and-brown blotching that camouflages them superbly in bushland. Nobody knew much about them. Were the males smaller or larger than females? Did males ignore each other, or battle like reptilian gladiators? I couldn't see any way to answer those questions, because Perth is a long way from Sydney (about 4,000 km, or 2,000 miles).

But in 1993, I was invited to join such a study. David Pearson, a wildlife researcher employed by the Western Australian government, wanted to look into the conservation status of Carpet Pythons for his Ph.D. By the time I met David, he had already had a decade of field experience in remote areas, and knew a lot more than I did about the fauna of Western Australia. But he needed a supervisor who knew something about snakes, and such people were scarce, so he approached me about playing that role. David's interest was motivated by conservation concerns, including the one that had stimulated Dave Slip's work on Diamond Pythons ("pet-keepers are poaching all the wild snakes"). But other issues were distinctive to Western Australia ("forest clearing has eradicated the snake's habitat"). The Conservation Department would support the project with David's salary, as well as other assistance and equipment. So—would I take on a student with extensive field experience, and guaranteed support from the government, on a project I had wanted to do for years? And that would give me an excuse to visit one of the most beautiful parts of Australia? Hell yeah!

David's most spectacular study site was Dryandra Forest, an ecological cathedral that resembles the manicured lawns of a country estate in England. Stately Wandoo trees dot the open meadows, much tidier than the tangled jumble of bushes that dominate most Australian forests. But it's home to rare marsupials like Woylies and Numbats. One of my favorite memories comes from my initial trip to Dryandra when (with my family in tow) I first encountered the pythons of Western Australia. The best way to find snakes in this area is to drive slowly along dirt roads at night. Although a 2-meter python is invisible by day in foliage, the same snake is easy to see as it crosses the road at night and is illuminated by a car's headlights. The tracks through Dryandra were flat, wide, and well-maintained, so I convinced Terri that it was safe for our sons Mac and Ben (12 and 10 years old, respectively) to climb up onto the field vehicle's roof-rack with me. As extra inducement, I offered the boys a reward (a dollar each) if they spotted a python for David. We rolled quietly through the open forest, cuddling up in the evening chill as we spotlighted scurrying marsupials, sleeping birds, and occasional snakes. Both boys were asleep in my arms within an hour—but it was the best $4 I ever spent.

David studied Carpet Pythons in other spectacular places as well, including islands off the coast of southwestern Australia. Islands have a special place in the hearts of biologists, enshrined by the role of the Galapagos Archipelago in kick-starting Charles Darwin's thinking about evolution. Darwin realized that each island had its own distinctive type of finch and giant tortoise, a situation that surely arose because the original colonists had adapted to local conditions. Isolated on separate islands, each population follows its own unique evolutionary path. Small differences in climate, landforms, and fauna drive initial colonists along different trajectories, creating evolutionary divergences. That provides an outstanding opportunity for scientists to compare organisms from nearby areas but with separate evolutionary histories. But there's a downside to islands as well: they are graveyards for biologists who don't take the right precautions. Many researchers have been lucky to escape with their lives, after an accident such as snakebite, or storms at sea. Some island researchers never make it back home.

But David's first island site didn't involve such perils. Fifty kilometers (30 miles) south of Perth, Garden Island is connected to the mainland by a bridge. Most of the island is bushland, with plenty of snakes. It didn't take David long to answer the first question I asked him—which sex grows bigger, male or female? The adult female pythons he captured were huge heavyset animals, whereas the males were puny. A typical Garden Island male python was a meter (3 feet) long and weighed less than 250 g (half a pound). In comparison, adult females averaged more than 2 meters (7 feet) long and weighed 5 kg (11 lb). This sex difference was far greater than in Diamond Pythons or in east-coast Carpet Snakes. The sex difference in adult body weight was around 10% in the east-coast animals versus 90% in their western relatives.

The next question was: do the males fight with each other? As we expected from their dwarf size, the Western Australian pythons were tolerant souls. No combat bouts, no battle scars. Courtship involved a giant female surrounded by pygmy suitors. The Western Australian Carpet Pythons were like Diamond Pythons in that respect, but even more extreme. The adult male pythons on Garden Island were so small that I thought they were juveniles when I first saw them, whereas some females were bigger than any Diamond Python I'd ever encountered.

The story became even more interesting when we looked in other places. On one trip I accompanied David to the Recherche Archipelago, south of Esperance. In this almost-undisturbed region, the islands are remote and the seas are rough. Getting onto an island from the government boat was like a Boy's Own Adventure: the inflatable dinghy we used for the transfer was liable to be flipped over by the powerful waves that propelled us headlong into the beach. We were thoroughly drenched by the time we clambered onto shore. All of our equipment and spare clothing was stored in (allegedly) watertight drums, that sometimes made it to shore and sometimes did not. But the scene when I poked my head out of the sleeping bag next morning was worth the discomfort. Sandy beaches, thick scrub, and the opportunity to look for snakes in a place that few people had ever visited.

Although the habitat on the Recherche Islands looked nothing like Dryandra, Carpet Pythons were common. And intriguingly, the body sizes of males and females varied dramatically among islands. Female pythons always grew larger than males, but the difference between the sexes varied. The females were far larger than the males on some islands, whereas sexual near-equality was the rule in others. What was going on? The absence of male combat explained why females grew larger than males overall, but why were the sex differences greater in some areas than in others? Did the mating system also vary, sometimes giving an evolutionary advantage to ultra-miniature males?

No. The relative sizes of males and females depend on food not sex. Dryandra and the larger islands contain a diversity of edible-sized mammals, so that any snake can find a mammal that is just the right size to eat. Juvenile pythons start out on mice, then move up through rats, possums, and Woylies. A python switches from smaller to larger prey as it grows older. As we saw with Diamond Pythons in Sydney (where again, there are many different types of mammals), the female pythons at Dryandra grow only a little bigger than males. But in contrast, a small island contains very few kinds of mammals. Garden Island provides only two sizes of menu items for a Carpet Python—and those two types are very different in size. Male pythons specialize on House-mice and geckoes, and that's why the larger males have a lean and hungry look. They eat mice and small lizards as fast as they can, but it's impossible for a meter-long snake to grow any larger if all it eats is tiny rodents and lizards.

The other edible mammal on Garden Island is much bigger. Adult Tammar Wallabies weigh around 5 kg, and it would be a brave snake indeed that tried to overpower one. However, younger wallabies are small enough for a python to eat. That leaves a massive gap in prey sizes—no mice or lizards bigger than 20 g, and no wallabies smaller than 2 kg. If you can cross the gap, you have access to plentiful food, and you can grow to a monstrous size. Only female pythons manage that feat. They are half-starved at intermediate sizes, just like the males, but they keep on trying. Only a few succeed. The end result is a horde of small emaciated male pythons eating mice and geckoes, and a few enormous well-fed female pythons dining on wallabies. As a result, the difference in male versus female body sizes is far more dramatic than in other sites.

David tracked dozens of snakes for long periods, and he found many of the same patterns as revealed by our earlier studies on the other side of the continent. For example, the importance of "personality". I remember two medium-sized pythons at Dryandra Forest that lived in adjacent trees. They were the same sex and size, caught at the same time. One of them stayed in its home tree, rarely moving more than a few hundred meters away. The other snake was seized by seasonal wanderlust and went on major excursions—sometimes meandering for kilometers. The snake would have been in trouble in those open sites if a hungry fox appeared, but fortunately, poison-baiting at Dryandra (to protect marsupials) keeps foxes scarce.

Another result from David's radio-tracking was so bizarre that some of my scientific colleagues refused to believe it. If I hadn't seen it with my own eyes, I would have been skeptical too. When he went out to locate a snake, David sometimes found just the transmitter, sitting on the ground. His first thought was that a predator had killed the snake and eaten it, leaving the transmitter behind. But when that happens, we find bits of carcass, and teeth-marks on the transmitter. These transmitters were unmarked, with no sign of a dead python. And tellingly, the transmitter was inside a pile of python feces. Had the snake somehow crapped it out, and then wandered off? That seemed impossible; the units were placed into the body cavity, not the gut. These are totally separate compartments. But sure enough, David later recaught snakes that had lost their transmitters, and they were perfectly healthy. They had indeed passed the transmitter in feces. But how did they do it?

A python's ability to transfer a transmitter from their body cavity into their digestive tract, then poo it out, is impressive because the gut is full of microbes, whereas the body cavity is sterile. If I had even a small hole in my gut, I would develop a raging infection (septicemia) as bacteria spread into my body cavity. Without antibiotics, I would die. How do snakes avoid that fate? The answer relates to the way that pythons save energy between meals. Fasting pythons downregulate their digestive machinery in the weeks (or months) between dinners. The stomach shrinks, and its lining becomes thinner. When the snake finally catches another possum, the snake's gut swells enormously as it reactivates. And as it swells, the stomach folds around that transmitter that is lying in the body cavity, and transfers the irritating unit into the gut. After that, it's easy to jettison the transmitter. Dave Slip and I had seen Diamond Pythons do the same thing, but we needed David's larger sample size to publish the story in a scientific journal. My colleagues were skeptical until American research revealed the same ability in other snakes.

David's pythons made him unpopular with colleagues who were radio-tracking wallabies at the same sites. More than once, their study animals ended up inside his snakes. But David's biggest problem during his Ph.D. was more substantial than a few frowns in the tearoom. It involved a chainsaw applied to his groin while he was 4 meters (12 feet) up a gumtree. A snake was due for transmitter replacement, because the batteries were about to run out. She was nestled inside a hollow limb in a big tree, so David climbed up and began trimming away branches so he could open a hole in the limb. It didn't go as planned. The first branch that he cut swung away, hit another branch, and pushed it down against the chainsaw—pressing the whirling blade into his upper thigh. As David describes the incident, two world records were broken. One was for how much blood could spill out of a leg in a few seconds, and the other was how quickly a man could drop a chainsaw and climb 4 meters back down a ladder. Fortunately, the government department for which he works has strict safety protocols. Rules and training can't prevent a freak accident, but they put procedures in place for emergencies. David was soon in a plane on the way to hospital. Months of physiotherapy later, he was back on his feet, with an impressive scar running perilously close to a critical part of his anatomy.

The results from David's research were encouraging in terms of conservation. The Carpet Python in Western Australia is doing better than people had feared, at least in parts of its range. It's not all good news, because the wheatbelt area inland from Perth has been extensively cleared for broad-acre cropping, leaving only a few large bushland reserves. Over large areas, granite outcrops offer the only hiding places for a large ambush predator or the species it feeds upon. But in the bushlands and forests, and even on the fringes of suburbia, the Western Australian subspecies is achieving the same unlikely success as its relatives in the leafy suburbs of northern Sydney and the fruit-and-nut orchards of northeastern New South Wales. In retrospect, the concerns that had stimulated my studies into Australian pythons proved not to be major issues. But those concerns encouraged me to look closely at these magnificent animals.

Although the python work was a great success, I failed to put the results into a broader context. In retrospect, I can see opportunities that I failed to follow up: aspects of python ecology that should have led to Eureka moments. For example, Dave Slip's pythons were finding and exploiting resource hot spots—backyards—within an otherwise inhospitable landscape, where meals were few and far between. Likewise, Mark Fitzgerald's Carpet Snakes were taking advantage of a locally abundant resource created by human activities—feral rats in a macadamia orchard. David Pearson's Western Australian pythons were flexibly modifying their body sizes in light of prey availability. But I didn't realize the importance of those observations at the time. Everything we found was new, and general patterns were hidden amid the maze of details. It was only years later, as I saw many types of snakes zeroing in on scattered patches of high-quality habitat, and I saw massive variation through space and time in almost every trait that I measured, that I appreciated the importance of flexibility. It's difficult for a predator to find food in the Australian bush; and success depends on exploiting every opportunity that arises. Blinkered to think in terms of averages not variation, it took me years to recognize the central role of flexibility in the ecology of snakes.

From my own selfish perspective, there was an important lesson for my career. Increasingly, I was confident that the English postdoc who harangued me in the Canberra pub was wrong. Very wrong. I didn't have to choose between studying snakes versus doing good science. Despite what that postdoc had argued, snakes can offer superb opportunities for research on fundamental questions in evolution and ecology. Opportunities that had been neglected. Our papers describing ecological and reproductive effects on sex differences in body size were published in good journals, and frequently cited by other researchers. The future looked rosy for a young researcher determined to make his mark.

6 Between a Rock and a Hard Place

Most times when death stares us in the face, it's a quick glimpse. The car skids on the icy road, the roof of the house collapses, your heart stops beating, or a blood vessel in your brain bursts. It's all over before you can think about it. But sometimes, it's a much more drawn-out situation. That's where I found myself one sunny winter afternoon in 1969, on the cliffs above Tianjara Falls on the south coast of New South Wales.

The falls are spectacular—an amphitheater of steep cliffs around a stream that hurls itself over the ridge to fall in a vertical sheet of water 40 meters (120 feet) to the gully-bottom below. The valley is covered in lush forests, from which the vertical sandstone cliffs emerge into the sunlight. But as a young snake enthusiast, partway through my undergraduate degree, I was less interested in the view than in turning over thin sun-warmed rocks to find snakes. The snakes prefer rocks very close to the cliff edge, so I concentrated on those. A single misstep would have me plunging over the edge, and it would take me the rest of my life to reach the ground below.

But I was young and athletic, and I was having a great time. I found more Small-Eyed Snakes than I had ever seen before. And excitement led to recklessness . . . I took more chances. Looking under rocks that were closer and closer to the cliff edge, and without any forethought. I would crawl my way out to a good-looking rock, then realize that it was impossible to turn over that rock without losing my balance and toppling over the edge. Or discover that jumping down onto a ledge was easier than getting back up to the clifftop again.

My really, truly, amazingly stupid decision, though, was to investigate some flat rocks on a smooth rock ledge that sloped downward toward the abyss. I crept out on hands and knees and reached the first few rocks without difficulty, but I had to crawl on my belly to get to the rest. And then, as I reached a steeper part of the ledge, gravity overcame the friction of my body on the rock, and I began to slide slowly toward the chasm. I instantly gave up all thoughts of further rock-checking, and tried to crawl back up toward safety. But I only managed to scramble a short distance before I began sliding backward. And to my horror, I couldn't stop. My feet were perilously close to the edge, and the only thing that stopped me going all the way was the friction from my belt buckle.

In my memory, the cliff was 100 meters (300 feet) high, my legs went over the edge, and I lay there for hours. In truth, I suspect that the cliff was only half that high, no part of me actually dangled over the edge, and the episode lasted only 10 minutes. But 10 minutes is an eternity when any slight movement—even a breath—makes you slide downslope toward death in the ravine below. I could hear my friends talking

DOI: 10.1201/b22815-6

only a few meters away, but I couldn't call to them. If I had tried to fill my lungs, I would have slid over the cliff.

Nobody has hugged their lover as passionately as I embraced that rock. As I slowly moved my hands around, my fingers encountered minute bumps and hollows that gave me just a fraction more purchase—until those small bumps broke off. My fingernails were torn, my fingertips were bleeding, and my arms were aching. I doubted that I could escape from this predicament.

Eventually, though, I managed to inch my way upward, with intermittent slides backward. Overall, I was gaining ground. And a few minutes later, I could get to my hands and knees, stumble a few more meters, then collapse in a quivering heap in the bushes behind the cliff edge. And swear to myself that next time I would think about what I was doing before *I did it. I didn't always remember, of course. After all, I was 19 years old. But it was a sobering lesson.*

One day in 1980, I read a scientific article that changed my view of reptiles as much as George Williams' book had revolutionized my view of natural selection several years earlier. Written by a taciturn American biologist, Harvey Pough, the article was titled "The Advantages of Ectothermy for Tetrapods". It took aim at the classical view of ectothermy ("cold-bloodedness") as primitive and inefficient. Harvey made a powerful case for the opposite conclusion: that ectothermy is a sophisticated tactic to liberate reptiles and amphibians from the tyranny of high energy requirements. Mammals and birds ("endotherms") rely upon metabolic heat production to keep themselves warmer than the world around them. That imposes a huge cost. First, an endotherm has to eat almost constantly to fuel that furnace. Second, an endotherm has to be covered in insulation (fur or feathers) to stop that expensive heat escaping from its body. And third, it can't afford to have a large surface area relative to its body volume. More surface area allows more heat to escape, so a mammal or bird can't afford to be small or to be elongate. The only way to maintain a small surface area relative to volume is to be big and round. In contrast, ectothermy allows animals to be almost any size and shape, and to exploit niches where an endotherm would starve. Harvey said "The way of life of amphibians and reptiles, in contrast to that of birds and mammals, is based on low energy flow".

By the late 1980s, I was an enthusiastic convert to this "new view" of snake biology. My studies on Blacksnakes and pythons had convinced me that these are sophisticated creatures, intricately adapted to the harsh Australian environment. But I was struggling to translate that insight into real outcomes for conservation. My research was going well, and my publications list was expanding. But I was frustrated. I wanted to build this new view of ectothermic biology into a novel framework for research and management—one that incorporated the fundamental attributes of snakes, in ways that had been ignored by earlier workers. To forge a new path, rather than follow the approaches developed by conservation biologists who studied mammals and birds, and whose ways of thinking reflected the traits of their study organisms. It was time for a fresh start. But I didn't know how to do it.

Somehow or other, the way forward must lie in focusing on how ectothermic reptiles differ from endothermic mammals and birds. I wanted to incorporate those differences into the way I did my science. But how could I bridge the gap between "academic" studies on topics such as reptiles behaviorally regulating their body temperatures, and population-level issues such as the conservation of endangered species? Fortunately, a project arose that pushed me in the right direction.

The catalyst was my star student Jonno Webb. After he completed his Honors work on Blindsnakes, Jonno decided to continue his studies for a Ph.D. Snakes have always been Jonno's grand passion, so his preferred study animal was sure to have many scales and no legs. But what species should he focus on? There were a lot to choose from! One day in 1992, Jonno came to me with an ambitious proposal: to study one of Australia's most endangered snakes.

I was wary. To a pragmatic researcher, endangered species are a two-edged sword. The payoff could be huge—your information might bring a species back from the brink of extinction. And more cynically, endangered species are iconic, which makes it easier to publish papers about them in top journals. Although the anonymous reviewers for those journals try to be objective, they will be more excited about a story on Koalas than on fruit flies. But there's a downside to research on endangered species as well: by definition, they are scarce. It's more time-effective to work with common species than to chase near-extinct creatures, no matter how charismatic they are.

But Jonno had a killer blow to his argument. Although the snake species in question had disappeared from most of its former range, it was still abundant at one site on the southern coastline of New South Wales. There, at least, the snakes were still easy to find. And they were big enough—just big enough—for radio-tracking. I agreed that it was worth a try. Thus was born a project that is still going today, and has yielded one of the most detailed datasets ever gathered for any snake species worldwide.

To understand the ecology of these snakes, you first need to understand the rocks they depend upon. In the Sydney region, sandstone was laid down on the ocean floor during the Triassic period many millions of years ago. That sandstone is soft and porous, and easily eroded. The sandy soil that results has very few nutrients, but paradoxically supports a diverse and spectacular flora. To make a living in this harsh world, plants have evolved a range of distinctive forms, colors and ecologies. But it's a tenuous life—the magnificent forests of the New South Wales coast grow on a thin layer of sand. Whenever a windstorm follows heavy rain, the shallow roots fail, and huge trees are uprooted.

It's a magnificent landscape, because mineral inclusions in the sandstone create a kaleidoscope of colors. When the afternoon sun strikes the bare cliff edges, they glow golden in the waning light. The grandeur of these sandstone cliffs has attracted admiration from many people. In 1836, the great English naturalist Charles Darwin ventured to Blackheath, and wrote in his diary that it was one of the most spectacular chasms he had ever seen. And that says a lot, because Darwin wasn't impressed by Australia overall. By the time his ship (the Beagle) reached the island continent in 1836, Darwin was homesick and travel-weary. When he left Australia, he wrote in his diary "*Farewell, Australia! You are a rising child, and doubtless some day will reign a great princess in the south: but you are too great and ambitious for affection, yet not great enough for respect. I leave your shores without sorrow or regret*".

These words disappoint me (because I worship Charles) but I don't take it personally. He was feeling grumpy. Just a few weeks before, he had dismissed New Zealand even more abruptly ("*we were all glad to leave New Zealand. It is not a pleasant place . . . the greater part of the English are the very refuse of society. Neither is the country itself attractive*"). The All Black rugby team should remember those words whenever they play the English team.

Even in the depths of his homesickness, however, Darwin admired some parts of Australia. In a letter to his sister, Charles described the sandstone cliffs of the Jamison Valley as "magnificent" and "stupendous". But if Darwin had taken the trouble to turn over some of the loose rocks by the cliff edge, he would have found animals that (to my eyes at least) are even more magnificent and stupendous than the landscape around them. Show me a precipice, and I'll show you a high-class apartment for a discerning reptile. Clifftops attract snakes because of the unique accommodation that they provide.

Ectothermic animals don't have a single "preferred temperature"; sometimes they seek out heat (for example, when they are digesting prey, or are pregnant) and sometimes they need to be cool (perhaps to save energy, or to trigger seasonal cycles in reproductive hormones). Snake-keepers know that to keep their pet healthy, they need to provide it with a choice of temperatures. The same principle applies in nature. When choosing a home, reptiles look for a thermal mosaic where they can move around to choose whatever temperature they want, bask in a sunny spot, then retreat to a cool burrow. But if the species' ecology requires it to remain in one place—for example, if it relies on ambushing prey—then it needs a site that offers access to both warm and cool conditions within easy reach of each other. That requirement shrinks the real estate market dramatically.

How can a snake modify its body temperature without leaving the safety of its shelter? Most retreat sites don't offer that opportunity. In a dense forest, for example, everywhere at ground level is the same temperature because sunlight doesn't penetrate through the forest canopy. But add a rocky precipice, and the situation changes. A steep cliff doesn't provide any foothold (or more precisely, roothold) for trees, so the clifftop towers above the trees. As a result, the open ground along the cliff edge is exposed to full sunlight for at least part of the day. An experienced snake-catcher drools with delight at the sight of a thin flat rock on a sun-exposed clifftop. That rock will be a magnet for snakes from the surrounding area. Best of all are thin sheets of rock that create tight-fitting crevices between themselves and the bedrock. These "onion-skin" rocks or "exfoliations" are formed when the thin surface layer swells as it heats (by day) and shrinks as it cools (by night). Over thousands of years, the daily cycle of expansion and contraction causes the surface layer to break off, forming a snug-fitting crevice beneath. Bushfires can speed up the process by heating the surface rock.

The end result is nirvana for reptiles. In a sandstone landscape, the sun-exposed cliff edges—especially the western side, baked by the afternoon sun—provide perfect shelter sites. Inaccessible to hungry predators, a crevice provides a wide range of thermal conditions. In the late afternoon, a sun-exposed rock is much hotter than a shaded one. Lift one of those sun-exposed thin rocks, turn it upside down, and put

FIGURE 6.1 The sandstone escarpments that house Broadheaded Snakes are spectacular, but they require a constant focus on safety. A single step backward off the cliff could be your last. Photograph by Terri Shine.

your hand against the undersurface you have exposed. It's hot—around 40°C (104°F), the same as the rock's upper (sun-exposed) surface—because heat has flowed all the way through the thin rock. But now, put your hand on the patch of bedrock that the loose rock had been shading. It's cool—the same as under the shade of a tree. A reptile inside that crevice has a choice between the cool ground and the underside of the sun-heated rock above. An ideal apartment for an ectotherm: a sauna on the top floor and a walk-in coolroom in the basement. And rooms in between, with intermediate temperatures. Choose any temperature you like. Avoid the scorching temperatures at midday by keeping your head down. As the evening cools, warm yourself up by pressing against the underside of the top rock.

Why go to all this trouble? Because heat matters. Temperature determines how a reptile's body works, and even a small difference is vital. A hot snake or lizard thinks and acts more quickly than a cold one. That can translate into the difference between catching a prey item versus missing it. Or between escaping from a predator versus being killed. Or between digesting your prey versus having it decompose in your stomach. Or (if you are pregnant), having your embryos develop quickly and successfully versus being born small, slow, stupid, and too late in the season to feed before hibernating.

Staying warm isn't a problem for a tropical reptile. It's always hot, even at night. But the cool coastal habitat south of Sydney offers far greater challenges. The species of lizards and snakes that live under sun-warmed rocks along the plateau edge are the southernmost pioneers of evolutionary lineages from warmer places. Because they evolved in the tropics, their physiology works best under warm conditions. They thrive under the rocks at Tianjara Falls only because the local geology creates tiny pockets of subtropical conditions even during winter. Warm crevices under thin exfoliated rocks along the edges of steep unshaded clifftops. Herpetological heaven.

The same process occurs all over the world. Any amphitheater of sun-exposed rocky cliffs offers an ideal winter retreat for the local serpents. In the Sydney region, Diamond Pythons and Brown Tree Snakes love spots like this. In the dense cool rainforests further north, Small-Eyed Snakes seek out the same kind of habitat. In the rocky ramparts of Kata Tjuta near Alice Springs, Centralian Carpet Pythons bask on sunny cliff edges. It happens even in the tropics. On the idyllic island of Lifou, in the Loyalties chain in eastern New Caledonia, the dense forest gives Pacific Tree Boas little opportunity to bask in the sun. And so, the boas move to the clifftops above the sea, on the western side of the island, to soak up a few extra rays.

On the chilly clifftops of southern New South Wales, the thermal smorgasbord offered by rocky crevices is enjoyed by many native animals. The most specialized are Flat-Rock Spiders that can squeeze into even the tiniest space. They look like a normal spider that has been stepped on, flattening it so much that it's virtually two-dimensional. During her research on these creatures, my student Claire Goldsbrough gave Flat-Rock Spiders a choice of shelters and confirmed that they prefer hotter rocks. And by setting up laboratory incubators at a range of temperatures, Claire showed that the spiders benefit from their warm little worlds. Hotter spiders grow faster. The only downside of Claire's project was that baby Flat-Rock Spiders can escape from almost any container—they can literally squeeze through the eye of a needle. During the year that Claire worked in my research group, baby spiders took over the lab.

Claire's escapees gave me the heebie-jeebies. People often assume that a snake enthusiast will be comfortable with spiders, but I'm scared of them. Small spiders induce a jolt of adrenalin, and big hairy spiders terrify me. And unfortunately, some of the cliff-edge rocks are home to huge spiders—real nightmare material. And under hot rocks, the spiders are turbo-charged. As soon as I turn over a rock, any spider hiding beneath it hurtles off at a pace that defies belief, looking for alternative shelter. A frightened spider may decide to head up my leg, inside my pants, at a speed that makes cheetahs look like snails.

If you've never had a large spider imperil your gonads, you probably can't imagine what it feels like. At a rational level, I understand that the spider is just looking for somewhere to hide—but that knowledge isn't as comforting as it sounds. My companions laugh as I leap around grasping my leg at thigh level, to preclude arachnid access to my genitals, then frantically disrobe. But the danger is real. On a cliff edge, one false step can turn entertainment into tragedy. Never stand between the rock and the cliff edge as you turn that rock over. A giant spider hurtling up your pants might cause a sudden involuntary leap backward. Plan an escape route before you turn over the rock.

But fascinating as the rock-dwelling invertebrates are, they weren't the magnet that attracted Jonno and me to the rock outcrops. We were, of course, looking for snakes. And not just any snakes. The species we were looking for is one of the most beautiful, but cantankerous, serpents in Australia.

The hero of the story (or the villain, if you give it half a chance to bite you) is the Broadheaded Snake. Growing to a meter (3 feet) long, the Broadhead is resplendent in a yellow-and-black lacework pattern, as elegant as a high-fashion model on a catwalk. Broadheads also have wide flattened heads (giving them their common name) and muscular bodies. But the first thing you notice about a Broadhead is its attitude. These are snakes that demand to be left alone, and they fizz with nervous energy if you disturb them. They shift from slumber to aggression faster than a junkyard dog, and instantly pull back into a striking pose.

The Broadhead's spectacular threat display and aggro personality, coupled with its magnificent colors and rarity, made it a high-profile "pet species" in the early days of snake-keeping in Sydney. Unfortunately, a Broadhead makes a terrible pet. First, it vanishes as soon as you put it into a cage. The only time you see it is when you lift the shelter to check that the snake hasn't escaped. Second, Broadheads take offence easily, and bite anyone who harasses them. The results will be unpleasant, because the venom is highly toxic. Mercifully, most pet-keepers came to their senses, and they turned to more sensible choices like Children's Pythons. That way, if young Stanley from the house next door wants to play with your pet snake, you don't have to phone the ambulance before opening the cage.

The Broadhead's scientific name is *Hoplocephalus bungaroides*. You might expect a snake as distinctive as this one—brightly colored, unusually shaped, with a chip on its shoulder, living in the Sydney suburbs—to have been among the first Australian species to be described by science. As a result, there wouldn't be any confusion about the scientific name of this iconic beast. But my friend Glenn Shea (who glories in the arcane intricacies of herpetological history) discovered a litany of mistakes by early workers. The scientific name *Naja bungaroides* was originally given to a juvenile King Cobra from India. The King Cobra had already been described, based on somber-colored adult specimens. But juveniles are brightly banded, and a biologist who

FIGURE 6.2 Beautiful but deadly, Broadheads believe that the best line of defense is attack. Photograph by Sylvain Dubey.

found one didn't realize that it belonged to the same species as the duller-hued adults. The original description of this juvenile King Cobra as a new species in 1837 was brief, and the specimen was lost. Early Australian scientists should have given a new name to the nasty-tempered rock-dweller from Sydney but instead, they mistakenly concluded that it was the same species as the earlier-described *Naja bungaroides*. And so the name *bungaroides* was attached to the Sydney snake. It's been applied for so long that it's best to retain it, to avoid further confusion.

***

The genus *Hoplocephalus* contains three species—all broad-headed, all tree-climbers, all grumpy, and all occurring in eastern Australia. The closest relative of that pugnacious trio is a poorly known species from woodland areas of southwestern Australia (the Lake Cronin Snake), a distribution which hints that the broad-headed lineage was once widespread across the continent. As the climate dried out a few million years ago and the forests were replaced by desert, the tree-climbing snakes died out with the trees. The three *Hoplocephalus* species that survived that great extinction event are the ones from the eastern coast, where dense forests remain.

If we expand the Broadhead's family tree still further, the next closest relatives are deadly species like Tigersnakes and Copperheads. Broadheads are smaller, but they pack a lot of punch. Their wide heads provide plenty of room for venom glands. Nobody has ever been killed by a Broadhead's bite, as far as I know, but I've seen some forlorn faces in hospital beds after a brush with these feisty little elapids. The head is flattened as well as wide, helping the snake to squeeze into crevices under close-fitting rocks. But there's more to a Broadhead than its unusual head shape, its powerful venom and its elegant color. Broadheads are at home in trees. That's almost unique among the Australian elapids. When you find a snake up a tree in Australia, it is almost certainly a python (like a Carpet Snake) or a harmless colubrid (like a Green Tree Snake). It won't be an elapid. Although elapids outnumber all other Australian snakes, they prefer to live on or under the ground. None of them are strongly aquatic, and only the Broadheads (and a close relative, the Rough-Scale Snake) are adept climbers.

The Broadhead species that Jonno chose to work on (*Hoplocephalus bungaroides*) certainly looks like it climbs trees for a living. Its body is semicircular in cross section because of its flattened belly; and it has a sharp ridge running along each side of the belly-scales. All over the world, snakes with this body shape are tree-climbers. But strange to relate, nobody had recorded Broadheaded Snakes in trees prior to Jonno's work—the snakes were always found under sun-warmed rocks on the edge of cliffs, as I've described above. People speculated that they might go up into trees as well, but we didn't know for sure until Jonno caught some snakes (under rocks), implanted radio transmitters, and released them. Jonno's snakes moved up into the trees as soon as the rocks became too warm in summer, returning to the outcrop when cooler weather arrived. His Broadheads were rock-snakes from late autumn through to early spring (when it was cool), and tree-snakes the rest of the year (when it was hot). Or at least, adult males and non-pregnant females showed this seasonal shift. Juveniles and pregnant females tended to hang around the rocks all year.

Although Jonno's radio-tracked snakes commuted from the rock outcrops to the forests, they weren't frequent flyers. Whether it was under a rock or up in a tree, a Broadhead that found a comfortable crevice stayed there for the next few weeks. Just like captive individuals, wild Broadheads pick a good spot and settle in for a long wait. That's a safe option—predators are unlikely to find you—but you have to solve the problem of finding a meal. Stay-at-home habits only work if you have access to a food-delivery service.

When a Broadhead chooses its rocky apartment, it picks one where an unwary lizard (probably a Velvet Gecko) is likely to wander by. The snake then curls up inside its crevice and waits. And waits. The geckos leave their home rocks at dusk, to look for food and to check out the neighbors. To do that, they need to look under those exfoliated boulders. A lizard wandering across the exposed rock outcrop soon cools down, reducing its ability to detect and evade a predatory snake. But under a rock that has been heated by the afternoon sun, a curled-up Broadhead remains warm for hours by pushing up against the sun-warmed rocky roof of its home. When the snake detects an incoming lizard, it strikes quickly and accurately. Goodbye Gecko.

When the snake abandons its winter home for its summer house, all that really changes is the view. Broadheads are as selective about their summer homes as they are about their wintertime retreats. After it enters the forests, a snake may travel hundreds of meters before climbing a tree. Broadheads prefer large trees festooned with hollows and dead branches—home sites for warm-blooded treats like small marsupials, mice and bats. And those crannies and crevices also allow a Broadhead to lie in wait for weeks, almost invisible. The snake can move within hollow branches, seeking out warmer places without revealing its presence. Just as it did under the rocks in winter, the snake waits for dinner to arrive.

These small scrappy serpents must have been common in sandstone ridges along the New South Wales coast for millions of years, but they didn't last long after Europeans arrived and began to build Australia's largest city. In one of the first books on Australian snakes, published in 1869, Gerard Krefft lamented that Broadheads were already disappearing from the Sydney suburbs. The reason was apparent even then: people were destroying the outcrops by removing thin flat sandstone rocks. Attractive colors, convenient shapes, perfect for landscaping their gardens. Ideal for a rockery around the ornamental pond, or a stony mound to show off the orchids. The removal of those rocks eliminated the thermally distinctive habitats that were the core of this unique ecosystem. As flat sun-warmed rocks were transferred from natural outcrops to suburban gardens, Broadheads were doomed.

Sadly, more outcrops are destroyed every year. As the suburbs expand, thousands of new backyards generate a massive demand for landscaping, creating a commercial incentive for rock-thieves. When jobs are hard to find, it's an easy way to make some cash. Drive into the national park, fill up the truck with rocks, and sell them to an unscrupulous garden-supply company. And tragically, the most valuable rocks for landscape-gardening are large flat thin ones, light enough to carry. Exactly the same ones that heat up and cool down fast enough to create a perfect shelter site for a Broadhead. The National Parks and Wildlife Service does what it can, sealing off roads and prosecuting miscreants—and just as importantly, educating people. Sandstone rocks in your backyard are a symbol of environmental vandalism. You wouldn't steal a painting from the Art Gallery to hang in your living room, so why support the theft of something equally precious from the natural world?

Suburban landscaping isn't the only risk to the Broadheads' favorite rocks. Children throw them off the cliff for fun, campers build a circle of boulders to

FIGURE 6.3 Broadheads have to be handled with care—note the heavy gauntlets! Photograph by Terri Shine.

contain their campfire, hikers erect rocky cairns to make a summit point. And sadly, reptile-fanciers are to blame also. If a shelter rock is moved only slightly out of its original position, it becomes worthless to snakes. Displacement leaves an open gap along part of the rock's edge; it no longer fits closely to the ground. Dirt and leaves blow in through that gap, changing the thermal properties of the crevice. Reptiles avoid it. One of the most frustrating aspects of our fieldwork on Broadheads was the fact that reptile-hobbyists often trashed our study area. Broken rocks, misplaced rocks. Sometimes, the crushed body of a reptile beneath a recently turned boulder. It was sickening. Casual entertainment for these knuckle-draggers had devastating consequences for the animals they pretended to care about. It takes thousands of years for an exfoliated rock to form, but only a few seconds to destroy that unique habitat. And that destruction is a death sentence for reptiles.

The more we learned, the more worried I became. The numbers of Broadheads were declining, and the survivors were concentrated in a few small patches. Year by year, the rock-thieves were encroaching further. Entire ridge-tops were being denuded of loose rock. It was awful to walk around these sites after they had been trashed by vandals. I could still see the "scars" on the bedrock where boulders had lain for millennia but were now gone forever. And it was difficult to keep the criminals out. When the National Parks and Wildlife Service tried to exclude vehicles from our study area, four-wheel-drive enthusiasts tore the new gates off their hinges in the name of "freedom of access".

The situation was precarious. If our study site contained the only remaining healthy population of Broadheads, a weeks' effort by illegal bushrock collectors might shatter the species' last bastion. I hoped that the remote ridgetops around our site also supported this endangered snake, but I wasn't optimistic. These snakes need a combination of habitat characteristics. A Broadhead needs a winter home (a rocky crevice with specific thermal properties) *plus* a summer home (a large tree with numerous hidey-holes), and the two places need to be close together. Without the right combination of habitat types, the snakes can't survive. We accompanied National Parks rangers on helicopter-based surveys to otherwise-inaccessible outcrops, in the hope of finding other snake populations. Although we saw some marvelous country, largely untouched by European hands, the Broadheads weren't there.

So, all of our Broadhead eggs were in one basket. This was a high-stakes game. A gang of rock-thieves could rapidly destroy our entire study area. There was no way to keep the vandals out, and as long as people pay money for bushrock, other people will steal it.

Jonno and I could see only one way out of this nightmare—we had to find a way to replace the rocks if they were stolen. If we could construct artificial rocks to serve as substitutes, we could repair the damage caused by rock-thieves in weeks not centuries. But critically, we needed to have this "insurance policy" in place *before* the natural rocks were stolen from our study area.

Designing an artificial alternative to replace those flat thin rocks wasn't a simple challenge. The new rock had to fit perfectly onto the bumps and hollows of the bedrock, to form a tight-fitting crevice with the same thermal characteristics as the original. Precast pavers or concrete blocks wouldn't work. Jonno and I agonized over designs for fake rocks, and asked colleagues for advice. A brainstorming session with John Weigel, the director of the Australian Reptile Park, was especially helpful. But every method we tried had drawbacks. For example, we tried pouring concrete onto a plastic sheet in situ, so that the bottom of the concrete cast conformed closely to the base rock. Take the plastic away, and we had a new rock—durable, heavy, and of no value to rock thieves. But it took too long to be feasible on a broad scale. We needed a rock that was cheap, could be constructed in bulk, and was adaptable to any surface.

The Australian Research Council funded me to develop a workable design, with support from Forests New South Wales, Zoos Victoria, and the Reptile Park. In the end, we found a simple answer. A tough but thin ferro-cement rock, smooth on top (molded from a real rock, and with pigment added to match the outcrop) but with a hollow etched into its underside, to form a crevice beneath it. To obtain that all-important close fit along the edges, we used a compressible rubber seal that closed off the perimeter except for a single small opening. Light enough to carry, fitting to the bedrock, offering a range of temperatures. A cosy replacement home for a rock-dwelling reptile.

But importantly, how big should we make the crevice beneath the rock? How thick, how wide, how long? Our student Ben Croak measured the natural crevices used by snakes and lizards, by injecting expandable foam into the space beneath a rock (after removing any lizards and snakes that lived there). This foam swells up to fill the space available, then solidifies. A day later Ben could lift off the top rock, and have a perfect foam cast of the crevice between the upper rock and the bedrock. By carefully measuring those casts, Ben identified the ideal dimensions for a reptile's home crevice.

Our ferro-cement rocks fitted snugly to almost any area and created a crevice that would (we hoped) attract rave reviews from Broadheaded Snakes, Velvet Geckos, and Flat-Rock Spiders. It looked like a real rock, so didn't irritate passing bush-walkers. And because it was concrete, it had no value to rock-thieves. Turning that idea into reality still took a lot of work. We paid a manufacturer to construct 200 rocks, then Ben deployed them in areas where Broadheads occurred (or used to)—especially, areas where rock-thieves had destroyed the habitat. It isn't easy to carry concrete rocks up steep cliffs. Some of the best sites were accessible only by leaping across chasms and clambering up towering cliffs through spiky bushes. And if anything goes wrong, it's difficult to get back out. On one trip, my wife Terri came along to photograph the research team. We were clambering along a steep cliff-face when Terri tripped, and her foot jammed between two boulders as she fell forward. A sickening crack told us that the ankle was broken as well as dislocated, and it took an hour to carry her down off the mountain. Terri claims that the worst ordeal came later, when she had to tolerate my cooking until she was off her crutches.

Placing all those fake rocks out into the right places was only the first step. Would snakes use them? The first field season was tantalizing. Spiders and lizards flocked to the rocks, but the snakes stayed away. We told ourselves that Broadheads only move around in autumn, so we wouldn't find them under our rocks until then. I'm not sure that we really believed it—but to our joy, that's what happened. As the weather cooled down, the Broadheads moved back to the outcrops—and soon, we were finding snakes under our artificial rocks.

I breathed a huge sigh of relief. Even if vandals stole the natural rocks from our study area, we could reverse the damage. So far, it hasn't been necessary. The National Parks rangers are doing a great job of protecting the area. Our research played a role there as well; through collaborating with us, the rangers became more aware of the importance of the problem. But if the worst comes to the worst, and

FIGURE 6.4 Jonno Webb checking under one of our fake rocks, hoping to find a Broadheaded Snake, while I watch on. Photograph by Terri Shine.

some idiot despoils our study site, we don't have to wave goodbye to the Broadheaded Snake. We have a simple way to restore critical habitat for an endangered species and the other wildlife species on which it depends.

But hands-on conservation biology is almost always frustrating: solving one problem reveals another threat, unseen until you deal with the first issue. Rock theft isn't the only peril confronting Broadheads. Even in sites where the rocks remained undisturbed, Jonno noticed a worrying pattern. Some of the best rocks—the ones with perfect crevices, exposed to the afternoon sun—lost their snake-attracting power as the years went by. Jonno's capture records revealed that more and more of our prime rocks were snakeless. And the reason was clear. Trees were growing up along the cliff edge, shading rocks so that they fell out of favor with our scaly subjects. Unless it was fully exposed to the afternoon sunlight, a rock was too cool. To test that interpretation, we trimmed some of the offending branches with garden clippers. Sure enough, the rocks warmed up, and the snakes began to use them again.

Shading of those preferred rocks wouldn't be a problem if new rocks were being exposed to sunlight at the same rate that others were being shaded. For example, new rocks sometimes become available when trees fall during a windstorm. But losses outweighed gains. Many of our prime rocks were no longer popular with snakes, and very few new snake-attractors had emerged. Was the forest becoming denser overall? To answer that question, we examined old aerial photographs of our study area to measure the extent of open area exposed to the afternoon sunlight. Sure enough, the area of bare rock had halved from 1941 to 2006. We don't know why. Perhaps it was because foxes exterminated the rock wallabies, whose grazing had prevented trees from growing on the rocky slopes. Perhaps it was a change in fire regimes, with the loss of Aboriginal "firestick farming" as Indigenous culture was disrupted by European colonists, and then exacerbated when many Aboriginal people left their cultural homelands. Perhaps climate change was at work; tree growth responds to rainfall, temperature and levels of carbon dioxide. But whatever the reason, more and more trees were shading the once-open clifftops—and if that trend continued, Broadheads would disappear. A fake rock is a waste of time if it isn't exposed to the afternoon sunlight.

What could we do? Forests are a lightning rod for the politics of conservation. Some of Australia's most passionate environmental debates have targeted forestry operations. The mantra of the green movement is "more trees are better", but that's not always true. For an endangered snake that relies on sun-exposed rocky outcrops, forest regrowth is a threat. To see if localized clearing could help the snakes, we asked the Park managers to let us cut down trees that were shading otherwise-suitable rocks. I admire the courage of those managers—imagine their problems if the newspaper ran a scare story about scientists chain-sawing the heart out of the wilderness. For his Ph.D., David Pike identified which trees to remove, and then spent a day with a professional tree surgeon who wielded his chainsaw as adeptly as a fine musician uses a violin.

For the next few years, David monitored wildlife numbers under the rocks on cleared sites, as well as on nearby "control" locations where we didn't cut down any trees. Before long, the reptiles moved back in to the newly cleared areas. Our prime rocks, deserted for years, reemerged as popular choices for Broadheads.

Should we scale up this study? Go ahead and thin the forest at many cliff edges, to protect Broadheads? I think we should. The woodlands in this region are vast, and the amount we cleared for David's project was less than one-tenth of 1% of the forest—but it doubled the habitat available for Broadheads. It would take political courage to intervene and "manage" a protected forest—many voters would view chainsaws in a National Park as an obscenity. But the alternative is to lose this remarkable Sydney ecosystem.

The same battle is occurring all around the world. Many well-intentioned environmentalists believe that natural ecosystems will take care of themselves if we just leave them alone, but that's a recipe for disaster. Forest thickening in southeastern New South Wales is due to processes such as the extinction of native herbivores in the jaws of feral foxes; climate change; and the loss of traditional Indigenous burning practices. Those issues won't be solved by locking the gate and walking away. Today's problems will grow into tomorrow's catastrophes.

****

Our study area for Broadheads is surrounded by wilderness, with a fantastic fauna that includes supersized reptiles. Diamond Pythons are common, creating a risk for anyone who relies on color pattern to identify snakes. The similarity between Broadheads and young Diamond Pythons has fooled several unwary snake-collectors. The heaviest local reptile, though, is a lizard not a snake. The black-and-yellow Lace Monitor or goanna grows up to 2 meters (6 feet) long and can weigh 10 kg (22 lb). It's one of the largest lizard species in the world—although not too long ago, Australia had much larger lizards. One of them was the awesome Komodo Dragon, that can grow more than 3 meters (9 feet) long and weigh more than 100 kg (220 lb). Sadly, the Komodo Dragon is now extinct in Australia, and is restricted to a few small islands in Indonesia. But even the mighty Komodo Dragon is far smaller than the monstrous goanna *Varanus prisca*, which roamed the Sydney basin 50,000 years ago. More than 6 meters (18 feet) long, weighing over 1,000 kg (2,200 lb). Now, *that's* a lizard. It must have added some excitement to the day-to-day lives of early Aboriginal colonists of the region.

A Lace Monitor is happy to eat just about anything that moves. And also, anything that once moved but has since died and is now a decomposing mess. Lace Monitors can often be seen beside the highway gorging on the putrid remains of a road-killed kangaroo. And Lace Monitors are bold, often entering tents and cabins in search of an unsecured pack of bacon. Jonno was lying in his tent one morning when a goanna snuck in, seized one of his socks, and then took off with its prize. After a week of fieldwork, the sock probably smelled like a dead kangaroo. Jonno needed that sock so he took off in hot pursuit, but it was to no avail. The goanna climbed a tree, bashed the sock against a branch a few times to make sure it was dead, and proceeded to swallow it. Jonno checked every pile of goanna poo he found for the next few weeks, but his sock had disappeared forever.

Despite their size, though, goannas aren't the apex predators in that ecosystem. The top dogs are dogs—or, more accurately, dingoes. We often saw their tracks on

FIGURE 6.5 Lace Monitors are one of the largest species of lizards in the world, and an important part of the sandstone ecosystem. Photograph by Sylvain Dubey.

dirt paths, but rarely saw the animals themselves. One day, however, Jonno flushed a large goanna as he walked through the bush—and had barely recovered from the shock of a giant lizard erupting in front of him when two dingoes materialized and tore the lizard to pieces. And there were other nerve-wracking episodes too, like realizing that you've blundered into a marijuana plantation in remote bushland and may be under close observation by people with firearms. And late at night, drunken four-wheel-drive enthusiasts who demand your help to pull their vehicle out of the creek where they have managed to bog it.

One of Jonno's most memorable experiences in the Broadhead research program involved an overseas colleague. Soon after Chinese scholar Weiguo Du joined my research group, he accompanied Jonno on a trip to survey Broadheads. Jonno explained that the snakes were dangerously venomous, and he gave Weiguo a thick welder's glove to wear on one hand. The protocol is simple—turn the rock with your ungloved hand and grab the snake with the gloved hand. Unfortunately, when Weiguo found his first Broadhead, he seized the snake with the wrong hand. And Broadheads being Broadheads, the outcome was inevitable. Any opportunity to bite you, and they will.

Initially, Weiguo wasn't worried. A Chinese snake that size wouldn't be too dangerous. But soon Weiguo felt light-headed, so he sat down and called Jonno over to tell him what had happened. Jonno insisted they head straight to hospital—which Weiguo thought was an overreaction until he stood up, and everything went black. He completely lost his vision. It was terrifying. In a strange country, in the middle of the bush, beside a steep cliff, unable to see, and with snake venom coursing through his body.

Jonno carried Weiguo back to the car, through the thick and spikey bush and across the uneven ground—even though Weiguo is bigger than Jonno. But they made it, and Weiguo was in a hospital bed in Nowra within 2 hours. He stayed there for a few days and made a full recovery. All's well that ends well, and Weiguo is now an eminent professor at the Chinese Academy of Sciences.

One advantage of long-term studies—and Jonno's work on Broadheads has been going on for more than 20 years—is that we can follow populations through time. That's taught us many things. For example, we had feared that wildfire might be devastating—our Broadheads would be trapped under the rocks and incinerated. But in fact, the Broadheads were unaffected when a huge forest-fire finally went through. Without vegetation to fuel it, the fire didn't burn as fiercely on the open rocks as it did in the forest. Reptiles in the undergrowth, like Small-Eyed Snakes, were in far more trouble. Broadheads that were sheltering in hollow limbs in the giant gum-trees survived also, because they were above flame height. And in the longer term, the fire benefitted the Broadheads by removing trees that had been shading otherwise-ideal rocks. For a year or two after the fire, though, the open canopy created a different threat. Foraging snakes were easy targets for predatory hawks. Fortunately, Broadheads rarely ventured out into those open areas. They stayed within their rocky homes while more mobile snake species were converted into bird-poo.

Any field biologist who works at the same site, year after year, develops a rich and complicated set of memories. Some years the forest is thick and green, after heavy rain. Other years it's dry and brown, as drought bites hard. If windstorms blew through the area before our annual Broadhead surveys, we had to chainsaw our way through the fallen trees that blocked the road. There were years when it seemed like there was a snake under every rock, and others when we ended up with a small tally after days of back-breaking work. The massive wildfire that swept through the site in 2001 changed the area dramatically, but left patches of unburned scrub amidst the devastation. Tiny accidents of wind speed and direction made the difference between black ash and green leaves. And through time, seeing the forest in all of its moods, I began to realize the huge importance of chance and change. A Broadhead Snake that lives for 20 years—as many of them do—doesn't experience a constant world. Year-to-year changes in the forest habitat have powerful impacts on the life of such a snake. How much prey is available, where that prey is to be found, how best to capture it. When to move from the rocks to the trees, and back again. How often to shift from one rock crevice or tree-limb to another. Whether or not to go ahead and breed this year. In an unpredictable world, the only tactic that pays off is flexibility. Assess what's going on, and change your behavior accordingly. No rigid rule will succeed. To thrive in the ecological lottery that is the Australian bush, you need to roll with the punches.

I have participated in the annual Broadhead surveys less often as the years have passed. Partly due to my age—turning rocks hour after hour, day after day, is brutally hard labor. But also, the pressures on my time increased. I needed to serve on more committees, take on more administrative tasks, and fill out an infinitely expanding array of forms. I found myself waving goodbye to students as they embarked for the field, then turning back to read the agenda papers for another meeting. My colleagues were trapped in the same way. People who became biologists because they loved fieldwork were dragged into a suffocating maelstrom of bureaucracy. Once irrepressibly enthusiastic, my friends began to talk of early retirement.

To escape that downward spiral, I refused requests that would take me away from research (which I was good at) into administration (at which I was patently incompetent). My selfishness saved my soul. And by maintaining my focus on research, I obtained fellowships that funded me to find out about snakes rather than to discuss the allocation of parking spaces behind the Biology Building. I was more useful to my institution as an enthusiastic researcher than I would have been as a reluctant bureaucrat, and it was a lot more fun.

But although research remained my professional focus, the nature of that research changed. My body was feeling the ravages of age, football, and parenthood. As I became less agile, I grew more strategic. I know where the best rocks are for Broadheads, but I can no longer clamber across to them—so I just point them out to my nimble students. Nonetheless, it's great fun to turn over a favorite rock and find the same snake under it as I did last year. A bit like meeting an old friend (although in this case, he/she immediately tries to punch me in the nose). I like Broadheaded Snakes. They are elegant, feisty, and now a bit less endangered because of our work. But no matter how much fun it is to catch a Broadhead, it's not worth hurtling over a precipice. I still remember slipping toward the edge of the cliff at Tianjara Falls, all those years ago.

Broadheads clarified my thinking about many issues in conservation science. For example, why does a species become endangered? The answer usually involves more than one factor and for Broadheads, it's a triple-whammy. Three factors have driven them to the brink of extinction. First, they live in the wrong place. Any animal that shares its home with five million people is bound to be in trouble. Urban development is encouraging rock theft, encroachment of weeds (that can shade otherwise-ideal rocks), and half-wild cats. Second, the Broadhead relies on a distinctive, scarce, non-renewable resource. It needs access to sun-warmed rocks in winter and large hollow-bearing trees in summer. That combination is difficult to find except where tall cliffs emerge above the shaded forest. And third, the Broadhead's reliance on ambushing its prey means that it feeds infrequently—and so it lives in the slow lane. A female Broadhead has only three or four babies in a litter, is pregnant only once every two or three years, and her offspring take six years to mature. That laid-back lifestyle is a recipe for disaster because a small increase in mortality rates can send the population into free-fall, and recovery is slow even if the threat disappears.

Ambush predation introduces another risk factor as well: low mobility. A snake that spends weeks at a time under a rock, waiting for a gecko, isn't likely to move very far. That may explain why Broadhead populations are so patchy, even in suitable habitat. If a local population is wiped out, immigrants from the ridge next door won't arrive any time soon. When Jo Sumner and Scott Keogh looked at the genetics of the Broadhead, they found that the species is composed of two different races—a northern one and a southern one. The gap between the two groups occurs in a rockless region south of Sydney. That no-go zone means that over evolutionary time, the snakes in northern and southern areas have rarely exchanged genes with each other. The resulting differences exacerbate the conservation challenge; if there are two distinctive types of snakes, then saving Broadheads in just one part of their range isn't enough.

Overall, then, ambush foraging lies at the heart of the Broadhead's problems. Other snakes in the same area search for their prey at night, while most of the edible lizards are sleeping. If times are good, a searching snake can find a lot of food. But an ambush forager doesn't have that flexibility—it has to wait for the prey to arrive. And so it eats less often, and grows slowly, and reproduces infrequently. Bob Reed and I found that in general, Australian snake species that rely on ambush foraging are the ones most likely to have become endangered. There are exceptions, like Carpet Pythons, but the general pattern is clear. When Europeans arrived in Australia two hundred years ago, native animals that depended on either a specialized foraging mode or a specialized habitat were in big trouble. Rely on both, and you're in even bigger trouble. Live near Sydney, and you are hit by a triple-whammy.

****

As Jonno's research shone a bright light on the ecology of Broadheaded Snakes, I began to wonder about their relatives—the two other species in the genus *Hoplocephalus*. And there was a good reason to investigate them, because all three *Hoplocephalus* species are becoming rare.

Endangered status often runs along family lines. If a species is endangered, so are its relatives. Why is that? What attributes put a species at risk? Are the other *Hoplocephalus* species also ambush foragers? Do they move seasonally between rocks and trees? Do they have "slow" lifestyles? Are their habitats vulnerable to human activities? Understanding those issues might explain why one species becomes endangered whereas another does not.

The man who answered those questions was Mark Fitzgerald, the banana-farmer from Mullumbimby who keeps a thesaurus beside his bed. I introduced Mark in an earlier chapter, as he radio-tracked Carpet Snakes in north-coastal New South Wales. My next collaboration with Mark took place nearby, when he decided to do a Ph.D. under my supervision. His project centered on a *Hoplocephalus* species that is a little larger than the southern Broadhead, is just as beautiful, and has just as big a chip on its metaphorical shoulders. The Stephens' Banded Snake is an animal of the thick wet forests, not the rock outcrops. Its hues are more subtle than those of its southern relative, but its pattern is bolder—cream or pale brown between black bands. Adults sometimes reach well over a meter (3 feet) in length. Living further north—and thus in a warmer climate—it doesn't need sun-warmed rocks. It can stay warm year-round inside the hollow branches of trees. Stephens' Banded Snakes occasionally shelter in rock outcrops among the forest, but they spend most of their time in trees. The best way to find them is to cruise slowly through the forest at night and look for snakes crossing the road.

For his Ph.D. Mark cruised forest roads, caught 16 snakes, and put transmitters in them. His radio-tracking revealed a species that is similar in some ways to its southern relative, but different in others. It's a secretive beast, spending long periods waiting in tree-hollows to ambush small mammals. It only comes down to the ground to move from one tree to the next. But for a few weeks in autumn, when young bush-rats are leaving their mothers to fend for themselves, the snakes spend more time on terra firma. Baby rodents are less wary than their parents, and so are sitting rats for a snake in ambush in the low sedges. Like Broadheads, Stephens' Banded Snakes stay hidden. They keep warm by moving within hollow branches to a place where sunlight hits the branch, rather than by doing anything as extrovert as basking. As a result, Mark rarely saw his radio-tracked snakes after their initial capture. All he'd get is the beep-beep-beep of a radio signal coming from a tall tree. Not much emotional reward for 2 hours clambering through dense forest, with lawyer vines tripping you over in every gully.

Rainforest looks attractive in a photograph, but it's a nightmare as a workplace. Without radio transmitters, it would be impossible to study rainforest reptiles. But Mark is tenacious. Bit by bit, he assembled a comprehensive picture of the lives of these mysterious animals. Parts of the story were strikingly similar to Jonno's results on Broadheads. The snake's ambush lifestyle restricts their food intake, and thus their rates of growth and reproduction. Reliant on a specialized ecological niche, with no margin for error, any deterioration in their habitat has devastating effects. Stephens' Banded Snakes depend on really large, really old trees with abundant hollows. A dead tree can remain standing for decades, with more hollows forming every year. As a result, elderly trees are critical habitat for Stephens' Banded Snakes (as they are for Broadheads, further south). Unfortunately, big old trees get in the way of timber production, so commercial operators knock them down to make way for healthy young replacements. And even when regulations protect the forest giants, it's difficult to convince foresters to leave enough middle-aged trees to replace the

FIGURE 6.6 Shy inhabitants of thick forest, Stephens' Banded Snakes spend most of their lives high in the treetops. The only time we can catch them is when the snake leaves one tree and meanders across the ground to find another. Photograph by Stephen Mahony.

oldsters when they finally fall. Saving the Stephens' Banded Snake is about maintaining a wide age range of trees within the forest. Easy to say, but hard to achieve when many of the forests are protected in national parks, and the foresters are intensively harvesting the only ones that remain accessible to them.

The third *Hoplocephalus* species is the Pale-Headed Snake, a smaller species that extends into hotter drier parts of the continent. Its body lacks the patterning of its coastal relatives—most of the body is pale brown to grey, without bands or flecks—but reflecting its links to more colorful relatives, it has a cream nape and a pale grey head. The biggest Pale-Heads I've seen have been around 80 cm (30 inches) long. We know less about the Pale-Headed Snake than its larger counterparts, but my dissections of preserved Pale-Heads in museums were not encouraging: the species showed that same combination of slow growth, infrequent reproduction, and small litters that render the other *Hoplocephalus* so vulnerable to habitat deterioration. To fill in the gaps, Mark collaborated with snake-keeper Brian Lazell to radio-track Pale-Headed Snakes in the Pilliga Scrub, beside the Namoi River. It's sun-drenched open country, similar in many ways to the Macquarie Marshes where I had studied Blacksnakes many years before. The real Australia, far from the equable coastal lands. Searching with flashlights at night among iconic River Red Gums and Coolibah trees, Mark and Brian found 17 snakes. All of them were close to the river, and most of the snakes were up trees.

The radio-tracked Pale-Heads sat inside tree hollows for long periods, with one female staying put for two months. They were ambush predators. Occasionally a snake moved from one tree to the next, usually about 50 meters (50 yards) away, but later returned to the original tree. Reflecting their smaller body size, Pale-Heads feed mostly on frogs and lizards, not mammals. But sadly, Pale-Heads are in as much peril as their relatives. Although these secretive little snakes live far from cities, the semiarid zone has been hammered by commercial agriculture. The trees have been cut down to make way for crops, and the rivers have been starved of water.

The plight of the genus *Hoplocephalus* symbolizes the challenges facing Australian wildlife. In the south, Broadheaded Snakes are under attack from urban development, and the theft of surface rocks. Forest thickening has added to their problems, whereas ironically, the other *Hoplocephalus* species have the reverse predicament. Trees that house Stephens' Banded Snakes are being cut down for forestry, and trees that house Pale-Headed Snakes are being bulldozed for agriculture. By degrading the habitats around them, the European colonists of Australia are dooming these elegant snakes.

But knowledge is power. Research by Jonno, Mark, and Brian has told us a great deal about the biology of the *Hoplocephalus* species—enough to save them from extinction, if we have the political will to do so. Will that happen? The jury is still out because sadly, snakes rank near the bottom of the political priority list. Most Australians don't care about venomous snakes that spend their lives hidden under rocks or up in trees.

Our research on *Hoplocephalus* was motivated by the elegance of these hot-tempered serpents, by a desire to discover why their numbers were declining, and by a determination to do something about it. But those studies had another benefit as well; they shone a bright light on connections between reptile biology and conservation. The Broadheads forced us to investigate issues such as temperature profiles underneath rocks, three-dimensional characteristics of crevices, and changes in the availability of sun-exposed bare rock within a forest ecosystem. Mammal biologists and bird biologists would have focused on very different factors. But by looking at the Broadhead's world from an ectothermic perspective—with an emphasis on subtle variations in temperature through space and time—we had framed new ways to conserve these endangered animals.

In many respects, these elegant tree-snakes embody the ectothermic lifestyle. Energy-misers, they have evolved a lifestyle unthinkable for any warm-blooded animal. For a Broadhead, the rules are simple. Find a place where an edible lizard will eventually pass by, and wait for that to happen. Don't waste energy on moving around. If one lizard arrives per month, you're in profit. You can use that energy to grow and eventually, reproduce. If times are bad—it's two or three months between lizards—you just hunker down and wait. No growth, no reproduction, but you can survive until things improve. Snakes are tough. They have evolved to deal with drought and bushfire. But when humans enter the picture, the equation changes. The rocky crevices that once offered shelter are gone, due to rock theft or forest overgrowth. The tree-hollows are disappearing, as the forest is plundered for timber. In the shelters that remain, a snake can't select the temperatures that it prefers. The system is out of kilter.

In contrast, mammals and birds are high-energy systems. To stay alive, they need to eat. Constantly. The risk of starvation is never far away. A population of endotherms cannot persist if its food supply is cut—even briefly. And hiding in a safe shelter for weeks at a time isn't an option for an endotherm, except for specialist hibernators in very cold regions. An endotherm's high metabolic rate soon depletes its reserves. If a devastating new predator from another land arrives in the local ecosystem, the frenetic endotherm can't avoid an encounter. Mammals and birds only know one pace—full steam ahead—and their lives are dominated by the need to fuel a high metabolic rate, just to keep their bodies warm. In environments where they can meet those energy requirements, endotherms thrive. But if something goes wrong—perhaps due to the ecological vandalism of *Homo sapiens*—dependence on a high-income lifestyle can be fatal. The effects of humans can be just as devastating for an endotherm as an ectotherm, but those impacts play out in different ways and on different timescales.

That difference is a challenge, because the science of conservation biology was founded by Americans and Europeans who studied mammals and birds. As a result, we know how to maintain viable populations of mammals and birds in the stable, productive lands of the Northern Hemisphere. But if we want to conserve reptiles—especially in a harsh and unpredictable Australian landscape—those approaches will be as useful as a screen door in a submarine. We need to abandon the old ideas and think anew. *Hoplocephalus* set me on that trajectory.

7 Snakes in Need of a Defamation Lawyer

I never met a snake I didn't like, but some strike a deeper chord with me than others. If I had to nominate my personal favorites, I'd admit an inordinate fondness for the genus Pseudechis. *That's a group of about ten species, a few of which look so similar to each other that they haven't yet been formally described by science. The group is usually called "the Blacksnakes" in honor of their best-known member (the Red-Bellied Blacksnake), even though most* Pseudechis *species are brown not black. They are all fantastic creatures, but one in particular—found over most of Australia—epitomizes the majesty of really large, really venomous snakes. And the Fear Factor worked powerfully on me as a young man; there's something special about a snake that scares the living hell out of you.*

Because it is distributed so widely, the species in question has several common names. The beast that South Australians call the "Mulga Snake" is the "King Brown Snake" in the Top End. Australians like abbreviations, so this giant brown serpent is known as the "Kingie" by snake-fanciers. And it is indeed a king—a monarch among reptiles. It's big, it's beautiful, it's deadly, and it has attitude.

As I've already mentioned—several times—snakes are incredibly flexible. They adapt to local conditions, opportunities, and challenges—and on a long timescale (through evolutionary changes) as well as a short timescale (by adjusting their behavior to a drought or cool weather). The Kingie is found over a vast range of habitats, from tropical savannas through to the Red Centre's sandy deserts, and the agricultural lands of the southern continental interior. Southern Kingies are dark and small, whereas the pale-brown monsters from the wet-dry tropics grow to 3 meters (9 feet) in length, larger than most pythons. And Kingies are heavy-bodied, large-headed snakes rather than slender whip-like speedsters like Taipans or Brownsnakes. A Kingie oozes power not velocity. Imagine a Red-Bellied Blacksnake or Indigo Snake painted brown, with a cream edge to each body scale.

During my early career, I met King Brown Snakes occasionally during fieldwork in southern Queensland and western New South Wales. My most enduring memory is of a feisty midsized animal that resented my attentions and came straight at me to retaliate. So, I ran away. It was an inglorious retreat for a "snake expert" who had often asserted that snakes don't chase people. The snake soon tired of the entertainment, but I realized that a big Kingie might be a challenge to subdue.

That prediction was confirmed in the mid-1980s, during one of my first trips to Darwin. My dissections of museum specimens showed that although female Red-Bellied Blacksnakes give birth to live young, all of the other Pseudechis *species are egg-layers. Variation among closely related species is a terrific research opportunity—it lets you compare species that are similar in most ways but differ in the characteristic*

DOI: 10.1201/b22815-7

of interest. In this case, I realized that I could look at how the evolution of live-bearing had modified the process of embryonic development. All I had to do was set up some Red-Bellied Blacks and King Browns in the lab, breed them, and study the embryos.

The first step was to catch some snakes. Red-Bellies were easy, but Kingies were harder to find. When I applied to the Northern Territory wildlife authorities for permission to collect the snakes, the officer in charge of permits phoned me to offer an easier option. The department ran a small zoo at Yarrawonga on Darwin's outskirts, and they had more snakes than they wanted. "Just drop by on your way to the airport when you're heading home", the permit manager said, "and you can select a few to take back with you". And then, strangely, he laughed.

At the end of my next trip, I assembled a few bags and an insulated cool-box, and I arrived at Yarrawonga Zoo with a couple of hours to spare before my plane was due to leave. But things rapidly spiraled out of control. The Kingies were not in standard indoor cages, but in a large sunken pit outdoors—20 meters long, 5 meters wide, and 2 meters deep (60 × 15 × 6 feet). A meter-high fence around the pit prevented careless visitors from toppling in. Tall grass covered the bottom of the enclosure, so I couldn't see any snakes, but that wasn't a problem. I could jump down into the pit, grab some snakes, and be on my way to the airport.

The zoo staff gathered to watch the fun, with wry smiles on their faces. It was a typically hot Darwin day, and I began to wonder how big the snakes were, how frisky they were, and how many there were. And if there were quite a few, how I could keep out of their way as I was trying to whisk their friends away to an all-expenses-paid flight to Sydney?

And then my plans had to change. The head ranger asked me to take all *of the snakes out of the pen so that the zookeepers could mow the grass. That request demolished my hope of taking the smallest snakes and leaving the bigger ones alone. I had three reasons for that plan. First, my innate cowardice. Second, big snakes need large cages, difficult to fit into a small lab. And third, I knew from my museum dissections that females were the smaller sex in King Browns. By targeting tiddlers, I hoped to get more females (and thus, egg-producers) than males. Still, you can't look a gift snake in the mouth. "No problem", I said, nonchalant and professional. After all, how hard could this be? They were just snakes.*

My next discovery, though, made every sphincter in my body pucker up. "Where did these snakes come from?" I asked innocently. The zoo-keepers chuckled. The rangers in Kakadu National Park were competing with each other to see who could catch the biggest Kingies and donate them to Yarrawonga Zoo. I hadn't met the rangers involved, but I had heard of their extraordinary exploits in capturing giant venomous snakes. I began to suspect that when we lifted the cover off the snakes' shelter, I might be seeing the creatures of nightmare.

And so it proved. When we heaved the lid off the shelter, it revealed a writhing mass of glistening brown scales and thick muscular bodies. The snakes shot off in all directions, but there wasn't far to go inside the pen. So they soon stopped, and lay there glaring up at the idiot who had disturbed their midday slumbers. I counted a dozen snakes. One was not much more than a meter long—and a female, I thought, from the shape of her tail. But the others were giants, 2–3 meters (6–9 feet) in length.

And after a few months of duckling for dinner, they were in superb condition. You could see the muscles rippling under the skin. They were the largest venomous snakes I had ever seen. The rangers set up a row of empty plastic garbage bins outside the pit, then stepped back to watch.

I had no hope of controlling a snake that size if I grabbed it by the tail—the pen was an enclosed space, with too many other scaly participants ready to join in if a battle got underway. A bite would be catastrophic; these huge snakes pack a lethal punch. But I had a pair of snake-tongs—metal grabbers a meter (3 feet) long, like the ones that council-workers use to pick up rubbish. I might be able to stand inside the pen, seize a snake mid-body, swing it across, drop it into a bin, and slam the lid down. I started with the smallest snakes. It was hot and tiring work, with no room for error. Finally, all I had left were two monstrous male Kingies. I seized one of them mid-body and tried to swing it across to the bin, but it exploded into frenzy and it was too much snake for me. Too long, too heavy, too powerful. Even with a strength born of desperation, I couldn't restrain that snake in the tongs—it flexed its muscles and broke free. So, I tried again. Same result. And again. Same result. And the Kingie decided that enough is enough. The third time it broke free, it raced up to the far end of the pen, then turned, lifted its head a meter above the ground, opened its mouth, began champing its jaws, and hurtled back straight toward me.

What to do? There was machismo involved—the zoo-keepers and a few members of the general public were lined up behind the fence, watching the suicidal boffin from Sydney trying to out-muscle large deadly snakes. To impress my audience I needed to wait until the snake was close to me, then grab it with the tongs yet again.

Instead, I squealed like a schoolchild who had encountered a large hairy spider on his/her pillow, and I leapt straight out of the pen. The laws of physics say that I must have got a foothold somewhere, but all I recall is soaring through the air in a graceful arc to end up on the safe side of the fence—miraculously, on my feet—staring down at the snapping jaws beneath. Deciding that I would rather survive than look manly, I caught the last two snakes by leaning over the fence, hauling them up in the air with my tongs, and waiting until they calmed down before dropping them into the bins.

As I made my way to the airport, I was drenched in sweat and trembling, barely strong enough to carry the snakes to my car. I heard the zoo-keepers chuckle as they began to mow the grass.

Australian environments have changed dramatically over the last few thousand years—and especially over the last 200 years or so, since European colonists arrived. A species has three options when the world around it changes, imposing new threats and new opportunities. Adapt to those changes, go elsewhere, or go extinct. Various species of snakes in Australia have followed each of those trajectories, but perhaps the most impressive are venomous snakes that have shown an amazing ability to adapt to new challenges.

When people think about "Australian snakes", it's the big elapid species that come to mind—like Taipans, Tigersnakes, Brownsnakes, and Death Adders. Not just venomous, but deadly. In reality, these big bruisers are the tip of a very large and diverse snakeberg. Of the 200 or so species of snakes in Australia, only a dozen are capable of killing a human being. I've talked about the rest—the neglected majority—in

FIGURE 7.1 King Brown Snakes are mellow if left alone, but you interfere with these deadly giants at your peril. Photograph by Rick Shine.

other chapters. But now, I'll describe some of my research with the snakes that scare people: the ones that can kill you.

During my youth, the danger posed by deadly snakes was one of the things that attracted me to them. As a doctoral student, I enjoyed seeing myself as a daredevil, boldly confronting death. That's a common syndrome among snake-lovers, contemptuous of anyone who doesn't view dancing with deadly snakes as the ultimate sport. Every time I saw a snake, I couldn't resist the challenge of getting my hands on it. And so, for absolutely no sensible reason, I harassed hundreds of deadly snakes over the first few years of my professional career. I'd seize the snake by the tail, dance around until it tired, then pin it down with a stick so that I could get a safe grip behind the head. Then I'd sit down to get my breath back, gaze at the animal with adoration, then release it to hurtle away from the maniac that had just disturbed it.

Stupid? I agree—in hindsight. But an obsession to catch large venomous snakes is widespread among snake enthusiasts (especially males, I suspect). No matter that you have no real reason for doing it, or that you terrify some poor blameless animal, or that you lose the opportunity to watch the animal go about its daily activities. Or that you are risking your life. Some deep inner demon tells you that you *must* catch that snake. The adrenalin rush was one of my motivations for working with snakes.

But it's juvenile nonsense, of course—the same kind of idiocy that drives young men to drive too fast and jump off steep cliffs into shallow water. For me, the epiphany came one evening as I helped a friend from the National Parks service, Tony Press, to survey Saltwater Crocodiles on the Mary River, in the tropical Northern Territory. I wanted to see how rangers count crocs and I suspected there might be a bit of white-knuckle adventure involved. And there was. The tidal surge in those rivers is huge and travels for many kilometers; a massive wave that hurtles upstream at breakneck pace and raises the water level by meters in a few seconds. As the tidal surge approaches, many crocodiles (previously hidden among branches or underwater) emerge and gather midstream at the water's surface. So if you're silly enough, you can surf along in a small boat on the crest of the incoming high tide, shining a spotlight around to see and count the crocs. Once in a while the boat drops down in front of the wave front, or gets spun around by eddies, or is caught up in branches by the riverbank, or collides with large crocs that are slow to get out of the way. It was an adrenalin-charged few hours, and we saw hundreds of crocodiles.

But the life-changing event occurred earlier. As we drove toward the river, a big King Brown Snake crossed the road. I called for Tony to stop the car, jumped out, caught the snake by the tail, and danced around for several minutes to avoid the open mouth as it hurtled at me from every direction. Time passes slowly when you've got a huge, enraged venomous snake by the tail. You enter a zen universe where nothing exists except you and the snake. You can't take your eyes off that fast-moving head, even for a second. Twisting the tail to upset the snake's balance while you pirouette, holding a heavy creature with your arm at full stretch above your head, leaves your muscles screaming for mercy. Finally, though, the snake tires (ectotherms run out of energy quicker than endotherms), and you can pin down its head and get a safer grip. Victory achieved. Humans 1, Snakes 0. But then what? I didn't need a King Brown for any research project, so I let it go.

Tony enjoyed the entertainment, clearly thinking I was an idiot. A few kilometers further down the road, we encountered another large Kingie—and we went through the same process. And sure enough, yet another one appeared in the headlights just 20 minutes later. I caught it with difficulty, because it was unusually athletic and I was already exhausted. Wow, I thought, three Kingies in less than an hour. World record? But as I sat on a log holding the snake, with both of us panting heavily from our exertions, I was struck with remorse. What right did I have to interfere with this inoffensive animal? It had looked so magnificent there on the road—and now it was stressed and horrified by its encounter with me. Capturing that snake had endangered my life and scared the hell out of an innocent creature. For no reason. I felt like an imbecile.

That evening cured me of a life-long obsession. For almost 40 years I hadn't been able to resist the thrill of trying to catch every snake I saw, but suddenly that hunger was gone. Instead of hurling myself onto snakes, I stood back and observed them. It was a huge weight off my shoulders, and I learned more from watching animals than I ever did by harassing them. It was a genuine liberation, and in retrospect I apologize to all those snakes that I man-handled. Thanks for not biting me, and I promise to keep my hands off you in the future unless I have a good reason for disturbing your life.

I've only ever been hospitalized by snakebite once, and that was in the United States not Australia. I foolishly grabbed a small snake that I thought was a harmless

FIGURE 7.2 These days I am happy to watch King Brown Snakes, not try to catch them. Photograph by Greg Brown.

Watersnake but turned out to be a Cottonmouth Moccasin. I immediately realized my mistake and dropped it—but the snake struck as it fell and gave me a sharp reminder about the importance of correct species identification. The worst part of the experience was not my arm swelling up to several times its normal volume, the long days in hospital, or the months it took for my snake-bitten finger to regain its function. Instead, it was the embarrassment.

I had taken a day off during an international scientific conference in Oklahoma to visit a field site with students. The conference had attracted the world's top reptile experts, including many of my long-term friends. As the news of my foolishness spread, a stream of colleagues visited me in hospital. But I wasn't severely injured, and sympathy soon gave way to amusement. The Great Aussie Snake Expert had grabbed a Cottonmouth with his bare hands! A copy of the book *The Complete Idiot's Guide to Reptiles and Amphibians*, extensively annotated with advice on distinguishing between Cottonmouths versus Watersnakes, was presented to me as a memento of the experience. It's one of my most treasured possessions.

Just to rub it in, my 83-year-old mother-in-law was living in Florida at the time, and not long after my Cottonmouth bite, Bobbi mentioned she'd caught a small red-brown snake in her house. She knew nothing about snakes, but she threw a towel over the animal, carried it outside, and released it. It adopted a strange pose—mouth wide open, showing a white throat. Only one species does that. Yep, my infirm geriatric mother-in-law had nonchalantly caught and released the same species that had hospitalized her son-in-law, the expert.

These days, I only work with deadly snakes if there is a good reason to do so. Given a choice, I'd rather work on a harmless species with a forgiving temperament. I've had enough close calls to last a lifetime, and those near-catastrophes have obliterated my interest in life-threatening science. I don't regret all those adventures with feisty Brownsnakes and Tigersnakes, but neither do I want to repeat them. Nowadays I choose my study species more carefully, avoiding ones that might kill me. I'm unusual in that respect. Many of my colleagues get a real kick out of handling deadly snakes. To them, a harmless snake is like Mexican food without the hot sauce. But in my wimpy opinion, deadly snakes are best admired from a distance. I've disappointed many pet-keepers by declining their invitation to fondle a captive Taipan or Death Adder. Since my Ph.D. studies in Armidale, I have only conducted substantial projects on two highly venomous species—Brownsnakes and Tigersnakes. And in both cases, someone else did the hard work. I just helped out.

The first project was the brainchild of Pat Whitaker, an Englishman who is much fonder of deadly snakes than I am. Pat grew up in England, but fell in love with venomous serpents at an early age and made a beeline for Australia as soon as he was old enough to leave home. He worked in a range of zoos and universities, with one common theme: snakes, the deadlier the better. After finishing his Honors thesis on Tasmanian Tigersnakes, Pat talked me into supervising his Ph.D. on Eastern Brownsnakes. They were his second choice; I vetoed his initial idea to work on Taipans. I agreed to supervise the Brownsnake study because the species is interesting from several perspectives. For starters, it's the greatest modern success story among Australian snakes. In undisturbed areas, Eastern Brownsnakes have a typical Aussie-snake diet based on lizards and frogs. But in agricultural land, where the native vegetation has been removed so that farmers can grow crops and graze sheep, the Brownsnake comes into its own. Grains attract mice, and Brownsnakes are fast enough and smart enough to feed on rodents. While most species of Australian snakes are declining in numbers, Brownsnakes are booming. I wanted to understand how Brownsnakes have modified their ecology so rapidly, to exploit the furry mouse-burgers provided by farmers.

Eastern Brownsnakes are interesting in other ways as well. When I first traveled to the United States, I was struck by the similarity between Brownsnakes and American snakes like Coachwhips and Racers: long, brown, slender, fast, active in the open on warm days, quick to defend themselves if attacked. These American snakes are non-venomous colubrids not deadly elapids—but in appearance and behavior, they could be the Brownsnake's cousins. Based on a quick glimpse, even an expert would struggle to tell them apart (unlike Watersnakes versus Cottonmouths, which any idiot can distinguish). And later, as I visited places like France, South Africa, and Brazil, I encountered other Brownsnake lookalikes. A detailed study on Aussie Brownsnakes might illuminate a more general story, enabling me to compare our lethal speedsters to "whipsnakes" from other countries.

Although I'd reached a stage in my life when the idea of a deadly study species no longer aroused me, the Brownsnake's toll on human lives is another reason to look at it as well. Eastern Brownsnakes kill more people than do any other Australian snakes. Perhaps if we knew more about the snake's ecology, we could understand how and when people encounter snakes, and why the snakes react as they do. We might even be able to reduce the numbers of people killed by snakes, and the number of snakes killed by people. Peaceful coexistence between farmers and Brownsnakes would be a huge step forward.

By the time that Pat began his Brownsnake study, I had already seen many of these snakes in the wild, usually on hot days when the snakes were out in open country rather than close to shelter. I was confident that these agile snakes are great wanderers, traveling long distances through the paddocks. Their camouflage, speed, and deadly venom reduce their vulnerability to predators, their light brown color reduces the risk of overheating, and their slender body shape lets them escape into tiny holes. No need to be a homebody, like other snakes—a Brownsnake can wander fancy-free across the landscape.

Because I expected Brownsnakes to have huge home ranges, I worried about radio-tracking them. Open habitats simplify that challenge, though, and Pat found an ideal study site near the town of Leeton in a cotton-growing region of the Murrumbidgee Irrigation Area in central southern New South Wales. Flat paddocks as far as the eye can see, filled with crops and stock, and with irrigation channels running through the landscape. The banks of those channels are honeycombed by small holes, originally dug by mice but now harboring Brownsnakes as well.

Leeton is more than 500 km (300 miles) from Sydney, so commuting was out of the question. Pat moved his family to the town and set to work. A snake enthusiast since he left the cradle, Pat had no trouble catching Brownsnakes and fitting them with miniature transmitters. Straining my research budget to the limit, I bought powerful (and thus expensive) units. I was certain that once those snakes were released, they would head for the hills—or at least, the farm next door. I advised Pat to be

FIGURE 7.3 Eastern Brownsnakes are abundant, fast, nervous, and highly venomous—and as a result, kill more people every year than does any other species of Australian snake. Photograph by Kris Bell.

ready—as soon as he released a snake, he had to monitor it frequently so that he knew the direction it was going as it zoomed out of range. The area was crisscrossed by farm roads, and I expected Pat to spend most of his time with a telemetry antenna out the car window, relocating snakes that had left his study area to see the big wide world.

I was wrong—utterly, completely, totally wrong. When Pat released those snakes, they disappeared down holes near where he had caught them, and they stayed there. It was like preparing yourself to ride a bucking bronco at the rodeo, strapping yourself in and awaiting the explosion of leaping and bouncing as the gate opens . . . but instead, the bronco meanders out and starts chewing grass.

OK, I thought, the snakes are just recovering from the transmitter implantation surgery. In a couple of weeks, I told him, your animals will take off to far-flung pastures. But they never did. Each snake lived inside its mouse-hole, coming out for a few hours on hot days to lie in the sun. After it had exterminated the local mice, the snake moved to a similar spot about 150 meters (150 yards) away and did the same thing. And then did it again. And again.

That was a humbling experience, partly because of the look of pity in Pat's eyes every time he told me that (yet again) his telemetered snakes were staying close to their point of release. I had captured hundreds of Brownsnakes before Pat embarked on his radio-tracking project, so how could I have overestimated their movements so greatly? When I found a Brownsnake out in open pastures, I had assumed that

FIGURE 7.4 Eastern Brownsnakes thrive in the disturbed habitats of agricultural land in southeastern Australia. Photograph by Kris Bell.

it was just moving through. I had been 100% wrong. That snake lived right where I encountered it, and if I'd looked more closely I would have discovered a mouse-hole that served as the snake's shelter and larder. Most Brownsnakes I found were hot and lightning fast, so I expected them to spend their life at high temperatures. Again, I was wrong. The radio-tracked Brownsnakes kept cool most of the time, down inside their burrows. On the rare occasions when they came to the surface, it was in hot weather and the snakes were speedy. I had assumed that Brownsnakes spent most of their time aboveground, hot, and in the open, because that was how I found them. Rather than high-temperature nomads moving across the landscape, they were homebodies who had briefly left their cool living rooms to sunbathe in the backyard.

And what of the famously psychotic personality of these cold-blooded killers? In every pub in Australia, the locals tell and retell blood-curdling tales about "aggressive" Brownsnakes—heartless murderers, lurking beside the path, saliva drooling from their wicked lips as they dream of slaughtering innocent passers-by. To find out if that was true, Pat and I constructed a questionnaire to ask people about their encounters with snakes, and we distributed it to the parents of children who attended the high school in Leeton. We asked people where and when they last saw a Brownsnake, what the snake did, and so forth. None of our 138 respondents had been bitten, but about 10% claimed that snakes had attacked them without provocation.

Pat and I were skeptical. The Brownsnakes we met tried to escape rather than to bite us. There's no doubt that if it is cornered, a hot Eastern Brownsnake will retaliate—but the stories about unprovoked attacks violated common sense. Why would any snake attack a human being, unless in self-defense? Assailing a large opponent uses up energy, wastes venom, and is very risky—with no benefit to the snake.

My opinions about Brownsnake non-aggression were based on numerous encounters with these lithe ballistic missiles, over many years. Time after time, they had a chance to nail me, and they didn't. They'd lift their head and a few coils up into a striking pose, feint toward me, then content themselves with a stern look, slowly unravel, and slither off into the undergrowth. Sometimes I was young enough and foolish enough to harass Brownsnakes intentionally, and several times I mistook them for other species and seized them by mistake. One day I took a flying dive at what I thought was a harmless Green Tree Snake, achieving the unlikely trick of seizing it behind the head on the initial grab. Which was just as well, since the animal proved to be a large emaciated Brownsnake. Blind in one eye, the snake had struggled to find food (and so was as thin as a Green Tree Snake); and by good fortune, I had lunged at it from its blind side, so it never saw me coming. If I had approached it from the other side, I might not be writing this book.

The most compelling evidence for viewing Brownsnakes as non-aggressive, though, came from the final year of my own Ph.D. I broke my ankle in a football game but carried on with fieldwork regardless, stomping around with my right leg in a plaster cast. It was impervious to a snake's fangs, so whenever I saw a Brownsnake, I walked right up to it with my plaster-covered leg to the front. I was hoping for a full-blooded charge by the snake, so I would have heroic tales to tell of a snake chewing on my plaster cast as I nonchalantly captured it while sipping a martini. But no such luck. The snakes either sat still, hoping I hadn't seen them, or they tried to zip off into the nearest shelter.

FIGURE 7.5 As a Ph.D. student, I broke my leg playing football. As a result, I spent one season catching snakes while my leg was encased in a plaster cast. That enabled me to approach venomous snakes closely, to see if they would try to bite me—and they didn't.

I had been abysmally wrong about Brownsnake home ranges and movements; could I be equally wrong about how these snakes respond to people? Pat and I were convinced that Brownsnakes were not the savage serpents they were reputed to be, but how could we document that fact? Unless we could actually give snakes the opportunity to launch an unprovoked attack, our opinions were no better than the anecdotes told by the citizens of Leeton. We devised a plan. With transmitters in 40 Brownsnakes, we could locate each animal whenever we wanted to. The transmitters beeped faster when the snake was hotter, telling us when a snake was out and about rather than sheltering in its burrow. So, we conducted a simple study: one of us (usually Pat) walked toward an active snake in a straight line at a standard speed, stopping 50 cm (20 inches) from the snake's head. If the snake didn't react, we gave it a nudge. No plaster casts to protect our legs, but heavy boots and jeans provided reasonable security (especially since we knew exactly where the snake was). If a snake came at us, we could get out of the way faster than it could launch an attack. Despite hair-raising tales about lightning-fast strikes ("quicker than the eye can see"), the head of a striking Brownsnake travels at a meter (3 feet) per second—much slower than a human boxer's jab.

We ended up with data on 455 approaches, a very respectable sample size. The result was clear. The Leeton snakes were pacifists, like the ones I had stomped up to years before with my plaster-shrouded leg. Half the snakes retreated as we approached (especially if we walked quickly or if the snake had been moving around beforehand). The rest stayed still, relying on camouflage. Those motionless snakes were prepared to tolerate close proximity; on average, we were less than a meter from such a Brownsnake before it reacted to us (although a bit further in hot weather, in open country, or if the snake was a juvenile). If we gently nudged the snake's body with a boot, the animal went the other way as fast it could. Snakes only approached us aggressively three times, out of those 455 staged encounters. The only strike occurred when Pat accidentally trod on the snake's tail. You can't blame the snake for that one.

So why did the residents of Leeton report that they had been attacked by Brownsnakes 10% of the time, whereas we were attacked less than 1% of the time? The reason was that local residents misinterpreted the snake's threat display as a full-blown attack. When it believes it's in peril, a Brownsnake rears up, lifts its forebody into s-shaped coils, and faces the problem head-on. It may even strike out, but usually in bluff. Those stories that "10% of snakes attacked me" were about defensive displays, not real attacks. Our high-speed video-footage showed that these defensive strikes often fail to make contact even if the target is right in front of the snake. Brownsnakes don't waste their precious venom if they don't have to.

But thoughts about economy in venom usage aren't front and center in your brain if a large snake erupts beside your feet as you wander through the paddock. Farmers encounter snakes when they least expect them. It's a helluva shock. A large Brownsnake materializes in front of you, forms a striking coil aimed at your ankles, hisses, thrashes around, and then disappears into thin air. It's easy to believe that you have survived a deadly onslaught. But if you can read snake body language, you realize that the animal was just as nervous as you, and just as unwilling to escalate the encounter. Given a chance, it hightailed out of there as fast as it could go. Brownsnakes are high-strung, and they are deadly retaliators, but they aren't aggressive.

****

The second high-profile "villain" I worked on is another Aussie icon: the Tigersnake. This well-known species occurs across much of southern Australia, and until a few decades ago, textbooks divided these snakes into several species. Tigersnakes look very different in different locations. Some islands off South Australia have huge jet-black Tigersnakes more than 2 meters (6 feet) long, whereas others have copper-colored pygmies much less than half that size. In Tasmania, Tigersnakes are brightly hued from yellow through to black, often with starkly contrasting bands. Mainland Tigersnakes are subtly adorned with indistinct dark bands against a brown background. It made sense to classify them as different species, but color is only skin deep. Genetic analyses tell us that all of these heavyset snakes are closely related and belong to the same species. I'd studied mainland Tigersnakes as part of my Ph.D. on the New England tablelands, but it wasn't until many years later that I collaborated in a study of the Tigersnake's colonization of small offshore islands in southern Australia.

As for Carpet Pythons in the Recherche Archipelago, islands are natural laboratories for Tigersnake evolution. When a few animals become isolated on islands, there is nothing to prevent those colonists from evolving in a new direction. Confronted by unfamiliar opportunities (such as novel prey types) and challenges (such as different predators), a small population can deviate from its ancestral form. And it can do so very rapidly. The islanders can't interbreed with their mainland relatives, so gene frequencies change faster than they can in a large population. Evolutionary opportunities are especially great on chains of islands, isolated from each other but colonized by the same mainland ancestors. Exposed to conditions that differ from one island to the next, those pioneer beach-combers can evolve into animals that are very different from the ancestral form. It's Evolution in Action. As Charles Darwin saw in the Galapagos Archipelago, an initial form can give rise to an array of distinctive species on each island.

The islands off southern Australia are young geologically. Many were separated from the mainland only a few thousand years ago, as sea levels rose when glaciers melted after the last Ice Age. Not long enough to see new species forming, but plenty of time to witness evolutionary changes. And to my delight (because two of my strongest passions are snakes and evolution), the main players in that dynamic process have been snakes.

Several types of reptiles have been marooned on offshore islands and have adapted to their new homes. Western Australia is especially blessed with island specialties: Rottnest Island possesses its own Shingleback Lizard as well as its own Brownsnake. And in an earlier chapter, I talked about how Carpet Pythons have adapted to their island homes in the Recherche. But the master of Robinson Crusoe Ecology—the species that thrives on many islands—is the Tigersnake. Distinctive races of Tigersnakes have colonized small windswept islands all across southern Australia, from Bass Strait in the east to Esperance in the west.

How did that happen? Did some early Tigersnake end up marooned on an island, evolve to deal with the new challenges, and then spread to more and more offshore homes? If so, Tigersnakes on all of the southern islands will be more closely related to each other than they are to snakes on the adjacent mainland. Or did the switch from mainland to island life happen many times independently? In that latter

FIGURE 7.6 and **FIGURE 7.7** Common in low-lying coastal areas cut off by rising sea levels, Tigersnakes have adapted to a Robinson Crusoe lifestyle on many small islands. Some populations have banded individuals, resembling mainland specimens in this respect, whereas the snakes on other islands are jet-black. Photographs by Stephen Mahony and Rob Valentic.

scenario, snakes were trapped on offshore islands all across the southern oceans as sea levels rose, and every population evolved to deal with it in their own way. Recently, Scott Keogh and his colleagues answered that question by analyzing the genetic composition of scale samples from Tigersnakes. Scott has been conducting cutting-edge research on the evolutionary relationships of snakes ever since his Ph.D. Now a senior professor at the Australian National University, he has transformed our understanding of snake evolution.

Scott and his team discovered that Tigersnakes made the transition from mainland to island life many times. Tigersnakes on South Australian islands are descended from ancestors on the South Australian mainland; whereas for Tigersnakes on Western Australian islands, the family tree begins in mainland Western Australia. There is no single race of "island Tigersnake" that adapted to its oceanic home and then spread from one island to another. Instead, snakes were marooned on different specks of dry land as the sea level rose. Those recent evolutionary events generated Tigersnakes of diverse sizes, shapes, and colors, leading earlier researchers to mistakenly describe them as separate species.

The genetic comparisons also show that different islands were colonized at different times. For example, snakes found their way across the water to Williams Island (in South Australia) 9,000 years ago and have accumulated many genetic differences over the intervening period. But other colonization events are much younger; Trefoil Island in Tasmania received its pioneer Tigersnakes only 40 years ago. In most cases, the process was straightforward. Tigersnakes are common in coastal swamps, where they feed mostly on frogs. When sea levels rise, some of these coastal areas end up as islands. And each time that happens, a small group of Tigersnakes is left staring wistfully back toward shore. Because the topography of the coastline is variable, some islands were isolated much earlier than others, even though sea levels rose at the same rate everywhere.

But sea-level change isn't the only process that has founded island populations of snakes. People have played a role also, with one remarkable case in January 1928. "Rocky Vane" (real name Lindsay Vagne) was a showman who traveled around Australia exhibiting snakes in circus sideshows. Rocky had a troubled time with law-enforcement authorities across Australia, for all kinds of transgressions ranging from burglary to selling illegal liquor. Besides exhibiting snakes, he sold his own special antidote to snakebite ("Rocky Vane's Snakebite Cure") that was medically useless (as shown by later testing), but nonetheless was popular with the public. Rocky had developed substantial immunity from repeated bites and he often induced snakes to bite him as part of his show, to demonstrate the power of his antidote. However, his colleagues were less fortunate. In 1914, his business partner Sandy Rolfe was bitten and killed by a Tigersnake. Fifteen years later, Rocky's assistant William Melrose met the same fate. But the central tragedy for this story was the death of his wife. Dot was Rocky's assistant, playing the role of Cleopatra; and sadly, she was killed by a Tigersnake bite during a show in Perth in January 1928.

There are two versions about what happened next. Some anecdotes suggest that Rocky was devastated by Dot's death and vowed never to have anything to do with the species again. But he didn't want to kill his snakes, so he hired a rowboat and released his 40 snakes on tiny Carnac Island, 10 km (6.2 miles) off the coast near

Fremantle. But the evidence is unclear, and there is another and less flattering version. Just a few years later (in 1931), Rocky was in court again, accused of harassing his new wife with deadly snakes. In court, Rocky claimed that he had "gotten rid of the snakes"—apparently by dumping them on Carnac. Rocky probably planned to use Carnac as a short-term repository for his snakes—satisfying the authorities that he had disposed of the animals, but on an island so small that he could go back and recapture most of them any time he wanted to.

Other introductions of Tigersnakes to small islands may have equally murky motives. I've heard rumors that after Aboriginal land claims were approved on some Tasmanian islands, rednecks retaliated by releasing Tigersnakes there. The aim was to prevent Aboriginal people from living on the islands and collecting seabirds for food.

At first sight, being abandoned on a rocky or sandy island looks like a death sentence. All across mainland Australia, Tigersnakes live in swamps and feed on frogs. They also eat mice or lizards if they can catch them, and I have found them inside tree hollows devouring nestling birds. But even if Tigersnakes will eat almost anything, an offshore island doesn't provide many opportunities. Most of those islands are tiny, with a few scrubby trees and thickets of grass. How can the snakes survive? Pioneering studies by Terry Schwaner, an exuberant American from the South Australian Museum, revealed a fascinating evolutionary story. There *is* food on the islands, but not the kind that Tigersnakes are used to. And adjacent islands differ in their available cuisines.

On most islands, frogs are absent. Not enough freshwater. But every island has enough lizards to sustain juvenile Tigersnakes. On islands with no other dietary options, even the adult snakes have nothing to eat except small lizards—and as a result, the snakes are pygmies. The best evolutionary option is to stay small, instead of growing so large that you starve.

If lizards were the only prey on islands, all offshore Tigersnakes would be small. But some islands offer another cuisine for a hungry snake: marine birds that nest offshore to avoid foxes, cats and dogs. Some islands are honeycombed with the burrows of shearwaters ("muttonbirds", so called for their taste). Other islands have nesting colonies of seagulls. The gulls feed at rubbish tips on the mainland by day but return to the island to roost at night.

In general, Australian snakes rarely eat birds. When I examined preserved snakes in museum collections, I found very few birds inside their stomachs. Most birds are too difficult to catch and too large to swallow. But an island Tigersnake faces an unusual situation. Its home contains few other options for the evening meal, and many island birds nest on or under the ground. A hungry snake has access to a cafeteria of feathered prey without needing to climb.

But that still leaves the problem of prey size. Even a nestling bird, straight out of the egg, is a huge meal for a baby snake. And unlike a lizard (or a human being), a snake can't break that prey item down into smaller pieces. With minor exceptions (like mangrove snakes that tear the legs off crabs), snakes only eat prey items that can fit into their mouths. That constraint has favored some impressive tricks to enable a snake to get its jaws around large prey. A snake's skull is a mechanical marvel, with flexible joints allowing expansion in several directions. But ultimately, the only way to swallow a bigger prey item is to have a bigger head.

That doesn't sound too promising for a young island Tigersnake trying to feed on muttonbird chicks rather than lizards. But evolution has given snakes extraordinary flexibility (as I keep on saying). In a pioneering laboratory study, my colleague Xavier Bonnet showed that a snake's head size is not fixed by the genes it inherits from its parents. A young snake that is consistently offered very large prey—the maximum size that it can swallow—develops a larger set of jaws as it grows. That work was done in Xavier's lab in rural France. With harmless local snakes, you might think? No, that would be too boring for a dashing Gallic alpha-male biologist. Instead, Xavier conducted the work with a deadly African species—the Gaboon Viper. It has the longest fangs of any snake. Why did a scientist in France have a lab full of huge, deadly, African vipers? Xavier is a free spirit. A zoo offered them to him, and he thought "why not?". And using them, he was able to demonstrate this kind of plasticity for the first time.

The idea of flexibly increasing the size of your head sounds bizarre. Can exercise during an animal's lifetime change the shape and size of its bones? Yes it can. Courses in evolutionary biology often disparage the French biologist Lamarck, a forerunner of Darwin. Lamarck thought that changes in an animal, wrought by conditions that it experienced during its life, could be passed on to its progeny. That, he thought, was the mechanism of evolutionary change. Lamarck was wrong about acquired characters being inherited by the offspring; but he was right about the ability of your lifestyle to shape your body. For example, professional athletes develop thicker bones in parts of their body that they use more frequently. And the same flexibility allows young snakes to grow bigger jaws, and thus swallow larger food items. If a Tigersnake is born into a world where the available meals are almost too big for it to eat, it grows a larger head. In scientific jargon, this is called "phenotypic plasticity": the conditions that an individual encounters early in life modify its shape (or its size, or behavior).

That flexibility enables island Tigersnakes to overcome the challenge that they face in a seabird colony. Phenotypic plasticity gives the young snake a fighting chance to grow bigger jaws, and so swallow larger prey. But what will happen in later generations? Would every generation of Tigersnakes be born with small heads (like their ancestors) and have to develop larger jaws anew? Probably not, and we don't have to adopt Lamarck's discredited views to explain why. If it's always an advantage to have a larger head, generation after generation, this consistent selection pressure will gradually result in an animal that develops a big head even if it doesn't encounter large prey. This idea was first proposed in 1953 by the British biologist C.H. Waddington. According to his idea of "genetic assimilation", evolutionary changes begin with animals flexibly responding to the new situation, but then shifting across to a genetically based solution. That is, a new characteristic (like a larger head in baby Tigersnakes) first arises as a direct response to the environment. The snake's genes say "develop a larger head if (and only if) you encounter large food items". But if having a large head is always useful (because every generation encounters large prey), evolution eventually replaces that flexible response with a hard-wired solution that is simpler and produces the same result. The snakes end up with genes that say "develop a large head, regardless of what kind of food you encounter".

"Genetic assimilation" makes a straightforward prediction: a new challenge will be solved by flexibility, but eventually that condition will become hard-wired into the genes. "Genetic assimilation" slipped out of scientific popularity because it sounded uncomfortably like the discredited ideas of Lamarck. But in fact, "genetic assimilation" isn't Lamarckian at all: it just predicts that under some conditions, a complicated genetic instruction will be replaced by a simpler genetic instruction. Mutation can easily provide the raw material for that kind of change. The other reason why "genetic assimilation" fell out of favor was a lack of evidence—until the Tigersnakes loomed into view. To test the idea properly, we'd need a species that has encountered the same evolutionary challenge several times, in different places. Tigersnakes were perfect. They had spread independently from the mainland to many offshore islands, with large prey (seabirds) on some islands but not others. Each population of Tigersnakes provides a separate evolutionary "experiment". And because those colonizations occurred at different times, we can examine various stages of the evolutionary process.

The man who used Tigersnakes to explore these ideas was Fabien Aubret, a debonair French student of Xavier's who traveled to Australia for his Ph.D. Fab's passion for snakes is exceeded only by his obsession with beach-fishing, and it's no accident that he chose research sites beside the ocean. In most of the photographs I have of Fab, he is displaying a massive silver fish that he's just hauled out of the sea. My favorite photo, though, comes from a day when we went fishing and (miraculously) I caught the largest fish. Looking at the picture of Fab, me, and the fish, you can see him gritting his teeth. Working with competitive alpha-male Frenchman can be highly entertaining.

Xavier had shown that young Gaboon Vipers develop larger jaws if they are given very large prey. Can Tigersnakes do the same thing? Tigersnakes on bird-nesting islands face the same challenge as Xavier had imposed on his pet Gaboon Vipers in the lab. Lots of food, but all of it tantalizingly large—almost too big to swallow. To test the idea, Fab collected pregnant snakes from two sites. One was Carnac Island near Perth, the population established by Rocky Vane in 1928 after his wife was killed by a Tigersnake. On Carnac, seabird chicks are the most common edible item. The other site was Herdsman Lake, a mainland population near Perth where frogs are the snakes' staple diet. And as we had predicted, the island snakes had very plastic jaw sizes: when we gave baby snakes large prey, the snakes grew bigger jaws. The mainland snakes were far less flexible. No matter what size mice we gave them, their jaws stayed at the same proportion of their body size. The underlying evolutionary pressure was clear: on Carnac Island, the only snakes that survived through babyhood were those that were able to enlarge their jaws in response to encountering large prey. Less flexible young snakes starved to death. It was the long-awaited demonstration of Phase One in the set of ideas that Waddington had suggested decades ago—before Fab or Xavier was born, and while I was still a toddler. A population responds to a new challenge by increasing the level of plasticity in its responses. We published our paper in the high-profile journal *Nature*, confirming my opinion that snakes are terrific study subjects if you can find the right topic.

But what of Phase Two? Would that flexible response eventually become hard-wired, so that all of the snakes in older populations were born with large jaws rather than having to develop them anew? Fab obtained a postdoctoral fellowship to compare baby Tigersnakes from islands with different colonization histories: some very recent (like Carnac) and some much older (like Williams Island). This was a risky study. First, Fab had to catch pregnant female snakes from remote locations. Island Tigersnakes are friendlier than their mainland counterparts, so that's not quite as terrifying as it sounds. On Carnac, for example, these giant serpents are gentle—but you're a long way from home if you are bitten. And even if you manage to contact the outside world for assistance, it can take a long time for a boat to evacuate you to medical care. Many islands that I've worked on are only accessible at high tide. And if the weather turns rough, you may be isolated for days at a time. My colleague Joe Slowinski died in 2001 in the remote mountains of Burma, when bad weather prevented helicopters from airlifting him to a hospital where he could have received antivenom for the snake that had bitten him. Working with deadly snakes is risky enough. Doing so in places where you can't get help takes it to a whole new level.

Fab was careful, though, and completed his fieldwork without being bitten. But then he faced an even bigger risk: measuring and handling the hundreds of baby Tigersnakes that were produced by those pregnant females. Baby Tigersnakes are feistier than their parents, and smaller animals are more difficult to handle (especially to measure their head sizes, because your fingers end up beside the fangs). Worried about having hordes of deadly snakes in the lab, I was delighted when John Weigel offered us space at the Australian Reptile Park. We purchased a demountable building, and Fab kept his snakes there instead of at the Uni. That arrangement also allowed Fab to live closer to good fishing spots.

Fab's results supported the venerable concept of genetic assimilation. Just as Waddington had predicted, snakes from recently colonized islands were born with small heads relative to body size, but rapidly increased their jaw size if we gave them large prey. On the islands that had been colonized thousands of years ago, it was a different story. The young snakes were born with large heads, and so could eat large prey right from the start. Giving them even bigger prey didn't change the sizes of their heads. If C.H. Waddington had still been alive, he might have bought us both a beer.

Like any good project, the work on island Tigersnakes raised many new questions as well as answering some old ones. Before Fab entered the project, for example, Xavier began working on Carnac. He noticed that many of the snakes bore major scars on their heads. Several animals were blind in one eye, and some had lost both eyes. It wasn't hard to work out why. Seagulls nest on the ground, so their fluffy nestlings look like easy prey for a hungry Tigersnake. But the parent birds guard the nest ferociously, using their razor-sharp beaks to stab at any intruder. A snake's head is tough, and most of those blows just slice a few scales. A peck to the eye is, however, very destructive—and as they grow older, snakes accumulate more and more damage.

Surely, losing your eyes is a disaster? If a snake can't see where it's going, how can it find food and shelter, let alone sustain a healthy sex life? Remarkably, Xavier's recaptures of blind snakes showed that the animals are doing perfectly well. They survive just as well as snakes with 20:20 vision, they grow just as quickly, and they even manage to find partners and mate. Human intuition is misleading. We depend on our eyes for finding our way around, and life is difficult if we lose our sight. But for a snake, as for a dog, the most important source of information about the world is scent not sight. The forked tongues of snakes are sensitive chemical-detection systems, enabling a snake to find its way to the nearest shelter, bird nest, or female. In a mainland habitat with predators like hawks and foxes, blindness is a disability. But on predator-free Carnac, blindness is just an inconvenience.

Given that toughness and flexibility, plus their toxic venom, you might expect Tigersnakes to have fared well since Europeans settled Australia. But they haven't. Unlike Brownsnakes, Tigers can't exploit the rodent-rich farms produced by commercial agriculture. Heavyset and slow-moving, a Tigersnake has little chance of outrunning a mouse, or fitting down its burrow to catch the rodent unawares. Although Tigersnakes have displayed incredible adaptability—by evolving into large-headed bird-eaters on offshore islands, and by soldiering on when blinded by seagull stabs—there are limits to that versatility. Once abundant across the waterways of southern Australia, Tigersnakes have become rare. As the decades have passed, medical records show that more and more Brownsnake antivenom has been used, and less and less Tigersnake antivenom.

What's gone wrong for the mainland Tigersnakes? As is true for most endangered species, including the Broadheads I discussed earlier, a combination of factors is at play. Habitat loss is important: intensive agriculture destroyed much of the Tiger's streamside haunts, and water allocations to farmers destroyed the once-mighty inland rivers. But the Tigersnake's dietary dependence on frogs was its Achilles' heel. In a healthy river or swamp, frogs occur in incredible numbers. With each clutch containing hundreds or even thousands of eggs, good weather conditions create an amphibian population explosion. All a frog needs is bugs to eat, water to breed in, and moist shelter sites. Sadly, Australians have trashed those waterside ecosystems by clearing the land, draining the swamp, and spreading pesticides . . . and so frog numbers have plummeted. And then a fungal disease (brought into the country from overseas) spread through our native amphibians. Several species were wiped out completely.

You can't sustain predators without prey, and you can't sustain streamside species without water. When I was a child in the 1960s, I could see dozens of fat Tigersnakes basking in the sunlight if I walked (carefully!) through the blackberry thickets along the edge of Lake George near Canberra. A local farmer, Sil Smith, captured literally thousands of Tigersnakes on his property, in the pioneering days of snake antivenom development. Those days are gone. When Jim Bull and I went snake-hunting with Sil in the 1970s, we only found a handful of Tigersnakes in areas where he had once caught hundreds. Like so many other species, the flexible and resourceful Tigersnake has been swept aside by the tsunami of environmental degradation that has deluged the Australian continent.

8 Rough Characters in the Billabong

I saw my first Saltwater Crocodile in the wild in 1972, near the top of Cape York in tropical Queensland. Before then I had seen captive "Salties" in zoos, and marveled at their size and speed. Even with a stout fence between me and the giant reptile, I could feel a surge of fear. But that sensation was far stronger for this first sighting in the wild. It happened around midnight, and I was in a small inflatable dinghy in the middle of a very large and very remote billabong (an oxbow lake). Worryingly, the boat was leaking so badly that it was likely to sink before we could make it back to shore. We'd have to swim for it—a nerve-wracking prospect, given that I'm a poor swimmer. But that problem faded into insignificance when two glowing red eyes appeared on the surface of the water, just a few meters behind the boat. I and Hal Heatwole, my Ph.D. supervisor, were on an expedition into the then uncharted territory of northeastern Australia. It was the first year of my Ph.D., and I was starting to think that it might be the last year as well.

In retrospect, we were naive and poorly prepared. The inflatable boat had been borrowed from a friend, and we hadn't even checked it for leaks. None of us had been in that part of the world before, so we made mistakes. The wind was strong, blowing us out into the center of the waterbody as soon as we left the shore, and the boat was difficult to steer. And as the water began to lap into the boat, with the shore still far away, it looked as though either Hal or I might pay a high price for our ill-preparedness.

Hal is a cheerful man, but his voice was serious when he cleared his throat and made a little speech to me. It was very measured:

> *Rick, I doubt the boat will make it to shore. That croc is getting closer. You're my student, so it's only fair to tell you my plans. When the boat goes under, I intend to swim down to the bottom of the lagoon and stay there for as long as I can hold my breath, then sprint for shore. The croc will go for whoever is swimming on the surface. I'm an experienced diver, so you are at more risk than me. But at least I've told you what I'll be doing.*

It was an honorable speech. There's nothing like a big croc and a leaking boat to reveal your supervisor's true nature.

Fortunately, the wind blew the boat into shore, and as soon as we hit the beach, we leapt out and away up into the scrub like startled wallabies. I have no idea how close the croc came to us, or what he did when his meal evaporated. But those glowing red eyes remain a vivid memory.

Decades later, I set up my main research base in the floodplains outside Darwin, in the Northern Territory. This is crocodile heartland, but in the early days—the

DOI: 10.1201/b22815-8

FIGURE 8.1 On my first trip to the tropics in 1973, trying to look professional while holding a Freshwater Crocodile. A few days after this photograph was taken, a much larger Saltwater Crocodile terrified the hell out of me.

1980s—I rarely saw these giant reptiles. And when I did, they were heading in the other direction. That's no longer the case. Crocodile conservation in the Territory is one of the true triumphs of wildlife management worldwide—spearheaded by one of my companions on that early North Queensland trip, Grahame Webb. Big crocs are abundant today, and are unfazed by humans. That creates interesting situations.

My most thrilling croc encounter occurred at Fogg Dam, the longtime focus of my tropical research. My postdoc Greg Brown routinely walks up and down the dam wall after dark, counting frogs and capturing snakes, and I join him whenever I can. We see crocs every night, but the only ones that display any aggression are nest-guarding Freshwater Crocs. The big "Salties" aren't a problem—at least during the dry season.

But with the advent of the monsoons, things change. Crocs move into newly flooded areas and turn up everywhere—even in backyard swimming pools. Occasionally, people are eaten. If you jump into your favorite swimming hole after it has been swollen by wet-season rains, not realizing that a massive "Saltie" has moved in overnight, the result can be tragic.

My Fogg Dam encounter occurred at the peak of wet-season flooding in February 2003. Greg had discovered that Macleay's Watersnakes gather at the flooded fringes of the dam at this time of year, to feast on small fishes that sleep in the shallows at night. A great chance, I thought, to run experiments on watersnake foraging. How do the snakes detect fishes? What information triggers a watersnake's strike? Smell? Movement? Touch? Vibration? I used a fishing rod to dangle rubber lures and dead fishes in front of foraging snakes, while Greg kept an eye open for crocs, and my research assistant Melanie recorded the data as I called it out (loudly, to be heard above the buzzing of a thousand mosquitoes). We discovered that the most important stimulus was one I hadn't even thought of originally. A splash of water was the key. A snake that felt a splash beside his head exploded into feverish open-mouthed thrashing. The reason was simple—a fish that detects an approaching snake flips into the air, landing with a splash close by. And so, the snake's best strategy is to lunge at any splash. Equipped with a water-pistol from the Humpty Doo supermarket, I could drive a foraging snake into a frenzy by squirting water beside its head.

To conduct this study, we had to wade in shallow water around the margins of the dam, late at night. Exactly the sort of place that Saltwater Crocodiles hang out. Greg was cool as a cucumber, but I was unsettled at the possibility of a monster croc moving in from elsewhere, and displacing the well-behaved local reptiles. And sure enough, when we were calf-deep in water halfway along the dam wall, a massive head broke the surface not far away. Almost 5 meters (15 feet) long, weighing 500 kg (1,000 lb), that "Saltie" was capable of eating all three of us, and then looking around for dessert.

But the big croc's behavior was bizarre. If he had wanted a meal of scientist, he should have crept along quietly underwater until he was close enough to pounce. Instead, he swam on the water's surface, hurtling forward in our direction at full speed. The bow wave was impressive. Indeed, bowel-loosening. Mel sensibly stampeded for dry land, while Greg and I began wading slowly backward, keeping our eyes on the croc. He was heading straight for us, eyes blazing in the beams of our flashlights.

I was about to turn and run when Greg reached into his pocket, pulled out a rock, and fitted it to the slingshot that he pulled out of another pocket. He let fly, and his aim was true. Admittedly, it was a large target, and less than 10 meters (30 feet) away. My hand would have been shaking like a leaf, but Greg had spent his Canadian boyhood with a slingshot in his hand. The rock thundered into the side of the croc's head and the big reptile hurtled up into the air in shock, then spun around and headed off. Within seconds, there was just a diminishing ripple. We went back to recording snake behavior, but it wasn't quite such a carefree frolic as it had been earlier in the night.

The word "ecology" means different things to different people. Even for a researcher, "ecology" covers a very broad field. At one extreme lie detailed investigations of individual animals, and at the other extreme lie analyses of population-level processes like rates of growth, reproduction, and mortality. We need both types of studies to build an overall understanding—but most of what we know about snakes comes from the former kind of research not the latter. The reason is simple: population studies require a lot of animals and a long time frame. And to keep a project going for decades, you need sustained funding. That's a huge hurdle. It's difficult to convince anyone to hand over substantial amounts of cash, year after year, to explore the ecology of snakes.

Early in my research career, ambitious population-level projects were way out of my league. Up until the late 1980s, all of my fieldwork was based on modest numbers of animals. Blacksnakes, Broadheads, and Carpet Pythons were common enough for me to find 20 or 30 animals to radio-track. In turn, those sample sizes told us a great deal about snake movements, habitat use, and so forth. But I couldn't extend that work to the population as a whole. To measure rates of survival and reproduction, I would need hundreds of animals. I could never obtain those sample sizes with giant pythons or endangered elapids.

The tyranny of small sample sizes meant that I couldn't tackle the conservation threats that really matter: those that imperil populations. I desperately wanted to grapple with those issues. Frustratingly, I knew what to expect under Harvey Pough's revolutionary views about ectothermy. A population of ectotherms should spend much of the time doing very little—with low rates of feeding, growth, and reproduction—and then explode into action during brief weather-driven windows of resource availability. Was this happening with snakes? To answer that question, I needed to find a place where snakes were abundant, and then convince someone to fund my studies.

First problem first: a place where snakes were truly abundant. By the mid-1980s, I doubted that any such site occurred in the southern half of Australia. I visited many supposedly "snake-infested" areas, but none of them measured up. I would have to look further afield.

The obvious place to look was in the tropics. The southernmost part of Australia, the island of Tasmania, has a grand total of three species of snakes (all elapids). By comparison, the area around Darwin, in the Northern Territory, contains more than

50 species (8 types of blindsnakes, 5 pythons, 2 Filesnakes, 9 colubrids, 14 elapids, and 17 sea snakes). And some of them might be common enough for me to conduct a population-level project.

But I was just guessing, because tropical reptiles were rarely studied. Most universities are in cool climates and it's cheaper, safer, and easier to work close to home. A mismatch between the geographic distribution of researchers and the geographic distribution of snakes. Back in the 1980s, I was the only snake biologist outside northern Europe or North America. None of us were based in the tropics. Our research was conducted in a world of warm summers and cold winters.

That concentration of research left a huge knowledge gap. The tropics contain a vast number of snakes, including groups that are found nowhere else. But how can you work in the tropics if you live in Sydney—or Paris, or Vancouver? It's not impossible if you want to look at snake genetics, or physiology, or classification. You just need to collect specimens and then bring them home with you. But to study tropical *ecology*, there are no shortcuts: you need to spend long periods in a faraway field site. And even if you can get there, the obstacles are formidable: poor scientific facilities, labyrinthine permitting procedures, unstable political systems, ghastly diseases, and limited access to medical care. And although tropical habitats have great biodiversity, it's rare for any single species of animal to be common. Great for a tourist but not for a researcher.

But among the few tropical areas where fieldwork on snakes is possible, Australia sits at the top of the list. It offers a large chunk of tropical real estate with political stability, medical facilities, safe roads, and reliable electricity. A friend of mine, a very experienced fieldworker, has a simple rule when selecting field sites: "you can do science anywhere you can get cold beer". He doesn't actually drink beer himself, but the rule makes sense—the logistics that make cold beer available give you access to everything else you need. And not only do the Australian tropics have a diverse fauna, but a few of those species are very common. And the beer is admirably cold in those outback pubs.

My background hadn't prepared me to be a tropical ecologist. I spent my early years in southern Australia. My only experience with warmer climates came during Christmas holidays, when Mum and Dad threw the three kids into the back of the car and drove up to the subtropical hinterland of the Gold Coast, south of Brisbane. I loved the humidity, warmth, and exotic foliage. Plus, of course, the spectacular lizards. I spent hours each day walking along narrow paths though the scrubby forest in search of scaly creatures.

But the suburban fringes of southern Queensland were a far cry from the *real* tropics. I was an avid reader, and my childhood passions were fired up by books about intrepid hunters in search of man-eating tigers. I imagined myself stalking through the forests of Kumaon in northern India, rifle in hand, following the pugmarks of the man-eater. In high school, though, I discovered exciting landscapes closer to hand. Books about tropical Australia told of wild places and heroic pioneers. Despite a deplorable lack of man-eating tigers, the Australian tropics were fascinating. My favorite books, written by Ion Idriess, centered on the remote Kimberley escarpments and Northern Territory savannas. Hard-bitten men wrested a dangerous existence from an unforgiving land, and many died lonely deaths. Noble Aboriginal warriors like Nemarluk fought against the rising tide of European expansion. Pioneer women endured hardships that were unimaginable to a city boy like me. Reptiles didn't feature as often as I would have liked, except for enormous crocodiles lurking in the tidal rivers. But those books touched a deep chord. I vowed that one day I would experience that intoxicating tropicality firsthand.

My first real taste of tropical fieldwork was a month-long expedition to the tip of Cape York, the northeastern tip of Australia. Early in my Ph.D. at Armidale, I had become friends with another student, Grahame Webb. Grahame is a bear of a man, passionate about science as well as life in general. During a conversation one evening, we discovered that we shared a secret dream: to conduct fieldwork in far northern Queensland. And to my astonishment, Grahame said "let's do it!" He assembled a ragtag team of family members, friends, fellow students, and even my supervisor Hal Heatwole and Hal's son Miguel. One of Grahame's friends who sold used cars took the "For Sale" signs off four trucks, and off we went. Our aims were vague: see Cape York, collect reptiles and amphibians, go fishing, and have an adventure.

That trip taught me many things. The landscape was a spectacular mosaic of rainforest and dry woodland, often shifting abruptly from one habitat to another as a result of some ancient fire. And the wildlife was fantastic—crocodiles, pythons, Frillneck Lizards—creatures I had read about for years, and now could see for myself. But our naivety led us into blunders. We underestimated the distances involved and overestimated the quality of the roads, so we spent most of our time driving to the next site rather than collecting animals. Without clear objectives, conflicts arose. Some people wanted to stop and go fishing, while others wanted to hurtle forward to the tip of the Cape. And with everyone hot and tired, tempers flared. In fieldwork, the most common focus of friction is food.

The only food supplies we had brought with us were two huge boxes full of cans of preserved meat. Too tired to cook after driving all day, we would each grab a can and eat it cold, then fall into the swag or tent for a few hour's rest. One carton contained cans of corned beef. It was palatable. But the other option was awful—a gray slush that was well-nigh inedible. So the cans of corned beef were the first to be eaten and before long, stocks were low. People began to purloin the cans, to hide them for a later meal. If I woke at midnight, I often saw a flashlight beam shining inside one of the trucks, as someone searched for hidden cans, whisked them away, and hid them somewhere else.

There were adventures aplenty on that Cape York trip, and some amazing biology. But in truth, my most valuable lessons were about the challenges of maintaining working relationships under difficult conditions. Some friendships strengthened and others frayed, even leading to threats of violence. But more than anything, I learned about myself. To my chagrin, I discovered that I wasn't destined to be a bare-chested tough-as-nails heroic fieldworker. Without decent food and a good night's sleep, my brain stops working. That insight was critical when I finally began working in the tropics in earnest, a decade later: I looked for a home base with enough comfort to keep me functioning effectively. I admire people who can sleep on the ground, skip meals, and wrestle a crocodile before breakfast—but my own style of working requires more comfort.

The second lesson I learned was about planning. I love the romantic idea of being a free spirit, disappearing into the wilderness with a knife and a packet of matches to discover new and exciting things. But it doesn't work. Good fieldwork requires planning. The logistics have to be in place—for safety as well as productivity. You need a clear understanding of what you are trying to achieve, *before* you head out into the wilderness. When they join my research group, students are surprised by the military-level planning that precedes our fieldwork. But they soon realize just how important that forethought can be.

****

When I began looking for a tropical base for fieldwork in the early 1980s, I had three options: the east, west, and center of the Australian tropics. To the east, the rainforests of the Queensland wet tropics had a fantastic array of reptiles. But there was a new university in Townsville, already filling up with tropical ecologists. And anyway, rainforests are hell to work in. Thick undergrowth, stinging trees, and thorny vines mean that relocating a radio-tracked snake that has moved to the next gully can take you all day. Recalling the frustration of rainforest fieldwork with Stephens' Banded Snakes, I crossed the northeast off my list.

Across to the west of the continent, the Kimberley is magnificent—but again, the logistics are tough. It's very hot and very dry for most of the year, and very hot and very wet the rest of the time. Infrastructure is poor and roads that are destroyed by monsoonal flooding may not be repaired for years. And it's an expensive place to work—partly because of its isolation, and partly because an upswing in mineral prices fueled a mining boom that drove prices sky-high. A scientist looking to rent a room couldn't compete with well-paid miners.

That left the legendary Top End, smack in the center of northern Australia. In the coastal regions of the Northern Territory near Darwin, the climate is more seasonal than in the Queensland tropics, but less extreme than in the Kimberley. The "wet-dry" tropics take their name from the seasonal nature of rainfall. From May to September, rain is rare. Blue cloudless skies, hot days, cool nights. But from October onward, the clouds form, the humidity increases, and the comfort level drops. Dry heat is OK, but hot and swampy conditions are hell. As the "buildup" begins, the locals stay indoors and start drinking. The clouds promise rain, but they tease for weeks until finally, in December, the skies open in earnest.

The monsoonal rain in Darwin has to be experienced to be believed. It's like standing beneath an upturned bucket. Under that ferocious deluge, the landscape undergoes a metamorphosis as dramatic as a tadpole turning into a frog. Yesterday's dusty forest is transformed into a wonderland. Within a few days, green replaces brown. Dry creekbeds fill, the tired trees leap back to life, and the locals cut back on their alcohol consumption. And the numbers of reptiles and amphibians explode. Only a few species remain active throughout the dry season, but an avalanche of wildlife appears as soon as the first storms of the wet season soak into the soil.

Kakadu National Park, east of the Northern Territory's capital city of Darwin, is often described as "the jewel in the crown of Australia's national parks". It's one of the largest intact ecosystems on Earth, covering almost 20,000 km^2 (7,600 square miles), from rainforests to semiarid woodlands. It is the size of Wales, and about half the size of Switzerland. Sitting on the ancient sandstone rocks of Noorlangie at dusk and looking out over a vast expanse of magnificent country, there is barely a trace of any changes wrought by Europeans. But the older occupants of the land—the Gagadju people—have been modifying the landscape by "firestick farming" for at least 60,000 years. By lighting fires at specific times and places, Aboriginal Australians created the habitat mosaic we see today. And deep within rock shelters all across Kakadu, you can find the artworks created by the early inhabitants of this amazing land.

The wildlife of Kakadu is as spectacular as its sandstone gorges: around 300 species of birds, 80 mammals, 150 reptiles, and 10,000 insects. And everything is bigger and brighter in the tropics. The reptiles include monsters like Saltwater Crocodiles and Oenpelli Pythons, the waterbirds swarm in thousands on the drying floodplains, and wallabies line the roads at dusk. The first time I saw Kakadu, I felt like I had been transported to a different planet.

Although the Top End is a mecca for anyone with an interest in reptiles, I'd never been there. It was too far away and too expensive to visit. But by 1982, the economic, social, and political landscape of northern Australia was changing, in ways that created an opportunity for me. The federal government had declared a massive new national park—Kakadu—but had excised a small area in the heart of Kakadu for uranium mining. The fledgling mine soon hit a road-bump. Aboriginal people in the area still lived traditional lifestyles—including, harvesting wildlife. Would chemicals from the mine contaminate the "bush tucker"—animals like wallabies, snakes, and lizards—that Indigenous communities had been eating for millennia?

A government body with the awesomely bureaucratic title of "Office of the Supervising Scientist" had been set up to minimize the impacts of uranium mining. One of the organization's first actions was to hold a meeting in the town of Jabiru, to identify issues that concerned the public. Understandably, Aboriginal people at that meeting asked if uranium mining might contaminate their traditional food. The increasing political power of Indigenous Australians meant that their concerns couldn't be ignored, but there was no way to answer their questions. Western science knew almost nothing about the ecology of the lizards and snakes that were a food source for Aboriginal people. Until we understood what those reptiles eat, and what habitats they use, it was impossible to assess the risk of contamination from mining. But who could conduct such a study? Reptile ecologists were few and far between. One day in 1981, I got a phone call from a friend who had recently been employed by the Office of the Supervising Scientist. He offered funding for me to conduct a study in Reptile Wonderland.

The focus of Aboriginal concern—and thus of the contract on offer—was on two reptile species. Both are spectacular icons of tropical Australia. One was a giant monitor lizard, the Yellow-Spotted Goanna. Growing to more than 2 meters (6 feet) in length and 7 kg (15 lb) in weight, this is an impressive beast. I was happy to work on it, although a little unsure about my ability to outsmart and out-wrestle these formidable lizards. But from my snake-obsessed perspective, the jewel in the crown was the other species—the Arafura Filesnake. An adult Filesnake can be well over a meter (3 feet) in length. And they are weird. They not only look different from other snakes, but they feel different. Most snakes are muscular. When you handle them, you feel the power within that sleek body. But Filesnakes are thickset dumpy creatures, with loose skin and sagging muscles. When I handle a python, the snake feels like an athlete. When I handle a Filesnake, I feel like I'm kneading dough. Filesnakes are related only distantly to other kinds of snakes, and look like no other living species, so their ecology was likely to be unusual also.

Back in 1981, nobody knew much about either species; indeed, both the Yellow-Spotted Goanna and the Arafura Filesnake had been formally given their own scientific names only a year earlier. Until then, they had been viewed as regional varieties of widespread species—which amazed me as soon as I saw them. They looked nothing like the species with which they had been confused. A 2-meter lizard and a bizarre snake, neither of them recognized as separate species until 1980.

I read everything I could find about Filesnakes, but it wasn't much help. Only three Filesnake species are known—a freshwater form in Asia, a saltwater form in the oceans between Asia and Australia, and a freshwater form in Australia and southern New Guinea. This third one was my target, and it was a creature of mystery. The only well-established facts were that the snakes are entirely aquatic, and that they produce live young rather than laying eggs. In all other respects, the family Acrochordidae was a black hole. For example, one oft-repeated statement was that Asian Filesnakes "eat soft fruit". That would be astonishing if it was true, because one of the only generalizations you can make about snakes (apart from their lack of legs) is that they are carnivorous. There are no vegetarian serpents, apart from the apple-salesman in the Garden of Eden. I tried to track down the source of the "soft fruit" record through the older scientific literature, and finally found it in an early German book that described developing embryos in the female's uterus as "looking like soft fruit". When an English-language version of that book was being prepared, an inept translator assumed that this referred to food in the stomach.

So, how could I find Filesnakes in tropical Australia? My only clue came from John Legler, the American turtle-expert who had taught me how to find my way around inside the innards of dead snakes during my Ph.D. John had been influential in my later career as well, arranging my postdoc studies in the United States. Now he offered some more information. John had traveled through Kakadu several years earlier, placing funnel-traps in deep water to catch turtles. He used sardines as bait, and several times he found Filesnakes in his traps. Maybe that method would work for me also.

With funding from the Supervising Scientist, I didn't have to rely on volunteers to help me. I employed my first research assistant—Rob Lambeck, a tall, slender, quietly spoken man who had recently completed his undergraduate degree. Rob and I flew up to Jabiru in 1982, borrowed a four-wheel-drive vehicle and a small boat from the Supervising Scientist lab, and set off to do tropical ecology. We hauled the boat down to nearby Magela Creek and set the same types of traps as John Legler had used, in the same places as he had set them. Our hopes were high as we checked the traps the next morning, but they were empty. And so it continued. We checked our traps every day but over the next two weeks, we caught nothing.

Goannas were easy to find but difficult to capture. Most times, the giant lizards disappeared into impenetrable thickets or dived into the water at our approach. But

FIGURE 8.2 Surely one of the most bizarre species of snakes, the Arafura Filesnake is a personal favorite of mine. Photograph by Etienne Littlefair.

some lizards were braver or more foolish than others, enabling us to extend a long fishing rod toward them and slip a noose around their necks. Once we had the lizard in hand, we could measure and weigh it, check its sex, and encourage it to defecate or vomit up its prey so that we could identify prey items. And then, release the animal where we had caught it. We persisted and refined our methods through trial-and-error, and before too long we were accumulating information about goannas as fast as we were accumulating scratches on our arms from the lizards' powerful claws. But Filesnakes were a different story. We were failing dismally.

Although I asked everyone I met about Filesnakes, all I heard were vague rumors. Nobody knew how to catch these mysterious animals. I began to dread the cheery question "Caught any Filesnakes yet?" from the locals. The breakthrough came one Friday evening when Rob and I took time off to have a drink and dinner at the Jabiru Social Club. It was a rough and ready establishment but it was the only place in town where you could buy a meal that you didn't cook yourself. As always, the conversation turned to the elusive Filesnakes. Everyone who turned up at the bar asked me if I'd caught any yet, and I tried to find new and more convincing explanations as to why I hadn't. But this time, my excuses brought forth a new story. One of the National Parks rangers was sitting nearby. I'd never met him, but he was kind enough to sound apologetic as he delivered a bombshell. "I saw a Filesnake last night".

We listened avidly. Not long after dusk, the ranger had been driving home across the shallow ford on Magela Creek, less than 30-minute drive from town. Halfway across the creek, his headlights illuminated a large snake in the shallow water. Only a quick glimpse of a dark body, but he thought it was a Filesnake. And as soon as I heard the story, I felt like an idiot. Watersnakes of other species move around at night in shallow water, where they can be found by searching with a flashlight. Why didn't I think of this before? I had caught watersnakes in North America with this method. It hadn't occurred to me to try it here, because all of the stories I had heard about Filesnakes involved groups of them living in deep-water holes, relying on ambushing fish, and thus, rarely moving about.

Suddenly, we had a new option: to look for Filesnakes at night in shallow water, with a flashlight. And no power on Earth could stop us from trying it—immediately. As soon as we finished our fish-and-chips, a ragtag convoy headed off to Magela Crossing to take a look. We were accompanied by most of the people in the Social Club that evening, because there wasn't a lot of other entertainment in town: mine-workers, accountants, mechanics, and even one of the bar-keepers. Some of them didn't know this was a snake-hunt; they just jumped into the cars when all of their mates did. It wasn't an elite team, but it was remarkably successful. As soon as we reached Magela Creek, people piled out of cars and those with flashlights began peering into the water. And almost immediately, a shout rang out—"Here's one!" And then another voice, a few meters upstream—"One here too!" And within a couple of minutes, my own flashlight beam illuminated a large female Filesnake tongue-flicking carefully as she probed small holes among the tree-roots in shallow water. Bingo! No Olympian athlete ever held aloft a gold medal with as much triumph as I carefully lifted this magnificent beast from the water.

Not everybody sees beauty in an Arafura Filesnake. The coarse file-like skin that gives the snake its common name droops in folds from the animal's body. A Filesnake seems like it is wearing its older sibling's skin, two sizes too large. One of my students, on seeing her first Filesnake, exclaimed that "it looks like a sausage inside a pantyhose!" Out of water, a Filesnake moves with painful slowness. But when the snake is submerged, that loose skin flattens out into a vertical paddle that pushes against the water. And the small spines on the center of every scale, responsible for that rough feel, have sensory hairs that tell a snake when it touches a fish. Rough skin also helps in the next step of the process, as the snake enfolds its fishy prey and constricts it. That's an impressive feat. Many snakes constrict their prey on land, but it's a lot easier to squeeze a dry mouse than a slippery fish. Those rough scales make it possible.

A Filesnake's crowning glory, though, is its head. Males have small sleek heads, but females have big bulldog-shaped heads. Square across the front of the mouth, with tiny eyes well-hidden. Valved nostrils are positioned high on the head so the snake can sneak up to the surface and take a quick breath without being seen. Filesnakes extrude their long forked tongues to sense chemicals in the water around them, but—like everything else—they do it in slow motion. Baby Filesnakes are brightly patterned but with age, the pale background color darkens to a rich chocolate. A network of irregular black bands breaks up the animal's outline. In murky water, a Filesnake is invisible.

The weirdness of Filesnakes isn't just skin-deep. They have many other bizarre characteristics. To help them remain underwater, Filesnakes have a very large lung (to hold lots of air) and a high volume of blood (so, a huge oxygen store). And dive time is increased even more by the snake's low rate of metabolism. Filesnakes need very little oxygen to stay alive. When they surface, they take several breaths in a row, then slide back down into the depths. Being able to hold your breath for 90 minutes helps if you're searching for fish.

The downside of that energy-efficiency is that Filesnakes are unathletic. They've taken energy-saving to such an extreme that they can't mobilize energy rapidly—regardless of the circumstances. A Filesnake that is lifted out of the water will convulse a few times in your hands, then lie there exhausted. They are the sloths of the serpent world. The only thing that a Filesnake does quickly is feed. The strike is fast, and the prey is swallowed at lightning speed—because the fish hastens its own demise. Once the snake has its prey item by the head, pointing in the right direction, the struggles of the fish propel it down into the snake's stomach.

A predator that eats fish needs long teeth to penetrate the fish's mucus, so a bite from a Filesnake would be painful. However, these gentle creatures don't try to defend themselves. When Aboriginal people collect Filesnakes for dinner, the hunters just throw the snakes up on the bank a meter (3 feet) or so from the water. The unfortunate reptiles wriggle a few times, then lie there and await their fate. Why don't they crawl straight back into the water? Because they can't. They're exhausted. Their low metabolic rates mean that Filesnakes are deplorably bad athletes. If all you're doing is poking slowly around looking for a sleeping fish, you don't need rippling muscles and great endurance.

As you might guess from these peculiar characteristics, the three species of Filesnakes are not closely related to any other snakes. Their ancestors diverged from the main line of snake evolution more than 20 million years ago. And despite their

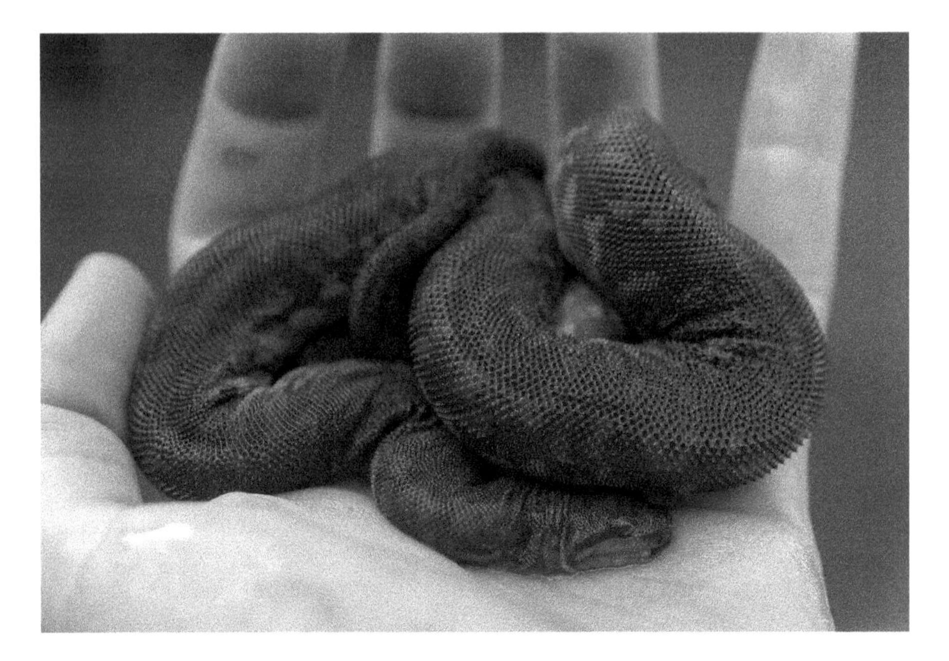

FIGURE 8.3 This juvenile Filesnake shows the distinctive features of the species: chunky build, loose spiky skin, and an overall floppiness that is very unlike most snakes. Photograph by David Nelson.

weird appearance, they are superbly adapted to their ecological niche. That low rate of energy use not only enables them to stay under water for a long time; it also means that they can survive months or even years without feeding, if local weather conditions decimate fish numbers. If Harvey Pough was correct—that ectothermy facilitates a sedate lifestyle that involves doing nothing for months at a time—the low metabolic rates of Filesnakes should make them a perfect example of that scenario.

I expected that this low metabolic rate also would force Filesnakes to stay close to home. A snake that is incapable of vigorous activity isn't going to wander far, so a Filesnake will spend its life sitting around waiting in ambush. Radio-tracking them would be boring. But at the same time, I had a nagging doubt. If a Filesnake is a couch potato, why does it move around at night in shallow water? Perhaps each snake spends the day in deep water, comes up to check its favorite stretch of shallows at night, then goes back home to recover for a week or two?

The way to find out was to put transmitters in Filesnakes and then release the snakes and see what they did. But there was a snag: radio signals don't travel as well through water as they do on land. Marine biologists use specially designed acoustic transmitters to study tuna and sharks—but I had no money for such extravagant technology. Could I just use those old transmitters in my desk drawer, the veterans of Blacksnake and Diamond Python research? Probably not, but it was my only option. So I replaced the batteries, brought the transmitters up to Jabiru, waded carefully out into a billabong, and deposited a couple of the units on the bottom in a meter of murky water. Then I went back and switched on the receiver and to my joy, the signals came in loud and clear.

So I anesthetized half a dozen of the Filesnakes we'd caught at Magela Crossing, surgically implanted my secondhand transmitters into their body cavities, then took the snakes back to the Crossing and waved goodbye. We brought the boat along when we went out to relocate them the next day, just in case one of the snakes had moved away. And I was pleased that we'd taken that precaution, because only one snake was close to where we had released it. All the rest had disappeared, although we picked up faint signals from two of them. We launched the boat and soon found those snakes, 100 meters (100 yards) away. Perhaps washed downstream with the current, I thought, before they could struggle into calmer water? But to my astonishment, other snakes had headed upstream—and my original female was 500 meters (550 yards) away from the release site. Surely too far for her to have gone under her own steam? Had she been eaten by a crocodile? We tethered the boat to a Pandanus palm beside the spot where the telemetry signal was loudest, and waited. And waited. The water was filthy, stained by mud washed in by monsoonal rains, so there was no way we could see a snake underwater. Thirty minutes later, the telemetry signal got louder, and Rob spotted a Filesnake's head come to the surface for a breath. This was definitely our female. She had performed a herculean feat by moving so far.

So, my newly released Filesnakes had traveled further than I had expected. I was confident that this was an aberration, induced by the stress of capture, handling, and surgery. They just wanted to get away from the place where they had been molested. Now they'll settle down. But they didn't. Every day when we went out to locate them, our snakes were hundreds of meters away from their previous sites. And when we watched Filesnakes in the shallows at night, they were perpetual motion machines. Slow but sure. Poking along the flooded edges of the creek, looking for sleeping fishes. We saw snakes catch fish, and I watched one large Filesnake consume half a dozen dead fish that an angler had discarded in the shallows earlier in the evening.

****

We rarely saw the animals that we were tracking. The telemetry signal would beep loud and strong, but it came from a snake hidden in murky water. To compensate for that frustration, though, Kakadu was a naturalist's paradise. The chop of a feeding Barramundi beside the boat; a wraith-like Golden Tree Snake disappearing into a tangle of branches; Mitchell's Water Monitors perched like miniature dragons on the Pandanus palms overhanging the water. Once, shockingly, a Saltwater Crocodile launched itself upward out of the water beside our boat to seize a Fruit-Bat that was clambering along a mangrove branch. Even the drive to and from Magela Crossing was a potpourri of wildlife sightings. On the day after rain, I'd often see Frill-Necked Lizards basking on the road. Large goannas on the roadside verge would raise themselves on their hind legs to peer intently at the intruder. At one dip in the road where water had pooled, Water Buffalos dug a wallow. Every time I drove that section, half a dozen massive bovines lumbered to their feet and galloped off into the savannah woodland.

In the 1980s, Water Buffalos were so abundant in Kakadu that they were a peril to life and limb. The pitch-black buffalo were almost invisible on the road at night, and many vehicles were destroyed by collisions with them. But for me, the biggest scares came when I blundered into buffalos while I was walking beside billabongs at night looking for Filesnakes. Several times, I almost touched a buff before I saw it. They always ran the other way, but Water Buffalos are ponderous thinkers, and there were a heart-stopping few seconds before they decided to retreat rather than flatten me. The most frightening encounter came late one night, as I was walking along a path near Magela Crossing with Rob Lambeck and Craig James. It was full moon, bright enough for us to see any Death Adders on the track, so we turned off our flashlights and walked along in companionable silence, enjoying the glow of the moonlight off the Paperbarks, and the sounds of a tropical night. But that meditative calm was abruptly shattered when a herd of 20 Buffalo erupted all around us. We had walked into the middle of a herd without seeing them; and when they noticed us it set off a full-bore stampede. Massive horns hurtled past on all sides, only inches away. Seconds later, it was all over and we were alone on the moonlit road, and luckily unperforated. Without saying a word, we all turned on our flashlights.

Those first trips were a steep learning curve for me, especially in terms of working with Indigenous people. Kakadu National Park is managed by the traditional owners of the land as well as by government-employed rangers. If we wanted to study Filesnakes, we needed permission from the Indigenous owners. I enjoyed spending time with people who have such a deep connection to the land, but communication between cultures wasn't always straightforward. One difficult situation arose when an older man asked to accompany us as we looked for Filesnakes, to see what we were doing. That was perfectly reasonable (it was *his* country, after all), so he came with us in the boat the next day as we relocated radio-tracked snakes. He didn't speak much English but he watched closely as we selected the correct radio frequency on our telemetry receiver, swung the antenna around until we picked up a signal, then followed the signal to where it was loudest. The Filesnake we were following was in very shallow water, and to demonstrate what we were looking for, I reached into the water, felt the snake's rough scales, and pulled it into the boat. We showed him the snake, then released it. He smiled. To make sure he understood what we were doing, I caught the next snake as well after we tracked it down to a similarly shallow site.

The problem came later when we returned the Aboriginal elder to his camp. The old man wanted our telemetry receiver. He liked this new way to find Filesnakes and he could use it himself to obtain snakes for dinner. We tried to explain that the snakes we found were ones we had already caught, and had implanted with radio transmitters, but he looked skeptical. Why would someone catch a Filesnake, put something inside it, then let it go again, only to go out and look for it every day? Whitefellas do stupid things, but this was absurd. He didn't believe us, but he allowed us to work on his land anyway.

The Whitefellas were forgiving as well. Some of them were straight out of the pages of the Ion Idriess adventure stories that I had devoured when I was at school. Men and women who had come up to the Territory to escape arrest warrants down south, or in search of a place where they could be themselves. Who thought nothing

FIGURE 8.4 Filesnakes spend their entire lives in water. Photograph by Steve Wilson.

of working all day in the tropical sun. People with gruff exteriors but warm hearts and observant eyes, willing to offer sage advice to an ignorant young scientist.

By the time I began working in Kakadu in the 1980s, the area was transforming from a remote wilderness to a tourist destination. Driving to the supermarket on bitumen-sealed roads, it was easy to forget how rough-and-tumble the region had been just a decade earlier. That historical perspective was brought home forcefully to me when I approached one of the pioneers of Kakadu for advice about catching Filesnakes.

Looking in the shallows at night had yielded those original Filesnakes, enabling me to get the radio-tracking underway. But unlike Magela Creek, most waterbodies in Kakadu either didn't have extensive shallow edges, or else they contained large Saltwater Crocodiles. Splashing through the margins of a billabong at night would be suicidal because that is where the crocs wait for a wallaby to come down to drink. And frustratingly, almost all of the Filesnakes I found in shallow water at night were males. I needed more snakes, especially females.

How else could we catch Filesnakes? One of the people I asked was Hal Cogger. He had played a pivotal role in my career already, allowing me to examine specimens in the Australian Museum; and now he came to my rescue again. Hal had conducted field surveys in the Kakadu region several years before, and he gave me the name of a local bushman who had worked with him: Dave Lindner, a larger-than-life pioneer who helped to establish Kakadu National Park.

Dave epitomized the rough-and-ready characters from Kakadu's early days. A huge man, fiercely independent, he had a long history of conflict with bureaucracy and officious administrators. Soon after Dave was first employed as a ranger in the

FIGURE 8.5 Rob Lambeck and I measuring a recently captured Filesnake before releasing the animal back where we had caught it.

1970s, he concluded that the Northern Territory wildlife-management system was corrupt. Authorities ignored the illicit poaching of fish and crocodiles, and Dave suspected that some of the "managers" were in cahoots with the poachers. Determined to smash that comfortable villainy, Dave pursued the criminals and corrupt authorities with a determination that bordered on obsession. Poachers went to jail, and contract killers were offered cash rewards to eliminate that troublesome new ranger. In the end, Dave triumphed.

On the day I first met Dave Lindner, Rob and I drove out at dawn to meet him at his home—the idyllic Patonga homestead, deep in Kakadu, beside a magnificent billabong festooned with waterlilies. My first impression of Dave in the half-light was of a giant—an oversized and immensely powerful man, with enormously muscular arms and hands. It was easy to imagine him wrestling crocodiles into submission, or walking unarmed into a poacher's camp to arrest miscreants.

By the time we dropped Dave back at Patonga at midnight, we had driven a lot of kilometers, and caught a lot of Filesnakes. But my most powerful memory is of Dave's nonstop narrative. A born raconteur, the stories he told about "the old days" made my own life look very humdrum. Many of those stories centered on his decades-long battle with a poacher named Roy Wright (known as "Wrighty" to one and all). Corrupt officials tipped Wrighty off about Dave's movements, enabling the poacher to avoid arrest, so Dave developed convoluted strategies—pretending to head off to one area but then taking back-roads to turn up in unexpected places. A year before our Filesnake trip, the coyote finally caught the roadrunner. Dave found an illegal gill-net and was hiding in the nearby bush when Wrighty arrived and attached the net to his truck, to haul out a few hundred Barramundi. Dave came roaring out but Wrighty didn't wait around; instead he took off at full speed, dragging the net behind him. The ranger chased the poacher out to the Arnhem Highway, where the pursuit continued. Two battered four-wheel-drive trucks hurtling along at breakneck speed, one of them dragging a net full of the mangled remains of huge fish; and the drivers shooting at each other with pistols out the window. It was an epic tale of crime and punishment, against the backdrop of a wild and unforgiving land. Ion Idriess should have written a book about it.

On that day in 1982, Dave took us to Alligator Heads, a remote floodplain on the Mary River that had been Wrighty's favorite site for poaching. Indeed, it was the place where Dave had caught him red-handed a year earlier, leading to that dramatic chase along the Arnhem Highway. The case went to court, Dave gave evidence, and Wrighty went to jail. Dave was telling us the story as we bumped our way across the floodplain toward the river—and when he pointed out the spot where it had all happened, we were astonished to see two vehicles beside the billabong. "Barramundi poachers!", said Dave, and drove toward them. A look of concern flitted briefly across his face as we came close. "That's Wrighty and his boys. Didn't realize he was already out of the slammer". No illegal nets, but a hard-looking group of men sporting jail tattoos. Dave drove up to them, stopped the car, and wound down his window. I saw a rifle leaning against a tree, and I wondered if my fledgling scientific career was about to be cut short.

Wrighty snarled at Dave, walked up to his open car window, turned around, and let rip with the loudest, smelliest, and longest-lasting fart I have ever witnessed. I swear it blistered the paintwork. Wrighty then turned around and screamed abuse at Dave, focusing on the trial that saw him convicted. "No way you could see it was me, you bastard—I was in the bushes by the time you got close enough". "Yeah, but I knew it had to be you, you bastard". I sat quietly in the front seat, pretending that I often encountered enraged poachers foaming at the mouth, furiously punching the front of my car, and swearing that they would subject the man beside me to a range of anatomically implausible experiences.

Dave kept calm, and Wrighty's passion eventually subsided. To change the subject, I asked him about Filesnakes. Yep, they often got caught up in the Barramundi nets. A bloke once offered him money to collect some for their skins, so Wrighty filled a 44-gallon drum with them. He handed it over but the man never paid for them. And as far as Wrighty knew, nobody ever managed to set up a commercial skin industry based on the snakes. It was an affable conversation, with a man who 5 minutes earlier had been screaming in outrage. I felt like Alice in Wonderland, a wimpy researcher transported into a strange frontier world where alpha-male enemies oscillated between obscenities and chit-chat.

That wasn't the end of the day's excitement, though. Dave took us to a deep billabong fringed by Pandanus palms. The only way into the water was a small track, where wallabies came down to drink in the evening. This looked like an ideal spot for a giant Saltie to lie in wait for an edible mammal (for example, an academic from Sydney), but Dave laughed and said confidently "I know this place—no Salties here", and promptly took off his clothes and jumped into the dirty water. Buoyed by Dave's confidence, we followed him in.

We swam out to the fringing Pandanus on the other side of the billabong, and groped around below water level. I was expecting a set of giant jaws to clamp onto my legs, but we didn't see any crocs. And sure enough, the Filesnakes were there—immediately recognizable by their rough skins. We each caught several, and clumsily swam back to the path with arms-full of snakes. We set about weighing and measuring them in the shade of the car. After we had finished, about 30 minutes later, I took the snakes back to release them. A huge Saltie was lying by the water's edge, right on the path. When I walked back and stuttered this awful news, Dave laughed. "Yep, I know that fellow, he's a good croc".

By the time we said our farewells to Dave back at Patonga, my brain was reeling. We had caught dozens of Filesnakes, some of them far larger than any we had seen at Magela Crossing. Dave had told us fascinating stories about Filesnake ecology. And more than anything else, I had been given a glimpse into how different life had been in the Top End just a decade or two earlier, compared to my bland existence in suburbia. And I was glad, and surprised, to be still alive.

****

Dave had learned the tricks of Filesnake-catching from people whose ancestors had been doing it for thousands of years: the Indigenous hunters. I was keen to see Aboriginal methods firsthand, but accessing that traditional knowledge wasn't easy. I employed young Aboriginal men to help me, but none of them had ever been Filesnake-catching. They didn't know where, when, or how to do it. Naively, I had assumed that all Indigenous people were expert about wildlife—but Aboriginal society is like whitefella society in this respect. Some people know about animals, whereas others couldn't care less. In Indigenous communities, the expert Filesnake-catchers are older women. And even among that group, not everyone fancies snake for dinner. Only two Indigenous families in Kakadu still harvested Filesnakes; the other families preferred buffalo-burgers. It took months of negotiation to organize a Filesnake-hunting trip, but finally it fell into place. The snake-catching ladies needed transport to their collecting sites, and I wanted to see Indigenous hunters in action: a win-win situation.

On the day of the Great Indigenous Snake-hunt, we turned up bright and early at the Aboriginal camp outside Jabiru. None of the snake-hunters were there. OK, things work slowly in the Territory. An hour later, an old man told us that the snake-catchers were waiting for us at another camp 30-minute drive away. And since we were going there, could we give a few people a lift? Of course. So our trucks were filled with Indigenous kids, whom we dropped off at the other camp. Where, it turned out, there was only one lady who was going snake-hunting. The others were back at the camp we had just come from. And 20 people needed a lift in that direction. So once again we set off. That story was repeated another three or four times, as our vehicles criss-crossed the local area redistributing Aboriginal people from one camp to another. Finally, though, we were ready to go. We had a large team of snake-catchers: four elderly ladies, and a group of about a dozen younger women.

One of the older ladies directed us through various back-roads to a beautiful Pandanus-lined billabong. But the plans changed again. This billabong wasn't any good for Filesnakes. We'd come here because the younger women wanted to look for turtles by probing the damp mud with iron bars. So, they climbed down from the vehicles, and told us to pick them up when we were heading home. Team Filesnake shrunk with every stop and eventually, only the four older ladies remained. We drove to another billabong where one of the ladies waded out, felt around under the water for 30 seconds, and then returned to the car. "No good", she said. I suspected that this was the end of the Filesnake hunt; we'd been hoodwinked to provide a free taxi service.

But I was wrong. Off we went to another billabong, the same lady got out, and this time she had a large Filesnake in her hands within a few seconds of entering the water. Soon all of us were knee-deep in the water groping around under floating grass mats and logs for the distinctive rough skin of a Filesnake. And we found plenty, of all sizes. Before long we had 45 snakes lying on the shore of the billabong. The hunters killed some of the snakes as soon as they were collected, by biting down on the snake's head. Other Filesnakes were thrown up onto the bank to await their fate.

Although I was reassured by the expertise of my hunting companions, this was classic Crocodile Country. I constantly scanned the water surface for any signs of hungry crocs, and I tried to keep at least one of the Indigenous hunters between myself and deep water. Unfortunately, the ladies realized that I was nervous, and decided to have a bit of fun. Every few minutes one of them would turn around, stare

FIGURE 8.6 Aboriginal hunters catching Filesnakes. One woman is killing a recently captured snake by biting down on its neck. Photograph by Rick Shine.

intently at a spot behind me, and splash the water with her hands as if to ward off an approaching croc. And when I jumped or squealed, they all roared with laughter.

When we had finished, the hunters gathered the snakes into a pile on the bank of the billabong. Conscious of conservation issues like harvest sustainability, I had intended to document which sizes and sexes of snakes the hunters kept and which they released. But that was a whitefella assumption. Why throw away a meal? All of the snakes were kept except for one emaciated female that aroused their sympathy—and anyway, she wouldn't have been very tasty. She went back into the water and the other snakes went into a bag. And then it was back into the car to drive back to town—circuitously. We spent another 2 hours picking up Aboriginal people from various billabongs and dropping them off at various camps—where we picked up more people and took them to other camps. But now that we knew the rules of the "Whitefella taxi" game, it wasn't so frustrating—and we had been amply repaid by the opportunity to see those expert snake-catchers in action.

For a year after that trip, our main collecting method was the one we had learned from Dave Lindner and the Indigenous hunters: to wade into the shallows and grope for Filesnakes under floating grassmats and Pandanus logs. The local scientists assured me that there were no big Saltwater Crocodiles in the upstream waterbodies where we worked. "The big blokes are all in the downstream billabongs", they claimed. And like an idiot, I believed them. We swam around the edges of billabongs, feeling for the rough scales of a Filesnake. We even set up a hookah system like pearl-divers, with an air compressor on the boat, so that we could spend more time around the steeper edges of the waterhole feeling up under the banks. Once a large fish hiding among the mangrove roots came hurtling out and crashed into my chest, almost precipitating a heart attack. And one day while I was feeling around, my hand encountered a soft area surrounded by a row of small sharp objects. I had stuck my hand into the open mouth of a small crocodile. I realized this when the reptile terminated the dental examination by slamming its jaws shut. Fortunately, it then let go and swam off. Blood dripping from my fingers, I decided that there must be a better way to catch Filesnakes.

We needed a method that kept us out of the water. Traps were the obvious answer, but John Legler's technique (baited drum-nests) had been a dismal failure. I decided to try a different kind of trap. If snakes were moving through shallow water at night, I could intercept them with a vertical wall of netting that projected straight out from the bank into the water. A funnel trap attached to one end of that mesh wall (technically, a "fyke net") would catch any snake that tried to go around the wall through deeper water. That setup didn't eliminate the risk from crocs, because we still had to wade around to attach the vertical wall of mesh firmly to the muddy bottom. But we didn't have to spend as much time in the water, and we could do it during daylight hours when I was braver and the crocodiles were less active. Once in a while a crocodile became entangled in the leader and had to be extracted, but they were little fellows. If a big Saltie took a dislike to one of our traps, all we found the next morning was bent poles and torn mesh.

The traps were a huge success. We often caught several Filesnakes per trap per night, plus assorted fishes. The traps were in the shallows, with their tops above water level so that the snakes could easily come up to breathe. I was confident that we had invented a safe effective way to catch Filesnakes—but my smugness was premature. One morning when we checked the traps, we discovered a catastrophe. A trap had caught so many snakes—20 of them—that their struggles had pulled it into deeper water. Unable to reach the water surface, all of the snakes had drowned. I had never killed a Filesnake up to that point, and these creatures are dear to my heart. I was sickened. From then on, we tethered our nets firmly to the bank, and added flotation devices (empty plastic bottles) inside each trap to keep part of the mesh above the water's surface, even if the trap was dragged away from its original site. We never drowned another snake.

In the meantime, though, what could I do with a pile of 20 dead Filesnakes? Guilt-stricken, I needed to find some positive outcome from my blunder. Museums would be grateful for specimens of these poorly known animals, and so I preserved all of the casualties. And before I shipped them off to the museums, I dissected the dead snakes to look at their stomach contents, reproductive organs, and the like. I noticed many strange features—Filesnakes are as weird on the inside as they are on the outside—but one aspect was especially thought-provoking. A few of the females were pregnant (making me feel even more like a mass-murderer), with embryos strung along the oviducts like a row of large yolky beads (or like soft fruit, according to that old German book). In every other species of snake I'd ever dissected, the eggs or embryos fill up the entire rear half of a pregnant female (from halfway down her body, all the way to the base of her tail). Strangely, though, all of the Filesnake embryos were in the upper and middle parts of the female's oviducts. The lower oviducts were empty.

Why would a pregnant female Filesnake carry her litter so far forward? Perhaps it's because a swollen rear end is too much of a burden for an aquatic snake? A snake swims by arcing its lower body from side to side, pushing against the water. The rear of the snake exerts the most pressure, and that's why sea snakes have paddle-like tails. So now I had a new hypothesis: carrying babies in the lower part of the body compromises a female's ability to swim, favoring an evolutionary redistribution of those embryos. And if that hypothesis is true, we should see the same pattern in other aquatic species. That is, the litter should be held further forward in the female's body in a watersnake than in a land-snake. As soon as I was back in Sydney, I went to the Australian Museum and looked inside their preserved sea snakes—and found that their embryos were held in the middle part of the body, just as in a Filesnake. This evolutionary adjustment to swimming has arisen independently in every group of snakes that has colonized the water. So, I had discovered an ecological cost of the switch from a life on land to water. The rear part of your body is a no-go zone, because eggs there would ruin your swimming style. As a result, aquatic snakes have to settle for smaller litters. That conclusion didn't make up for drowning the Filesnakes, but it meant that something positive had come out of the disaster.

When I dissected those accidentally killed Filesnakes, I also noticed that their intestines were looped into complex folds, rather than being a straight tube between mouth and anus as in other snakes. Carnivores typically have short, straight, and simple guts because meat is easy to digest. In contrast, plant tissue is difficult to break down, so animals that eat plants have long guts that provide a large surface area and a slow passage time. To fit that long gut into the body, it needs to be folded back on itself and coiled around. Looking inside a Filesnake is like looking inside a cow.

Some of those Filesnakes had eaten recently, and their intestines were swollen with digestive gases. I've never seen such bloating in any other snake. It may be due to bacterial activity that enhances digestive efficiency. But how do Filesnakes manage the buoyancy challenge of an intestine filled with large bubbles of gas? Unless the animal hangs on to a log, it would float to the surface. And imagine the Filesnake Flatulence Fiesta when that gas passes through!

Both the convoluted gut and the bacterial bloating may reflect other as-yet-unstudied adaptations related to energy-efficiency. A Filesnake needs to extract every possible nutrient out of every prey item that it catches. Nobody has measured digestive efficiency in Filesnakes, but I suspect that it's very high. Very little material comes out of the rear ends of these strange flabby serpents. Unlike almost every other snake I've ever handled, Filesnakes don't poo on me when I catch them. If I keep Filesnakes in a fish tank, I rarely need to change the water. If I put a sea snake in the same tank and feed it the same fish, next week I'll be cleaning excreted fish scales, and even fish eyes, out of the filter.

By 1983, we knew enough about the ecology of Filesnakes and goannas to conclude that they wouldn't be affected by contamination from uranium mining. The concern expressed by Aboriginal people was reasonable, but unfounded. Polluted water from mining activities was stored in settling ponds until it could be released into the river at the height of wet-season flooding. The vast amounts of water heading out to sea (measured in units of "sydharbs", defined as the volume of water required to fill Sydney Harbour) assured instant dilution of those chemicals to near-homeopathic levels. Even just a few meters from the outlet, sensitive assays could barely detect the chemicals from mining. And during the wet season, when the pollutants were released, Filesnakes and goannas were scattered across the landscape, far from the main channels that carry pollutants to the ocean. We could assure the local people that their "bush tucker" wouldn't be affected by mining.

When the contract finished in 1983, I no longer had any funding to continue the Filesnake studies. I couldn't turn my back on the Top End, though; those first trips to Jabiru were a powerful drug, and I was hooked. I was the first reptile-ecology researcher to work in the Australian wet-dry tropics, and (as our US cousins would say) I felt like a kid in a candy store. Spectacular animals, all of them unstudied. I wanted to keep investigating the fauna of tropical Australia, but I didn't know how to do it.

In 1985, a diffident young man named Darryl Houston provided me with an opportunity to launch back into tropical fieldwork. Darryl had recently completed his Honors degree at my old alma mater, the University of New England, and wanted to continue for a Ph.D.—preferably, on the ecology of Filesnakes. An encounter with a Filesnake in northern Queensland a few years before had struck a visceral chord, motivating Darryl to find out more about this bizarre reptile. And Darryl was well-suited to the challenge. He is not a man to rush headlong into any situation; he prefers to mull things over before deciding on a course of action. Filesnakes function the same way, so it was a marriage made in heaven. A researcher and his study animals both traversing the path less traveled, at a slow and steady pace. The timing was perfect as well. I had contacts in Kakadu, and I knew where Darryl could find abundant snakes and free accommodation. So began the second phase of the work on Filesnakes.

Darryl focused on DjaDja Billabong, downstream from my earlier studies but on the same floodplain. This huge waterbody was full of Saltwater Crocs, but our fyke-net technology made it safe to work there. Well, fairly safe. And Filesnakes were far more abundant in DjaDja than in the upstream billabongs where I had worked earlier. I dared to hope that this system would allow a genuine population-level study of snakes. Sure enough, snakes poured into Darryl's fyke-nets. His capture rates were phenomenal—more than 4,000 Filesnakes over the course of his Ph.D.

The abundance of Filesnakes in DjaDja is made possible by their incredibly low rates of metabolism. It requires only a few fish dinners per annum to keep a Filesnake alive, so a billabong can support a vast number of these superefficient predators. In a herculean effort under tough conditions, Darryl captured thousands of snakes, marked them individually, and then released them. Recaptures of those marked animals told us not only how far Filesnakes moved, but also how quickly they grew, and how often the females reproduced. The usual way to mark a snake is to clip a few of its enlarged belly-scales, so you can recognize the animal by counting the number of scales between the vent and the belly-shields that you clipped. But Filesnakes don't have enlarged belly-scales, so that method wasn't feasible. Instead, we lightly branded numbers (or more accurately, geometric symbols that translated into numbers) on the snake's sides. The symbols were Dave Slip's idea, based on a method used to brand racehorses. It pays to have students with a broad knowledge of life outside science.

******

The peculiarity of Filesnakes extends to their sex lives. Females grow larger than males, as in most snakes, but Filesnakes take things to extremes. Females grow vastly larger than their male companions. Groping around in the Mary River with Dave Lindner, I caught one mating pair where the male weighed 400 g (less than 1 lb) and the female weighed 4 kg (9 lb)—exactly ten times heavier than her lover. If the Guinness Book of World Records ever includes the category "greatest size disparity between mating animals", those snakes might win. But that inequality in size is just the beginning. Even at the same body length, males are slender-bodied and small-headed, whereas the females are heavy-bodied and have large square heads.

As an evolutionary biologist, I was intrigued as to *why* males and females have different-sized heads. The same kind of difference between the sexes occurs in many types of animals, but generally in the reverse direction. Males have larger heads than females in many lizards, for example, and the reason is clear. A more powerful bite helps when you're fighting your neighbor for access to mating opportunities. More matings by big-headed boys results in more baby lizards with genes that code for large head size in males. But advantages in sexual rivalry can't explain the situation where the female snake not the male has a larger head—because as far as we know, female snakes never fight with each other.

So why do female Filesnakes have larger heads than males? Apart from bite-force in battles, the most likely benefit of having a bigger head is that it enables a snake to swallow larger prey items. For that process to affect head sizes differently in males and females, however, the sexes would have to feed on different types of prey. Is that feasible? Yes it is. David Pearson's studies on Carpet Pythons in Western Australia had shown exactly that kind of divergence; at the extreme case (on Garden

FIGURE 8.7 Female Filesnakes (above) grow larger than males (below) and are much more heavy-bodied and have much larger heads. Photograph by Rick Shine.

Island), male pythons eat mice and female pythons eat wallabies. And by squeezing our trapped Filesnakes gently, to encourage them to regurgitate their prey, we soon assembled a dataset on the food habits of these peculiar aquatic snakes. Most of the snakes had empty stomachs, but Darryl's huge sample size enabled us to answer the questions we were asking. Sure enough, male and female Filesnakes feed on different prey. Females specialize on larger fish, and so they need big heads to swallow them. Males eat small fish, and so they don't benefit from having large heads. And Filesnakes take this sex-based divergence in ecology to an extreme; in many ways, the two sexes act like two separate species.

This sexual disparity is possible only because of the Filesnake's superefficient low-cost lifestyle. Their metabolic rate is so low that a month without food is no problem. A Filesnake burns up very little energy while it waits for Nature to restock the sushi bar. The capacity to survive for months between meals gives the Filesnake access to a remarkable evolutionary option: specializations that increase an individual's ability to catch one particular type of prey, rather than exploiting everything edible that it encounters. And Filesnakes have taken up that option. Females have evolved to be large, with supersized heads, and to specialize in ambushing large fish—like Barramundi, Catfish, and Sleepy Cod—in deep water. Males have evolved to use a different ecological niche. They cruise the shallows at night to find sleeping Rainbowfish and Gudgeons. A small head has hydrodynamic advantages—when striking at a fish, a smaller head moves through the water much faster. The disadvantage of a small head—an inability to swallow big prey—doesn't matter, because male Filesnakes ignore larger fish.

The end result of that niche partitioning is that male Filesnakes sneak along in shallow water looking for tiddlers, and female Filesnakes lurk in deeper water waiting for lunkers. The sexes even differ in the way they detect their prey. Males rely on scent, locating immobile and well-camouflaged sleeping fish by following the trail of chemicals in the water. In contrast, female Filesnakes look for movement—the best cue if you are lying in ambush for a mobile prey item.

Because male and female Filesnakes live different lives, and forage at different depths, they prefer different types of waterbodies. As the floodplain dries out at the end of the wet season, most of the male Filesnakes move into shallow billabongs, whereas most of the females end up in deeper steep-sided ones. That assures a good food supply for each sex during the long dry season, but it introduces a challenge for the snakes' love life. Filesnakes mate when the water levels are at their lowest, just before the monsoonal rains begin late in the year. Male Filesnakes are confronted by a choice between food versus sex—they can either spend the dry season well-fed but sexually frustrated, or starved but with plenty of female company.

Large female Filesnakes can eat huge fish. Most types of snakes have flexible heads, with joints between the jawbones that allow a wide gape, but few take this as far as a female Filesnake. And the billabongs of Kakadu teem with large fish. Large Barramundi occasionally forced their way into our traps, to feed on smaller fish already inside. As a result, our daily trap-checking sometimes provided a Barra for dinner, saving us the long drive back into town to buy supplies. On one trip when we'd run out of food, we were delighted to find a large fish in one of the traps—but unfortunately, we were a trifle late. A female Filesnake in the same trap had found the Barra before we did, and was swallowing it. Our dinner plans were in ruins until the snake took offence at being handled, and regurgitated its meal. That pre-marinated fish fed three of us.

Given the enthusiasm of (some) Aboriginal people for Filesnake Fritters, I have often wondered how a Filesnake would taste. But I still don't know. The only reptile I've ever eaten was a goanna (out of necessity; I was marooned on an island at the time) and the flavor was underwhelming. Nonetheless, the Aboriginal hunters who showed us how to catch Filesnakes were enthusiastic about the flavor. The recipe was simple. "Throw them on the coals, and wait until the skin starts to pop, just like popcorn", one Indigenous woman told me. This was early in the Filesnake research, when I was still deploying baited drum-nets in deep water. It was miserable work. Worried that any captured snakes might drown, I decided to pull up the nets at hourly intervals. But catch rates were low, and the nets were almost always empty. The work was OK during daylight hours, but became increasingly tedious as the night wore on. Sleep was out of the question because it was hot, we were soaked in sweat and insect repellant, and every noise might herald the approach of a crocodile or Water Buffalo. But it took substantial self-discipline to clamber bleary-eyed into the boat once an hour, and haul up empty traps. A Saltwater Croc helped us stay awake, by approaching whenever we stopped to check a trap; his sudden appearance beside the boat startled us back into wakefulness many times. We'd abandon the regular checks at around 3 A.M., try to get some sleep, and then check again at daylight. We decided that if we found a drowned snake in a trap in the morning, we'd dissect the animal then and there, record the data, and have Filesnake for breakfast over the campfire. It never happened, and so it was sausages instead of Filesnakes for the morning meal.

The early days of the Filesnake research were challenging, as I learned how to conduct fieldwork in the tropics. Running a research program in a remote area is complicated. The obvious problems to solve are the "known unknowns" like how to find, catch, and handle your study animals. But less obviously, you need to ask the right questions. What really matters to a Filesnake? Off-the-shelf ideas from North America or southern Australia are likely to fail. I needed to immerse myself in this strange new land. Even grasping the annual cycle of weather was a challenge—the Gagadju people recognize six seasons (based mostly on rainfall), not the temperature-based (winter-spring-summer-autumn) division I'd grown up with. Fortunately for me, the local inhabitants—both European and Aboriginal—generously shared their knowledge. Anecdotes from local naturalists often suggested new perspectives, or new places to look for animals. Practical issues were vital—like how to avert the potential catastrophes (heatstroke, mosquito-borne diseases, buffalo, crocodiles) that lay in wait as I drove down the road or walked beside the pond. Political dangers lurked in the background—who decides where we can go, and what we can do? Alienating the park managers or the traditional owners would have brought an abrupt end to the research program.

Not all the locals were supportive. My worst experience came late one night after I stumbled into a hole in the ground beside DjaDja Billabong, cutting my leg to the bone. My fault. I should have been using a flashlight. Rob and Pete drove me into

FIGURE 8.8 Female Arafura Filesnakes have large chunky heads that enable them to swallow remarkably large fishes. Photograph by Scott Eipper.

Jabiru so that the after-hours nurse at the clinic could patch me up. The pain as she stitched up the gash in my leg (without anesthetic) was less dire than the nurse's commentary as she did so. She had forgotten that she was on duty that night, and had already gone home and had more than a few drinks. She wasn't happy about being called back to the clinic at midnight. Maybe it was the alcohol that launched her into a vitriolic tirade against all Aboriginal people ("why don't they stop drinking?"), snakes ("kill them all!"), and scientists ("parasites, egg-heads"). It was a moral dilemma. I wanted to disagree with her, but whenever I did, she dug the needle in deeper. In the end, I sat quietly as she sewed me up.

But she was the exception. As I talked with other researchers and residents of the Top End, and spent my days and nights searching for reptiles, my perspective on snake ecology morphed into a more general view. I realized that the snakes I had studied in southern Australia live in a thermal prison. For much of the year, a snake in New South Wales has to seek out pockets of warmth before it can function. That quest for heat determines where and when snakes are active, and it constrains what they can do. As a result, thermal biology is central to the day-to-day lives of snakes like Broadheads. But that thermal dictator is trivial in the tropics. A reptile in Kakadu can get as hot as it likes whenever it likes. Liberated from that cold-climate straight-jacket, tropical snakes are free to evolve new and different ways of making a living.

Our research on Filesnakes reinforced Harvey Pough's key insight about ectothermy: the low energy needs of a snake or lizard enable exquisite flexibility in lifestyles. A month without food is a catastrophe for a mammal or a bird. In bad times emaciated dingoes scavenge along the roads and in the rubbish-dumps, while wallabies and waterbirds head off to greener pastures. But the Filesnakes wait it out. Our best evidence came from follow-up studies with Thomas Madsen and Bea Ujvari in a billabong close to our main tropical field site at Fogg Dam. With ten years of data, the patterns became clear. In a year when the wet-season rains are prolonged, life is good: the floodplain is inundated, the fish breed, and food is abundant. The Filesnakes feast. Their babies survive and grow, and the females replenish their energy stores and reproduce again next year. If the world stays like this for a few years in succession, a female Filesnake matures when she is 3 years old and produces a litter of babies every year after that. But if the wet-season rains stop early, water levels fall and fish are scarce. The baby Filesnakes starve and the adults cease growing. Females postpone reproduction. A female Filesnake who encounters a succession of poor (low rainfall) wet seasons may not mature until she is 10 years old, and produce offspring only once every seven or eight years thereafter.

That flexibility shocked me. I knew that the "pace of life" varies among snake species. For example, North American gartersnakes grow fast and die young, whereas rattlesnakes grow slowly, and reproduce late and infrequently. But I had viewed these as mutually exclusive alternatives, hard-wired by evolution. For the Filesnakes of Kakadu, though, these extremes are just the ends of a spectrum. Filesnakes seamlessly switch from gartersnake to rattlesnake tactics in response to rainfall patterns. In this dynamic ecosystem, where resource levels bounce around unpredictably, Filesnakes flourish because they adjust to whatever the monsoons throw at them.

The Filesnake's flexibility challenged a hidden assumption in my view of the world: the importance of "averages". As a scientist, I have spent a lot of my time finding out if one thing averages bigger or smaller than another. But it is meaningless to talk about the "average" age at maturity for Filesnakes, or their "average" reproductive frequency. That "average" is driven by local conditions—Filesnakes just go with the flow. It's a difficult idea for us to understand, because we rely on constant resources. Our warm-blooded bodies hurtle along at a fast, constant speed. Until we can visualize a life in which some of us mature at 5 years of age and others at 75, or where our wage fluctuates from a million dollars one week to zero dollars the next, we will fail to understand the life of a reptile. We have much to learn from Filesnakes, the paragons of plasticity.

9 Snakes, Rats, and Rainfall

The climate in Darwin is wonderful for ten months every year. The long dry season offers warm days and cool nights, with low humidity and no rain. And the brief wet season is refreshingly stormy and dramatic, as the ecosystem comes to life around you. But those slices of heaven are separated by a month of hell. Nobody in their right mind visits Darwin in November. During the "build-up", as the dry season transforms into the wet season, the weather is awful. The skies fill with billowing clouds that trap the heat, maintaining noontime temperatures to midnight. The humidity increases until every pore of your body erupts like a miniature geyser. There's no evaporation to cool you down. You pray for a cooling shower but the clouds just tease, building up then drifting away. By the time the monsoons begin in December, you are drenched in frustration as well as sweat.

Sensible residents of Darwin flee to southern Australia in November, or they stay inside air-conditioned rooms and cars. But those weren't options for me in 1986, when I set out to discover how tropical pythons care for their eggs. All of the textbooks claimed that female pythons were role models of Devoted Parenthood. After laying her eggs, the doting mother coils around them and shivers her powerful muscles to keep her offspring warm. Or at least, that's what pythons do in zoos, and that's what Diamond Pythons do in the Sydney suburbs. But what happens in the tropics, where most pythons live?

After Dave Slip finished his thesis on the Diamond Pythons of Sydney, I employed him to conduct a similar study on tropical pythons. But where should we do the work? Pythons are difficult to find in most parts of the world ("as rare as rocking-horse droppings", as a colleague once remarked to me) but there is one fabled site, near Darwin, where pythons are abundant—ludicrously abundant. The species is the Water Python, medium-sized as far as pythons go. A big adult female is 2 meters (6 feet) long; males are smaller. A Water Python is drab brown in poor lighting, but microscopic grooves on its glossy scales reflect light and make the snake sparkle in sunlight (or a flashlight beam). This iridescent species might have stimulated the Aboriginal "dreamtime" mythology of the rainbow serpent, the supernatural creator of life.

In October 1986, Dave and I traveled to Fogg Dam, not far from Darwin, to study Water Pythons. This was the breeding season, and some of the female pythons that we found crossing the road at night were grossly distended with eggs. We implanted radio transmitters in those reproductive animals and one of those snakes—#240, named for her transmitter frequency—soon disappeared into a goanna burrow on the side of the dam wall: a perfect egg-laying site. When she emerged from the burrow she was emaciated, a striking contrast to her pre-laying plumpness. Clearly, she had laid her clutch in that burrow.

DOI: 10.1201/b22815-9

But just a week later, to my astonishment, the daily telemetry check showed that female #240 had moved. Her eggs weren't due to hatch for another month or two, so what was going on? The textbooks said that female pythons brood their eggs until hatching, but #240 hadn't read those books. Had she been eaten by a predator?

The signal from her transmitter was coming from the floodplain below the dam wall, a flat swampy expanse leading down toward the Adelaide River. Early wet-season rain had turned the area into a soggy mess, and wading knee-deep through the black soil wasn't fun. Especially, with the air temperature at 37°C (almost 100°F) and the humidity at 99%.

Dave and I clambered through the mud toward a patch of reeds 50 meters (50 yards) away, toward the telemetry signal. Mud sucking against our boots with every step, we moved at glacial speed. I was reflecting on our vulnerability—if anything risky eventuated, like a Saltwater Croc hiding in the mud, we'd take an eternity to get back to dry ground. And just at that moment a huge Water Buffalo that had been lying down in the shade of the reedbed leapt to its feet, snorted, and lowered its head as if it was about to charge. Unlike humans, Water Buffalos can move rapidly through thick mud. If he decided to attack, there wasn't anything we could do about it. But then he wheeled around and ploughed off in the opposite direction, leaving Dave and I staring at each other with dazed expressions.

When we finally reached the reedbed, we found that the snake was indeed there, sheltering under grass. She was thin but uninjured, with no apparent reason to have abandoned her motherly duties. When we dug up her old burrow the next day, the eggs were sitting there unattended. Her behavior remained a puzzle for another few years, until we conducted a larger study. Those additional data showed that #240's behavior was common. Water Pythons are fickle mothers.

By the late 1980s, my opinions about the lifestyles of ectotherms had moved beyond speculation. Our Filesnakes had confirmed that a population of ectotherms can track climate-driven fluctuations in food resources. But Filesnakes were a special case, because of their extraordinarily low metabolic rates. Did other snakes also wait out the bad times and exploit brief and unpredictable resource pulses? If that was true, climatic fluctuations from year to year should drive variation in ecological traits of other snakes also.

Understanding how weather affects wildlife populations was just an intellectual puzzle for me when I first began studying that topic, but now the topic has taken on more significant and more somber overtones. The reality and consequences of climate change are beyond dispute. Anyone who has looked at the evidence agrees that the world's climate is changing, and that air pollution is the prime culprit. The burning of fossil fuels is causing hotter and more variable weather. Tragically, though, climate change has become a political football, with vested interests delaying any sensible response.

Climate is important to all animals (including our own species), but organisms that depend on environmental conditions to regulate their body temperatures are especially sensitive. Many types of reptiles bask in the sunlight to warm up, and they move to a shaded place to cool down. A tiny difference in environmental temperature can be important, by rendering particular times or places too cold or too hot for activity.

FIGURE 9.1 Fogg Dam, partway between Darwin and Kakadu National Park, has been the focus of my snake research for many years. Photographs by Ruchira Somaweera.

Even a rabid denier of climate change has to acknowledge that weather differs from one year to the next; and that this variation affects the landscape and thus the plants and animals. If you don't agree with that last sentence, you've never experienced a drought. It's heartbreaking to see a green, lush area turn into a dustbowl when the rains fail to come.

We live in a variable world. Weather changes, and wildlife responds. One central challenge for conservation planning, then, is to understand how that impact plays out. How do weather conditions influence wildlife? At first sight, that question sounds like it will be simple to answer. We just need long-term monitoring to see how a population is affected by wet years, dry years, and so forth. The problem is that long-term studies are rare, especially for "neglected" animals like reptiles, and especially in the tropics. And in turn, the reason is straightforward. Long-term studies require long-term funding.

Buoyed by the success of the Filesnake work, I looked around for another system that might show me how weather conditions affect the ecology of ectotherms. I needed a dense population of snakes, consuming prey whose numbers fluctuated from one year to the next depending on rainfall. And I found an ideal site not far from where we had studied Filesnakes—but this new site contained a very different kind of animal, and one that could be captured on land rather than in crocodile-infested waters. I was determined to expand my work to that new system, if only I could get funding to support the work.

My plans centered on Fogg Dam, partway between Darwin and Kakadu National Park. When I began planning my studies on Filesnakes, I asked all of the experts about possible study sites in Kakadu. And most of them, after pointing out potential sites, finished with "and of course, you need to take a look at Fogg Dam as well". The eminent Hal Cogger, curator at the Australian Museum, was insistent. For a snake ecologist, Hal said, this place is El Dorado. With Hal's endorsement ringing in my ears, I arranged my tropical visits so that I could stay overnight in Darwin on my way to Jabiru. It was a short drive from Darwin to Fogg Dam, and everything that Hal had told me was true. The abundance of Water Pythons at Fogg Dam was extraordinary. Whenever I drove along the dam wall at night, large iridescent bodies crawled across the road in front of me. And not just pythons, but other snake species as well.

It takes an hour to drive from Darwin to Fogg Dam, and the road is bordered, as far as your eye can see, by highly disturbed habitats. First suburbia, then farms devoted to monocultures of bananas, mangos, and melons. Finally, some patches of open woodland, and then pastures crammed with livestock—Water Buffalo, cattle, and feral pigs. Nowhere that a snake could find a place to hide, let alone to live. But Fogg Dam itself lies within a nature reserve, where agricultural development is forbidden; and it is an oasis of biodiversity within a shattered landscape. As water levels shrink during the long dry season, thousands of waterbirds gather around the remaining ponds. If I was a romantic soul, with a desire to work in pristine country, I would have opted for somewhere in Kakadu, among the magnificent sandstone outcrops. But I'm a pragmatist, and I decided that despite being hugely disturbed by human activities, Fogg Dam would be ideal for a long-term study.

I couldn't do it on my own. My job and my family were in Sydney. I had spent enough time in the tropics to realize that brief expeditions could never reveal the true story. The only way to understand this complicated system was to be there day after day, month after month, year after year. To achieve that continuity, I needed to employ someone to live there full-time. And to fund that person's salary, I needed a major grant.

But external funding was difficult to obtain. Nobody would give me money just to look at natural history; I'd have to frame my grant application around new and exciting ideas. And research on snake ecology wasn't mainstream; some assessors would shy away from it as a result. But if I could make a strong case that the pythons of Fogg Dam provided a unique opportunity to explore some general concept (like climatic effects on predator–prey systems), I might succeed. And luckily, I had background information to support my arguments. Fogg Dam had been a focus of research by the CSIRO (a government research organization) soon after the dam was built for commercial rice-growing in the 1950s. When government scientists surveyed the abundance of native rats and Magpie Geese, they found strong year-to-year variations related to annual rainfall. In particular, Cyclone Tracy—a fierce storm that devastated Darwin in 1974—virtually wiped out native rats from the floodplain. Breeding success of Magpie Geese likewise depended upon water levels in the floodplain, where the geese built their nests.

In my grant application, I confidently asserted (based on flimsy evidence) that the Water Pythons at Fogg Dam eat rats in the dry season, and goose-eggs in the wet season. And thus, I argued, annual variation in rainfall changes the abundance of prey (rats and geese) that in turn drive the population biology of predators (Water

Pythons). Very much like the story that was emerging from the Filesnake research, but the Water Python system might enable us to gather far more detail. In 1954, the pioneering ecologists Herbert Andrewartha and Charles Birch had argued that unpredictable environmental fluctuations drive dynamic responses in Australian wildlife. The two Adelaide-based scientists had based their conclusions on regular counts of small insects (thrips) on suburban rose-bushes. In 1980, Harvey Pough had argued that reptile populations thrive on low and variable rates of food availability. I could combine those ideas, and apply them to a new palette: rats, geese, and snakes on a floodplain in tropical Australia.

To make a strong case, though, I needed to confirm snake diets and demonstrate that my methods were feasible. So every year I made month-long trips to Fogg Dam with Rob Lambeck, Russell Hore, Peter Harlow, and Dave Slip to get the basics organized. The snakes were doing what I expected, and our methods worked smoothly.

And of course, we made other discoveries, starting with an answer to the most basic question of all: how many pythons live around Fogg Dam? I already knew that the snakes were very common, because I routinely found six to ten pythons per night along the 2-km (1.2-mile) dam wall. That's why I had chosen Fogg Dam as my study site. But surely, most of those snakes were the same ones as I saw the night before? After all, how many large pythons could live in such a small area? That should be easy to find out. All I had to do was see how many of each night's captures were the same individuals that we had caught the night before.

In the mid-1980s, we began catching, marking, and then releasing those resident snakes. In the process, we obtained data on their body sizes, reproductive cycles, food habits, and so forth—the information I needed to beef up the grant application. But something strange happened. We kept catching snakes, scale-clipping them, releasing them, and catching more snakes. But we never got a recapture. Even after we had caught and released hundreds of pythons, we never saw any of them more than once. This was absurd. There were only three possible explanations: first, the marks we clipped on the snakes' belly-scales were healing so quickly that we couldn't recognize a marked individual. That was impossible. Second, the snakes that we were catching were dispersing to other places as soon as we released them, so it was like sampling a human population at an airport. Most individuals are there only briefly, and don't return. If so, we were sampling a population from a huge area. But why would Water Pythons use Fogg Dam as a transit lounge? Implausible. Third, the number of snakes in the area was huge. They weren't dispersing—they were simply so superabundant that our marked animals were only a tiny proportion of the local population. Not impossible, but it seemed *very* unlikely.

To see if snakes were dispersing away from the dam wall, I could radio-track them. But I had the same problem as with the Macquarie Marshes Blacksnakes: transmitter-implantation surgery would be stressful. By the time a snake recovered, I'd be on the plane back to Sydney. How else could I get a transmitter inside a python? I tried the method that had worked with Blacksnakes—putting transmitters inside dead rats and dangling them in front of foraging snakes—but the pythons at Fogg Dam were too nervous. When I dangled a mouse in front of a python at night, the snake fled instead of feeding. I decided that I'd have to catch snakes, feed them the

FIGURE 9.2 Fogg Dam and its surrounding habitat are home to an astonishing richness of reptiles. Photograph by Terri Shine.

FIGURE 9.3 At its peak, the abundance of Water Pythons around Fogg Dam was simply phenomenal. Photograph by Dane Trembath.

FIGURE 9.4 Dusky Rats are the main prey for Water Pythons at Fogg Dam. Photograph by David Nelson.

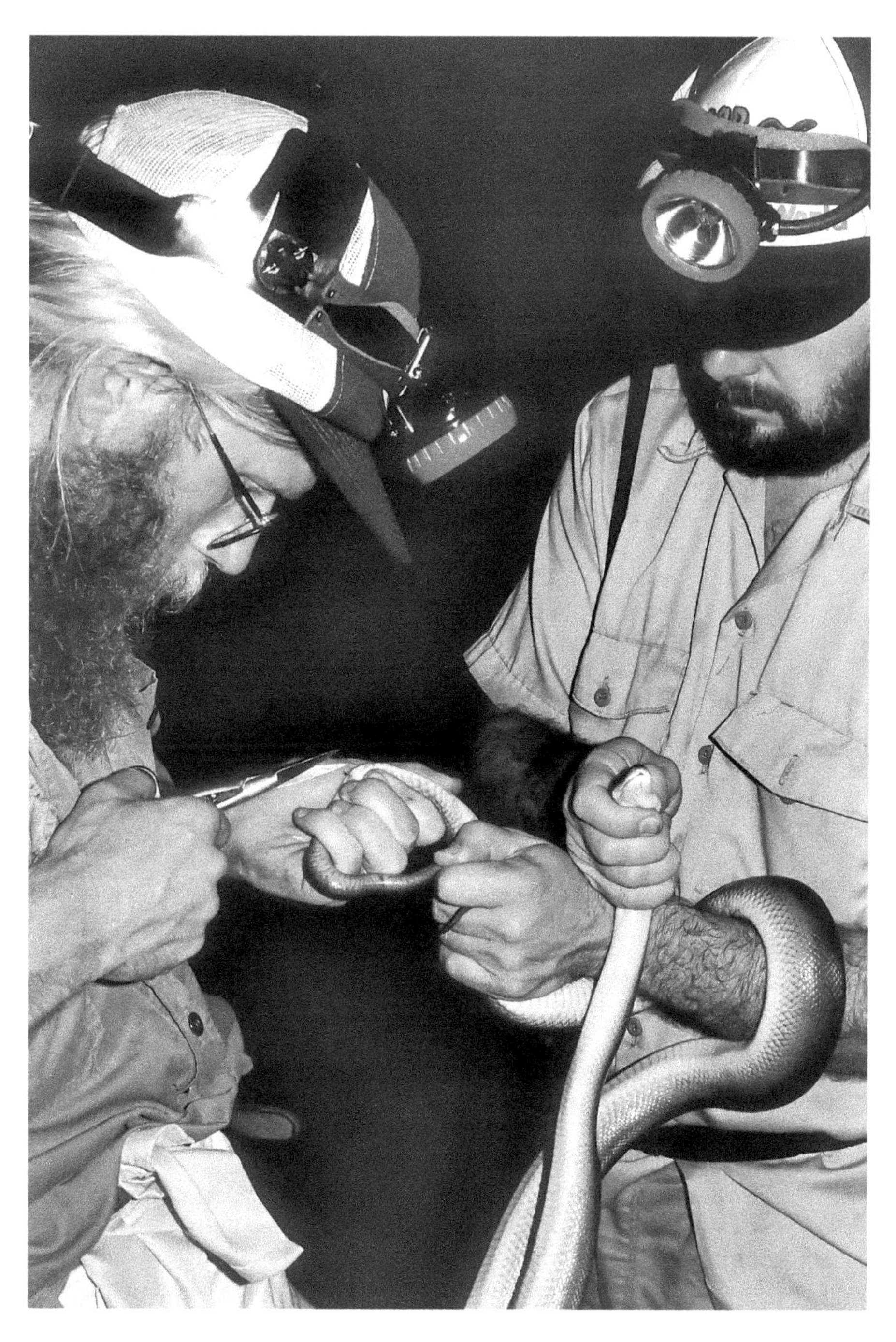

FIGURE 9.5 To find out how many Water Pythons lived near Fogg Dam, we gave each snake an individual scale-clip before releasing it.

electronic rodents in captivity, then release them. I didn't have any snake cages, but we were staying in a ranger's house at Middle Point Village, close to the dam. The kitchen cupboard had drawers for cutlery and plates, and the drawers looked escape-proof. So one night I put transmitters inside dead rats, then placed one electronically enhanced rodent in each cupboard drawer with a freshly captured snake. In the morning, I discovered that none of the rats had been eaten, and that all of the pythons had escaped. They had slithered out through gaps at the back of each drawer, that I had failed to notice. It took me hours to take the kitchen apart to recapture the (unhappy) pythons. And Water Pythons are snappy, so I lost a lot of blood in the process. In the end, I force-fed the electronic rats to the snakes, gently pushing lubricated rodents down into their stomachs.

After we released those snakes, they soon began to move about, their scale-clips were perfectly visible, and they weren't heading off to greener pastures. Why, then, wasn't I recapturing any marked pythons? As Sherlock Holmes said to Dr Watson: "*when you have eliminated the impossible, whatever remains, however improbable, must be the truth*". The inescapable conclusion was that the floodplain supported vastly more snakes than I had believed possible. Not dozens, not hundreds, but thousands. But how could that be true? It takes an enormous number of prey animals to feed an army of predators. Where were all those prey? How could the featureless floodplain contain enough food to support *thousands* of pythons? But when I looked more closely, I realized that the floodplain floor is crisscrossed with deep crevices that form when the wet-season floods subside. My radio-tracked snakes disappeared into those soil-cracks for weeks at a time—and if a soil crack can hide a python, it can hide a lot of rats.

I had blundered across a snake population that not only was easily accessible, but also was enormously abundant. I had the ammunition I needed to apply for funds to employ a full-time researcher. My university provided a few thousand dollars a year for my research—enough for short trips and basic equipment. But to conduct a serious study, I needed to find money elsewhere.

The only granting agency that might fund such a project was a government body called the ARC (the Australian Research Council). The ARC funds basic science, but less than 10% of applications succeed—and that figure drops to 5% for early-career researchers. A 1-in-20 chance of obtaining funds. I would be competing against experienced campaigners with long publication records and high scientific profiles. And, I suspected, networks of friends who might be evaluating those applications. The ARC had never funded anyone to work on snake ecology—but on the other hand, I was publishing papers in good journals, so I'd score OK on that criterion. And I had new ideas, effective methods, and a study population ideally suited to answer my questions. To my joy, my application in 1989 was approved. I could finally fulfill my dream of setting up a full-time research program in the tropics.

The next big decision was: who to employ? The project wouldn't be easy. I needed someone who was experienced in snake ecology, independent, had lived in remote areas, and was creative as well as hardworking. I didn't want to do copycat science, taking the standard approaches and measuring the same things as other people. I wanted us to ask new questions, embedded within a new view of population ecology. And I knew the perfect person for the job: a Swede named Thomas Madsen, about my own age. Thomas had just finished his Ph.D. on Swedish snakes. He was tall, thin, energetic, creative, and highly experienced. He had already studied snakes in sites from Africa to Sweden. But Thomas is also loud, confident, and argumentative, and his career prospects in Swedish science were poor because he had offended several senior figures.

Thomas accepted my offer, and he moved (with his wife Moni and two young sons Jens and Pedor) from their centuries-old apartment in Sweden to a small demountable building in Middle Point Village, just 2 km (1.2 miles) from Fogg Dam. It's difficult to imagine two places more different to each other. Thomas originally planned to stay for a year, then move back home. But more than 20 years later, if you drive down the Fogg Dam wall at night, you may encounter a tall Swede jumping out of his car to pick up Water Pythons. He's no longer at Sydney University, and his hair is gray not brown. But Thomas is still loud and argumentative, is still passionate about his research, and still coming up with new perspectives on biology.

Having a full-time presence up in the Northern Territory meant that we could restart the Filesnake studies as well. Darryl Houston had just completed his thesis research at DjaDja Billabong, so thousands of individually marked snakes were still out there to be recaught. Although DjaDja remained our focus, the project changed. The personnel shifted from an Australian introvert to a Swedish extrovert, our sampling sessions shrunk to once a year, and we upgraded our trap design. The original mesh fyke-nets often caught catfish, which were difficult to disentangle from the mesh. A trapped catfish erects its serrated dorsal spine, creating a tangle that is almost impossible to unravel without being stabbed by that spine. Such an injury is painful at worst for most people, but Thomas is so allergic to fish that a single stab from a catfish spine might be fatal. With Scandinavian ingenuity, Thomas developed a modified fyke-net with rigid wire mesh rather than catfish-entangling nylon. The trap still caught catfish, but we could dump them out without risk. Thomas caught more than 3,000 Filesnakes over the next few years, without being jabbed.

That first grant from the Australian Research Council supported Thomas for three years, and my application to renew it for a further three years was successful also. And it kept on going. I've had continuous grant funding for Fogg Dam research now for more than 30 years. That says a lot about our productivity. Only a small proportion of grant applications are funded. If our flow of scientific papers had stopped, even for a year, the river of money would have stopped as well. The Fogg Dam system was a goldmine. Thomas and I worked well together, talking non-stop as we spotlighted for snakes on the dam wall at night, processed pythons the following day at the village, and explored datasets with statistical software back in the demountable. A superb fieldworker, Thomas amassed datasets that I could never have gathered myself, and came up with many new ideas about the ecology of pythons. My skills were in translating that information, and our new ideas, into papers in high-impact journals. We achieved more together than either of us could have done alone. They were exciting times.

As a result of Thomas' flair and persistence, we now understand the ecology of Water Pythons at Fogg Dam as well as we understand any snake population in the world. For example, we now know why that female python (#240) abandoned her eggs, back in the days when Thomas was still a student in Sweden. Now that

FIGURE 9.6 Water Pythons were so common at Fogg Dam that we could just drive along the dam wall at night, stopping to pick up snakes. Photograph by Terri Shine.

we had funding for a full-time position and enough left over to buy transmitters, Thomas tracked many more female pythons. He discovered that the parental tactics of a female Water Python depend on where she lays her eggs. If she lays them in a cool area, like a burrow under the roots of a paperbark tree, the mother stays around throughout incubation. She guards them and keeps them warm, like the textbooks say that she should. But if she lays her eggs in a hot spot, like a goanna burrow, her eggs will stay warm even if she abandons them. And so, that's exactly what she does.

The mother's decision—to stay versus to desert her eggs—is a flexible one. Some female pythons switch from one nest type to another from one year to the next, and if they do, they change their mothering tactics also. That flexibility has a payoff. By incubating eggs in the laboratory, we know that warmer nests allow baby pythons to emerge sooner, and to be stronger and faster. So, if a female lays her eggs in a cool nest, staying around to keep them warm benefits the offspring. But if the eggs will stay warm anyway, inside a hot burrow, the mother's presence isn't going to make much difference. Motherhood is costly—not just because of the energy needed to produce and then brood the eggs, but also because a brooding female can't go out and catch a rat for dinner. Emaciated by the time the eggs hatch, she will need at least two years to build up enough energy stores to breed again. If she's unlucky, she'll be killed and eaten by a goanna before she recovers her strength. In contrast, a mother who deserts her eggs rather than brooding them is still fighting fit, can feed up, and can produce another clutch of eggs again next year. Some of those deserted eggs will dry out or be eaten by predators—but enough survive for the "flexible motherhood" strategy to succeed.

****

Like all field projects, our work at Fogg Dam generated a mix of insights—some of them only relevant to our specific system, but some much broader. That's one of the great joys of natural-history research, when glimpses of the private life of your study animal build into something bigger. If you're doing innovative research, the data often end up showing deep flaws in your original ideas. And your progress reports to the granting agency sound a trifle apologetic: "*we expected X, but we found Y*". In this case, though, the pythons proved that we were right.

My "big-picture" story for the Fogg Dam work was straightforward. I had argued that this tropical swamp could reveal fundamental truths about Australian ecosystems in general, and the ecology of ectothermic predators in particular. Borrowing from Andrewartha and Birch's insights many decades earlier, I contrasted the harsh and unpredictable ecosystem of Fogg Dam with the stable fertile lands of the Northern Hemisphere. Academics from Oxford, Oslo, and Ohio study a fauna and flora that evolved to deal with predictable winters and summers. Plentiful rain and rich soil promote plant growth and insect abundance. Mammals and birds dominate because they can fuel their high energy needs. Reptiles are scarce, and inactive for much of the year because of seasonally cold weather. Although some years are wetter or warmer than others, the variation is low. These stable and benevolent landscapes are ideally suited to high-energy organisms. Endotherms rule.

Australia isn't like that. It's a land of unpredictable climatic extremes. The volcanic activity and glaciation that created rich new soil in the Northern Hemisphere are just distant memories in Australia. Millions of years of geological stability mean that the soil hasn't been renewed, and nutrients have leached out of the surface layer. Rainfall is sparse and erratic, creating small pockets of favorable conditions in response to local rains. And that challenge has driven the evolution of the Australian wildlife. Kangaroos are incredibly mobile, moving hundreds of kilometers to any patch of land that receives a heavy shower. Even tiny marsupial predators move tens of kilometers to rain-drenched pockets in the landscape. But mobility isn't enough if there is no green oasis within hundreds of kilometers. Faced with long and unpredictable periods of food scarcity, Australian marsupials have evolved low metabolic rates to reduce their energy needs. Australia is a hard place to make a living.

But bad news for mammals and birds is good news for reptiles. As ectotherms, not wasting energy to keep themselves warm, the cold-blooded Aussies can wait out the bad times—the droughts, the cold weather, the hot weather—and exploit the good times when they finally arrive. Australia is a land of lizards and snakes: energy-misers, flexible, opportunistic, thriving in adversity. For an animal at Fogg Dam, every year is different. The floodplain is still moist and green in July after a good wet season; but it's a parched moonscape by March if the monsoons fail. A Water Python never knows whether next year will be Heaven or Hell.

If I were writing that initial grant application today, the term "climate change" would feature prominently. But in 1989, I'd never even heard of the idea. My application focused on a different justification. If Australian ecosystems work as I've described above—boom and bust cycles, driven by erratic variation in weather—then we need to understand what that means for conservation biology. We need a new science of ecology. Borrowing ideas from the Northern Hemisphere just won't cut the mustard. If things happen differently in the Wide Brown Land, we need a true-blue dinky-di vegemite-flavored Aussie paradigm of ecosystem function. Otherwise, we'll blunder when we try to manage Australian landscapes.

That message wasn't new but it had fallen out of favor. There are fashions in research just as there are in hairstyles, music, and clothing. In 1954, Andrewartha and Birch had argued that "ecological stability" was bunkum: plant and animal populations were pushed around by weather conditions too often to reach any equilibrium. Forty years later, Tim Flannery brought those ideas to public attention with his book *The Future Eaters*. Reading Tim's book is like taking off a pair of blinkers. Forget the old ideas. We need to look at Australia through new eyes, not view them through a Northern Hemisphere prism. Once you look at the Australian bush as a place of scarce nutrients and unpredictably harsh weather, many things begin to make sense. For example, I've always liked the Possum-Poo Hypothesis, to explain why gum trees have so many hollow branches. A tree-hollow provides an attractive home for possums that wander through the forest at night to feed, then return to sleep inside the hollow during the day. And inevitably, those possums deposit their (copious) poo inside the hollow or next to the tree—a subsidy of high-quality fertilizer. So, under this hypothesis, gum trees evolved to have hollow branches because those hollows attract fertilizers. It's a tactic that only makes sense in a land of poor resources.

What kinds of animals might triumph against the challenges I've described above? Low energy needs are critical, so "cold-blooded" (ectothermic) animals will do better than "warm-blooded" (endothermic) animals. That explains why there are more species of lizards in my Sydney backyard than in all of England, or France, or in the eastern half of North America. But if the harsh dynamics of tropical Australia are such bad news for mammals, why is the floodplain at Fogg Dam full of rats? The answer lies in two facts. One is that runoff from monsoonal downpours keeps the low-lying "backswamp" moist and well-fertilized because the rain carries nutrient-rich ash from dry-season fires. So the backswamp is watered and nourished like a suburban garden, encouraging abundant plant growth that provides food for rodents. The second fact is that the rat involved is no ordinary rodent. The Dusky Rat is an opportunist native species that erupts in huge plagues after good seasons but then dwindles in numbers until the next window of good times. The key to that ability is to live life in the superfast lane: the opposite tactic to that used by a snake. Dusky Rats breed at an amazing 4 weeks of age, and they produce large litters. When food is plentiful, a single pair of rats can produce a dozen offspring in a month, 70 grand-rats in two months, 400 great-grand-rats in three months, more than a thousand great-great-grand-rats in four months, and almost 10,000 rodents in six months. That's a lot of snake food.

To see if the Fogg Dam ecosystem really works this way, we had to study rats as well as pythons. But Thomas didn't want to, and our arguments raged for hours. His attempts to trap voles in Sweden hadn't succeeded, and he didn't want to repeat the experience. To describe Thomas as a man of strong opinions is like describing an elephant as a moderately large animal. Eventually, though, Thomas agreed to give it a try and fortunately the Australian rats were more cooperative than their Swedish relatives. Dusky Rats are readily enticed into a live-trap by a whiff of peanut butter and oatmeal. We could monitor rat abundance and breeding at the same times and in the same places as we were studying the pythons. We knew python feeding rates as well, because many snakes generously regurgitated rats all over us as we were measuring them.

I assisted Thomas with rat-trapping, but I never enjoyed it. The metal live-traps became dangerously hot for a trapped rodent after just an hour's exposure to sunlight, so we needed to start checking traps at dawn to ensure that our study animals didn't suffer. Not my favorite time of day, especially after a long night of snake-catching. And although Dusky Rats are smaller and more docile than many other rodents, they still have formidable teeth and lightning-fast reflexes. I'd rather handle a deadly snake than a rat. And lastly, rats are smelly. A rat inside a trap creates an aroma that even a human being can notice . . . a beacon to any King Brown Snake in the area. The snakes would lie unseen in the dry grass beside the trap, waiting to strike out at any movement—such as a hand reaching in to lift up the trap. We needed to be wary.

Dusky Rats drive the ecology of Water Pythons in space as well as time. When monsoonal rains inundate the floodplain in January, the rats abandon their soil-cracks and move to the higher drier levee banks that fringe the Adelaide River, a few kilometers downstream. And the pythons follow them. The snakes we were radio-tracking disappeared as the floodwaters rose; one by one we lost their signals. It was desperately frustrating, but the inundated floodplain contained too many Saltwater Crocodiles for us to venture out there on foot. A solution to our problems appeared in the form of a visiting helicopter pilot at the local pub one evening. We became friends over a few drinks, and the next day he gave us a lift to look for telemetry signals from the air. Thomas held the aerial out the open window of the chopper as we cruised over the floodplain, and the transmitter signals came booming in loud and clear as soon as we approached the levee banks. The pythons had followed the rats: a seasonal migration of predator and prey across the landscape, just like the lions of the Serengeti following herds of wildebeest. In both cases, rainfall drives the migration—but in Australia, the wildebeest are downsized to rats, and the role of lions is usurped by energy-efficient pythons.

A Dusky Rat's sex life is determined by rainfall late in the wet season. It doesn't matter how much rain falls earlier in the year. When the floodplain is already full, excess water just flows out to sea. But if the rain continues into April and May, the soil remains moist instead of drying out, and the rats keep feeding and breeding. An extra month of moist soil gives the rats time to produce thousands of furry snacks for hungry snakes. We could reliably predict the number of rats in our traps from rainfall the year before. Too little rain, and the ground dried out. Rats stopped breeding, and their numbers dropped. Too much rain, and the soil-cracks were flooded—so again, bad news for the rodents. But in a year with average rainfall, the rats kept breeding and their numbers exploded.

That year-to-year variation in rodent numbers regulates the world of Water Pythons. When rats are common, a predator's life is good: the pythons feed often, they grow fast, and their offspring survive. Adult females replenish their energy stores quickly and produce a clutch of eggs every year. The pythons are fat and sassy. In such years, it only took Thomas and me an hour to fill the car with bags of plump snakes when we cruised the dam wall at night. And those pythons regurgitated rats as soon as they were captured. When we hauled the pythons out of their bags the next morning to measure them, every snake was encased in a mash of half-digested rats. The stench was overpowering.

After a dry year, though, it was a different story. Our rat-traps were empty and the pythons were scrawny. Few females were reproductive. Young snakes were scarce because the hatchlings had starved to death. The survivors were tiny. It was stunning support for Andrewartha and Birch's view of Australian ecosystems as being in constant flux. There is no elegant equilibrium here—instead, the system is driven by weather. Just as in a desert, unpredictable rainfall events switch the system from bust to boom.

As Thomas and I continued to study the Water Pythons, we learned more about the links between rain, rats, and snakes. For example, one puzzle was that in some years very few young pythons survived even though adult rats were still common. The reason, Thomas realized, was that the rats had stopped breeding. A young python is too small to swallow a full-grown rat, so the hatchling snakes need bite-size morsels—baby rats. The young pythons are doomed if an early end to the monsoons stops the rats from breeding before the python eggs hatch. Surrounded by rats, the young snakes can't find any that are small enough to eat.

For a baby of any species, nutrition in the first few months of life can have long-term effects. Thomas uncovered two patterns that were driven by that fact. The first was unsurprising: in years when rats are common, the snakes grow faster. Just as for a pet snake, more food translates into faster growth; a young python that hatches in a "good" rat year grows rapidly.

But the second pattern is intriguing. It involves rates of python growth later in life. Even in the same year (and so, at the same rat abundance), some middle-aged snakes grow much faster than others. Is that variation among individuals caused solely by genetic factors? No. Instead, it's driven by the environment that a young snake encounters in the first few months of its life. The fast-growers are snakes that hatched out in times of plenty, when rats were easy to find. That early experience of good times sets them on an optimistic path in life; whenever they have the chance, they grow as fast as possible. Life has been good before, so they gamble that it will be good again; there is no need to store energy as insurance for a rat-less day. In contrast, young pythons that hatch out in bad years, when rats are few and far between, adopt a pessimistic approach. They allocate the energy from their food into storage (fat) not growth, so that they can survive if things turn bad again. They are shorter and chubbier than the fast-growers. It is a cautionary tale for an ecological researcher. To explain why some individuals grow faster than others, it's not enough to know how much food they are eating—you also have to know what kind of an environment they grew up in, years before. Ecology is complicated! In the harsh world of the wet-dry tropics, ectotherms flexibly adjust to the challenges they encounter.

Because we were catching enormous numbers of snakes, Thomas and I were able to ask and answer questions that wouldn't have been accessible in other study systems. That enabled us to publish papers in the top scientific journals and as a result, the grant money kept on coming. Fogg Dam found its way into ecology textbooks. Rainfall drives prey abundance, and prey abundance drives predator ecology. And as scientists began to realize that the world's climate is changing, the impact of climatic conditions on wildlife populations emerged as a critical conservation issue.

Besides yielding phenomenal data, the Water Python research was great fun. We were working in a place that snake-lovers dream about, watching exotic reptiles doing interesting things. And Thomas' exuberance was a constant tonic, helping me to forget the biting flies, mosquitoes and heat. Thomas' wife Moni was a genial host, happily accommodating my presence in the small demountable—and she even gave me lessons to improve my tennis game. During my trips to Fogg Dam, the Madsen family often invited local banana-farmers and buffalo-wranglers over for a Swedish Feast in their small demountable home. The evening would begin with ice-cold schnapps, followed by meatballs and potatoes, more alcohol, and finally a spirited group rendition of traditional Nordic drinking songs. The favorite song was "Heylan Gor", with the words (in Swedish) written out on a sheet of paper stuck to the refrigerator. The singing was notable for its volume rather than its musicality or accuracy of Swedish pronunciation.

Our flow of scientific papers on snake ecology at Fogg Dam alerted the outside world to the wildlife treasure-trove only an hour's drive from Darwin. Fogg Dam was ideal for the media not just because of its scenic beauty and vast flocks of waterbirds. Just as importantly, large pythons provide spectacular photo opportunities. And the snakes were harmless, so even a moderately intrepid TV presenter could be filmed with a giant snake draped across his or her shoulders for the closing footage of a news clip. The first TV cameras to arrive were from the local Darwin station, but soon we had crews from all over the world. Then it was the turn of the wildlife documentary makers to feature "the biggest snake-pit on Earth". Any vehicle rolling slowly along the dam wall after sunset was likely to encounter at least one or two pythons, so Fogg Dam was attractive for a filmmaker on a tight budget, or for a tourism operator looking for extra thrills for backpackers. For a few years, the serenity of Fogg Dam after dark was destroyed by tour-buses driving along the dam wall. Whenever his spotlight revealed a snake, the macho tour-guide would leap out and harass the animal amid dozens of flashing cameras. But eventually the craze wore out, and life on the dam wall returned to normal.

****

After several years working up in the Top End, the landscape and animals no longer seemed foreign to me. I felt as comfortable thinking about ecology in the tropics as in the cool climate regions where my research life had begun. Once every few months I flew up to Darwin, drove out to Fogg Dam, caught snakes with Thomas for a week or two, then flew home to Sydney. I knew as much about snake ecology in the Australian tropics as anyone else except Thomas, but that was not a major claim to fame, because we were the *only* snake ecologists in tropical Australia.

But what about the intervening habitats? Darwin and Sydney are a long way apart; how do the forests of southeastern Australia transition into the wet-dry tropics of the north? I couldn't make a mental connection between the landscapes of Darwin and Sydney—how did one turn into the other? I knew there was a very large desert in between, but what did the world look like as you moved out of that desert and into the tropical floodplains? In 1989, I decided to drive rather than fly up to Fogg Dam from Sydney and get a broader perspective and find out how my study sites fitted into the continent as a whole. My research assistant Peter Harlow was happy to join me in spending a month driving several thousand kilometers to check out the geography between Sydney and Darwin. We took our time, stopping at places that allegedly were "full of snakes" (I was always on the lookout for that next fabulous study site). We bounced around on rocky tracks, finding an occasional serpent, then heading back to the main road and continuing northward.

I enjoy desert landscapes, so the drive was never tedious. Well, not often. The biggest surprise was something that I could have learned from examining a climatic map before we took off on the trip. The wet-dry tropics are a thin belt of well-watered land near the coast; further inland, the monsoons run out of water after they dump it all on Darwin and Humpty Doo. So after a week of driving, the landscape didn't begin to look like the habitat around Fogg Dam until we were almost at our journey's end. Australia really *is* a dry continent, with its people and much of its biodiversity clustered around the coastline.

My second lesson from that trip was life-changing but far less pleasant. My lower spine gave way under the ceaseless thudding as the four-wheel-drive vehicle bounced across rock outcrops and zoomed down corrugated dirt roads. Finally, as we headed back south from Fogg Dam via Alice Springs, it reached crisis point. I heard a loud crunch in my lower back as we hit a bump, followed by the mother of all muscle spasms. I was in agony, and it just kept getting worse. My trip was over. I flew back to Sydney from Alice Springs, and Pete drove home by himself. My back refused to heal, and for the next five years my life became an endless round of appointments with lumbar specialists and physiotherapists. None of them could agree on what the problem was, or how to fix it. But after 40 years as an active person, I was horrified. Chronic pain can turn an optimist into a depressive. Some days it was difficult to get out of bed—emotionally as well as physically—because I knew that not long after I began moving around, I would be pole-axed by a stab of intense pain. A simple task, like tying up my shoelaces, was a frightening prospect. I stopped exercising, began to overeat, and moved closer to life as an invalid.

Something had to give, or my days as a field biologist were over. That motivated me to put real effort into my recovery. It worked. With exercising, yoga, and cutting back on sugar-rich foods, I lost 20 kg. I still have a sore back, and from time to time I have a very sore back. But it rarely flares up to the point of agony. Some kinds of fieldwork have been relegated to history. Hours spent lifting rocks looking for Broadheads make me feel as though someone is sliding a knife through my lower spine. But that still leaves many other activities. And fortunately, the Fogg Dam fieldwork—walking or driving along the dam wall—wouldn't challenge the athletic abilities of a healthy great-grandmother.

Thomas wasn't immune to the ravages of time either. His extensive fieldwork in Africa left him with major health problems. But as a true Viking, Thomas stoically feigned good health even when debilitated by a migraine or malaria. In contrast, I complained bitterly whenever my back went into spasms. But somehow or other Thomas and I worked well together, and we amassed incredible detail about the day-to-day lives of tropical snakes.

Because so little research had been conducted on tropical snakes, we were able to grab some low-hanging scientific fruit—stories that were obvious to anyone in the tropics, but not to scientists who worked in colder climates. For example, "behavioral thermoregulation" is a classic topic in reptile ecology. Many researchers have shown how European and North American lizards and snakes exploit patches of sunlight to bask, and thus stay warm even when the air is cold. In hundreds of papers, scientists have documented the advantages of careful thermoregulation in great detail. But staying warm isn't a problem if you live at Fogg Dam; a Water Python needs an air-conditioner not an overcoat. So, Thomas and I wrote a provocative challenge to the orthodox view that keeping warm was *the* central challenge in a reptile's life. Basking in the sunshine is a key aspect of reptile life in cold climates (where most scientists live), but it is irrelevant in the tropics (where most reptiles live).

Many reptile researchers disliked that paper, because it argued that their favorite research topic (thermal biology) was not as important as they had thought. They had overestimated the generality of their own work. But overall, the paper's reception wasn't as frosty as I had feared, and nobody abused me about it. My colleagues saw the paper as a challenge to conventional views, but not as an insult to the people holding those views. And more generally, I began to notice that my professional status was changing. My transition from "junior researcher studying peculiar animals" through to "respected mid-career scientist" and then to "expert" happened quickly. I was caught up in the day-to-day whirl of family life, teaching, and research, too busy to notice broader trends. But by the early 1990s, my career was heading to places I never expected it to go. I was publishing lots of papers, gaining grants, winning awards, and giving keynote talks at conferences.

Oftentimes I had a powerful case of "imposter syndrome"—feeling like it was all pretense. It was brutally apparent to me that many of my academic colleagues were smarter than me, that my research assistants could solve practical problems better than I could, and that my students found and caught more snakes than I did. I consoled myself with the thought that although I wasn't especially good at any of those things, I had my own set of skills. Somehow, it worked; yet every time I stood up on stage to receive an award or give a keynote speech, I felt like a charlatan who was taking credit for other people's work. But when I tried to say that to the people concerned, they shrugged it off. They knew that their contributions were appreciated, and we all benefitted from the team's achievements.

The digital revolution helped my career, because it enabled promotions committees and granting agencies to measure research productivity directly rather than relying on subjective evaluations. That subjectivity had created obstacles for a person like me, who was more comfortable with snakes than with my fellow human beings. My networking skills were rudimentary, and I wasn't the heir apparent to any Grand Old Professor. I had no political links to the academic gods of Oxford, Cambridge, or

Harvard. But hard evidence was now available about how often a researcher's work was cited in papers written by other scientists. That evidence began replacing the fallible impressions of esteemed professors. I scored better under the new numerical scoring system than I would have in the 1970s, when my prospects would have depended on who I knew.

And more generally, the old scientific hierarchy was crumbling—good news not only for an introverted Australian outsider like me, but more importantly for researchers whose gender, race, or sexuality had unfairly blighted their career options. Women were massively underrepresented in the world of snake research, and especially in the ranks of tenured university faculty. A few pioneers, like Joanna Burger from Rutgers University in New Jersey, accomplished terrific research—but any photograph from a conference on snake ecology in the 1980s and 1990s was dominated by white males. Nonetheless, change was in the air. Woman began taking more prominent roles in herpetology, initially through amphibian-focused studies, and before long the pendulum had swung to a healthier position. Perhaps the most spectacular example involves research on sea snakes. The previous generation of investigators—most of them now retired—were all male, and almost all white. But if you look at today's scientific articles about sea snakes, the dominant authors are women. And the same kind of transition has occurred with national identity—researchers from places like Brazil and China are now prominent contributors to knowledge. The field of snake ecology has benefitted enormously from that broadening of approaches and perspectives.

From the outset of my own career, my decisions about who to accept as students and collaborators were based on the individuals involved, not on their gender, nationality, or lifestyle. The resulting diversity of my research group has made it a far more dynamic team, as well as enriching our output immeasurably. A scientist's approach to research is fashioned by their life experiences as well as their education, and a diversity of backgrounds makes for a more exciting mix of ideas. The best science is unconstrained, so I didn't plan my research directions; I just selected projects that excited me, and I worked with enthusiastic and capable people. I looked for like-minded souls, not specific demographics.

Some experts pride themselves on their deep understanding of their field of study, but my own experience is different. It's easy to stay humble when you keep being wrong. No matter how many awards adorn my office walls, I continue to blunder, and to be puzzled by the animals I study. For example, during our studies on Water Pythons at Fogg Dam, Thomas and I discovered many things that didn't make sense. We realized that something interesting was going on, transforming an "unknown unknown" into a "known unknown". But we couldn't solve most of those mysteries.

As an example, here's one tantalizing issue that still frustrates me. We noticed a strong pattern—if we caught a Water Python on the dam wall at a particular place and particular time (say, at the eastern end of the dam, in April), we'd be likely to find that same snake in the same place almost exactly a year later. Our snakes weren't just meandering around the planet. They knew where they were going. A radio-tracked snake often made a bee-line for a specific grass clump that it had used the year before. It couldn't have been relying on scent, because the floodplain had been underwater for most of the intervening period. Snakes must have an internal compass, and an accurate one at that. But so far, nobody has worked out exactly how they do it. Or why they do it. I have no sensible explanation for why snakes travel along specific "highways".

More generally, any field biologist will admit how little they understand about the systems they have spent decades examining. Even if the rest of the world sees that person as an expert, he or she knows otherwise. The most famous exponent of that view was the great physicist Sir Isaac Newton, who wrote:

> I do not know what I may appear to the world, but to myself I seem to have been only like a boy playing on the sea-shore, and diverting myself in now and then finding a smoother pebble or a prettier shell than ordinary, whilst the great ocean of truth lay all undiscovered before me

Substitute "snake" for "shell", and Sir Isaac has explained my situation beautifully.

The most spectacular (and depressing) example of my fallibility came late in the study, after we had been studying Water Pythons for many years. Thomas had obtained his own funding to continue the study, so the data kept rolling in. But long-term datasets can be humbling; just when you begin to think that you truly understand an ecosystem, it turns around and bites you in the bum. By 2007, we had 15 years of detailed data on the Water Pythons of Fogg Dam. If we knew the wet-season rainfall in one year, we could reliably predict the abundance of rats, and thus snakes, the following year. We thought that we really *did* understand this system. But my smugness was washed away on the 3rd of March 2007, when a township 50 km (30 miles) upstream of our study site received a fierce but very local downpour—244 mm (10 inches) of rain in less than 24 hours. A massive surge of water—an inland tsunami—raced down the Adelaide River and hit Fogg Dam, inundating the floodplain for three days. When the flood receded, it left behind the bodies of drowned wallabies and rats. When Thomas trapped during the next dry season, he caught only 8 rats instead of the 100 or so we expected.

The deluge had a catastrophic impact. In a system that we had studied for 15 years, a single-day rain 50 km away had eliminated almost all of the snakes' food. I had been planning to find a mathematical modeler to look at our long-term data, to predict how the python population would respond to future climatic changes. All of a sudden, that looked like a waste of time. One day's heavy rain was a game changer—it had more impact on the rat population than any variation in monsoonal rainfall over the last 15 years.

Things got worse. With the rats gone, the pythons began to starve. They began to eat snakes and lizards instead of rats, but those alternative prey were scarce and difficult to catch. The days of easy living were over. Without food, a Water Python can live for months but not for years. The largest pythons were the first to die. It was awful to watch the snakes losing condition week by week. Some of them were just skin stretched over bones. It was the same nightmare that befell the giant Blacksnakes of the Macquarie Marshes more than 30 years before, when drought eradicated the frogs.

Not far from where the Water Pythons were starving to death, though, Dusky Rats were still abundant. On nearby Beatrice Hill, just 8 km (5 miles) away, the rats

FIGURE 9.7 My wife Terri and sons Mac and Ben accompanied me on several of the early trips to Fogg Dam. It added a wonderful extra dimension to fieldwork, experiencing this magnificent place through the eyes of my sons. Photograph by Terri Shine.

were fat and happy. They were feeding on pasture grasses, and the hill had saved them from drowning. Surely, I thought, the Water Pythons on the Fogg Dam wall will look elsewhere. These snakes are mobile animals; they travel several kilometers every year in their annual migration down onto the levee banks. All they had to do was go in the other direction. But they didn't. The pythons stayed, and they starved, and they died.

Water Pythons became scarcer and scarcer on the dam wall. The only survivors were small animals, growing slowly, but their extraordinary flexibility averted extinction. They would have died if they stuck with the old rules, such as delaying reproduction until they reached the "normal" threshold in body size and condition. But instead, they shifted those rules. Females matured at smaller sizes and produced smaller clutches (3–7 eggs versus 8–22 eggs in the Good Old Days). So, the python population persisted—but the snakes were small and scarce. If we had studied Water Pythons at Fogg Dam in those years only, we would have had an entirely different view of how big they grow, what they eat, and how common they are. If your study species is ecologically flexible, you need to stay with the system for a long time to capture the broad canvas of its life.

The bad times didn't last forever. Gradually, things improved. Rat numbers began to recover and in 2010, Thomas caught 60 rats in ten days of sampling. The pythons looked more like snakes and less like shoelaces. But tragedy struck again. In February 2011, Cyclone Carlos scored a direct hit on the Fogg Dam floodplain. More than 300 mm (12 inches) of rain poured down in 48 hours. The torrential downpour left no dry land. As had happened four years earlier with the Adelaide River tsunami, the rats were drowned. Once again, the pythons began to starve. By 2013, things were as bad as they had been in 2009. Then things improved . . . until January 2018, when another massive deluge wiped out the rats once again.

In time, the Water Python population at Fogg Dam will recover. But these three floods carry a terrifying message: even an abundant predator is vulnerable to a sudden brief climatic event. Mathematical models of climate change predict that such "extreme events" will be more frequent. Australia is already experiencing more heat waves, deluges, and droughts than it has for the last two centuries. If a bout of heavy rain can destroy the population of an ecologically dominant species like the Water Python—kings of the floodplain for 15 years—then the impact of extreme weather events on rare species may well be disastrous. Let's hope that other species deal with a novel climatic challenge better than Water Pythons do. If not, a catastrophe for Australian biodiversity is lurking just beyond the horizon. And that will apply even to ectotherms. Their flexibility will buffer the impacts of change, but there are limits to that resilience.

The water python project was the most intensive and long-running project I had ever initiated. I began to hope that I would encounter that English postdoc—the man in the pub in Canberra, who cast scorn on snakes as "boring inflexible robots". It would be fun to show him just how much we have learned about ecological flexibility, from organisms that make his beloved Great Tits look like zombies on autopilot.

10 Science on the Floodplain

Snakes have personalities. Some are timid, some are bold. Some stay close to home, whereas others are wanderers. Some don't retaliate even if provoked, whereas others are quick to go on the offensive.

Much of that variation occurs within a single species—sometimes, among siblings from a single clutch of eggs. But there are broader patterns as well. Some species are renowned for their docility (like the Sea Krait) and others are famed for their aggression (like the Black Mamba). Stories about snakes attacking people are almost always fanciful, but even a devout supporter of snakes has to admit that a few species are downright grumpy. The Slatey-Grey Snake sits at the top of the list.

Widespread through the Australian tropics, Slatey-Grey Snakes have the color and sheen of a gray roof-tile. They are capable tree-climbers with slender muscular bodies and glossy scales. And although the snake isn't colorful, it's formidable. Males can grow to 2 meters (6 feet) long, and their weapons are impressive. Slatey-Grey Snakes lack venom, but the teeth at the rear of the mouth are enlarged into razor-sharp blades. To slit open the lizard eggs that they eat, the snakes need to rip through a leathery eggshell.

Those same teeth can do a terrific demolition job on a human arm. The task is carried out with energy and enthusiasm; at the risk of anthropomorphism, I suspect that Slatey-Greys enjoy *biting me. And that joy is accompanied by explosive speed, remarkable power, and lightning-fast reflexes. When seized around the mid-body, most snakes take a few seconds before they decide they are under attack, and spin around to chomp down on the hand that's restraining them. Slatey-Greys aren't troubled by that namby-pamby indecision. As soon as you grab one, it bites. The snake's head moves too fast for you to see, but the blood welling out of those deep slashes on your arm tells you all you need to know.*

I met my first Slatey-Grey Snake a day after I arrived in the Northern Territory in 1982, excited about encountering tropical wildlife for the first time. I was nervous because I was supposed to be "the snake expert", whereas in fact I knew almost nothing about the tropical fauna. Late that night I was sitting with a roomful of experienced biologists at a research station in Jabiru, a monsoonal downpour drummed on the roof. I was hoping to learn from the locals, to avoid making a complete idiot of myself. But we'd barely started talking when Mike Cappo, a muscular young frog biologist, came into the room—drenched from the rain but proudly holding a cloth bag in his hand. "I caught a big Slatey-Grey Snake in the carpark", he said. And handed me the bag. Everybody waited for me to extract the serpent from the bag and show it around to the interested audience.

DOI: 10.1201/b22815-10

I'd never seen a Slatey-Grey Snake, but I knew they weren't venomous. Or at least, I was fairly sure that they weren't. And I had taken a lot of snakes out of bags, so I knew how to do it. The first step is to carefully feel the bag below the knot, to check that the snake isn't pushing upward with its snout against the knot. If it is, it will spring straight out as soon as the bag is opened. But the bag was the same thickness all the way down, so I assumed that the snake was at the bottom—first mistake. I didn't realize that Slatey-Grey Snakes are slender and muscular. The snake was pushing its head up against the knot.

Even then, it might have worked out OK if I had been more careful. But I was anxious to get my first look at a Slatey-Grey Snake, so I peered down into the bag as soon as I opened it. You can't blame the snake for what happened next. It saw something looming above it, blocking its escape route, so it did what comes naturally to a Slatey-Grey. It launched itself upward and bit onto that looming object, which happened to be my nose. And hung on like a bulldog.

It was an inauspicious beginning to my career as a tropical biologist. I was in front of a roomful of people I would be working with, and who I wanted to impress. And the first snake I encountered was a whirling dervish—2 meters (6 feet) of writhing fury, its body spinning around frantically as it hung from my nose. It must have been quite a sight. Human noses are well-supplied with nerves (thus, the bite was incredibly painful) and blood vessels (and thus my blood was pouring down the snake's body and spraying around the room as the snake gyrated).

The snake soon released its hold and I grabbed it mid-body as it tried to escape. That provided the group with an opportunity to compare rates of blood loss from a nose versus an arm. I finally got the snake under control by grasping it by the back of the neck, and then attempted to regain my prestige by pointing out various features of this fearsome beast to the group. Most of them didn't pay much attention. They were busy giggling and wiping blood off their clothing.

It was a rocky start to my relationship with Slatey-Grey Snakes, but you have to admire a snake that plays hardball. As long as you understand that interference with a Slatey-Grey Snake provokes instant retaliation, it can be entertaining. Catching a Slatey-Grey Snake is like sending a ballistic missile over the border into North Korea, or criticizing the All Black rugby team to a New Zealander. You can expect a rapid and dramatic escalation of the interaction. Whenever a new person joined us for fieldwork at Fogg Dam and seemed a little too cocksure of their snake-handling abilities, Thomas and I would suggest that they catch the first Slatey-Grey we found. The abashed newcomer inevitably returned to the car bleeding profusely, and without the snake.

One of my Darwin friends took it even further. He was out in the bush with a new companion whose ignorance of snakes was as great as his fear of them. So, when they encountered a Slatey-Grey Snake going about its business, my friend told the newcomer that the snake they had just found was a deadly "Territory Taipan". A bite would be fatal. He then seized the snake mid-body, it ripped into him, and he collapsed with a scream and played dead. You could use almost any kind of serpent for this performance, but few would play the part as convincingly as a Slatey-Grey Snake.

My friend and colleague Greg Brown has been based up at Fogg Dam, outside Darwin, for 20 years now. He has caught hundreds of Slatey-Grey Snakes and is the only person I know who unashamedly admits to liking *these psychopaths. Greg usually wears gloves when catching them, for protection, and recently upgraded from a bright blue rubber washing-up glove to a leather gauntlet. But an occasional snake turns up before Greg has a chance to slip the glove on, resulting in nice neat rows of Slatey-Grey tooth-marks on Greg's hand and wrist. Once when he had a particularly symmetrical and clear set of punctures, he contemplated buying some tattoo ink and rubbing it into the wounds to give himself a permanent reminder of his beloved Slatey-Grey Snakes. He hasn't got around to it yet, but he'll have many more opportunities.*

The tropics gave me a chance to shift from behavioral studies (on a few individual snakes) through to population-level analyses (on hundreds of animals). I wanted to combine those two approaches but frustratingly, the snakes that were abundant enough for population studies were poorly suited to behavioral work. The Filesnakes were deep in crocodile-infested billabongs and the Water Pythons were hiding in deep soil-cracks. What I needed was a small land-dwelling snake that would tolerate a close examination of its private life—and a researcher with more patience than me. In 1998, the pieces fell into place.

I've talked in the previous chapter about Fogg Dam, and how I chose it as my home base for studies on the ecology of tropical snakes. That decision was driven by the abundance of Water Pythons, and for the first few years they were my main focus. But pythons weren't the only snakes that I encountered as I walked along the dam wall at night. Two species of "harmless" snakes (colubrids, to give them their technical name) were common as well.

Contrary to the fears of English tourists, the Australian tropics are not littered with deadly serpents. You occasionally find an elapid such as a King Brown or Death Adder as you walk along the wall of Fogg Dam at night, but venomous snakes are scarce. Instead, most of the snakes you encounter in the tropics—whether on the road or in your bathroom—are non-venomous.

If you live anywhere except Australia, this won't seem strange. In Europe, Asia, Africa, and the Americas, most of the local snakes won't put you in hospital if they bite you. In evolutionary terms, the (mostly) non-venomous snakes of the superfamily Colubridae are a huge success story. Although they evolved quite recently, the colubrids have spread out into diverse ecological niches across the world. Well-known species like American Gartersnakes and European Grass Snakes are colubrids. Most snakes on the planet—in terms of individuals as well as species—are colubrids. Modern genetic methods are identifying tribes within that mega-Family, so the honorable old category "Colubridae" is being divided into smaller groups, each with their own name. Still, nobody disputes that these snakes all belong to a single lineage—and their success worldwide is amazing. Over the last several million years, few types of organisms have thrived and diversified as much as the non-venomous snakes.

Australia is the exception that proves the rule. Because the colubrids didn't evolve until after Australia separated from South America and Antarctica, there was no way for these new-fangled snakes to get to the Land of Oz. Instead, marine elapids colonized our shores, and evolved into the species (like Brownsnakes, Blacksnakes,

FIGURE 10.1 The two most common non-venomous snakes at Fogg Dam are Slatey-Grey Snakes (upper) and Keelbacks (lower). Slatey-Greys are "the thugs of the floodplain", always ready to retaliate if harassed. In contrast, Keelbacks rarely bite when captured. Photographs by David Nelson and Stephen Mahony.

and Tigersnakes) that still dominate southern Australia. About ten million years ago, though, Australia's slow drift northward brought it close enough to the Asian plate for a few "modern" snakes to make the move across. That contact zone, where tropical Australia meets Indonesia and Papua New Guinea, is clothed in rainforest and savanna woodland. The colonists, then, were snakes of the swamps, mangroves, and forests.

The Australian colubrids made that trip so recently that they still resemble their Asian relatives. Most Australian colubrid snakes are classified as the southernmost representatives of species that also occur in New Guinea. As genetic information accumulates, some of the Aussie snakes doubtless are being recognized as species in their own right; but however you look at it, they are recent arrivals.

Two colubrids are abundant at Fogg Dam, although they aren't as common as Water Pythons. We find those snakes the same way as we find Water Pythons—by walking along the wall of Fogg Dam at night, using flashlights to locate active snakes. On a good night, we can find several Keelbacks, a small (less than 1 meter [3 feet] long) water-loving snake. Keelbacks are the Australian cousins of American Gartersnakes and English Grass Snakes, but with rougher scales and a variable color. Most Keelbacks at Fogg Dam are a sedate brown with opalescent bellies; but in other parts of the species' range, their colors are more variable. Some Queensland Keelbacks are strawberry red and some are jet-black. Keelbacks feed mostly on frogs, and emerge from soil-cracks to forage for their prey on the Fogg Dam wall as soon as dusk settles in.

The other common colubrid on the dam wall is the Slatey-Grey Snake, larger and stronger than the Keelback, with a more diverse diet. One book on Northern Territory reptiles called the Slatey-Grey Snake "the thug of the floodplain", a phrase that perfectly catches their menacing bully-boy aura. There's not much that a Slatey-Grey Snake won't eat, from lizard eggs through to fishes, frogs, and small rats. I suspect that they prefer prey items that are capable of experiencing pain.

I encountered both of these colubrid species frequently during our Water Python fieldwork at Fogg Dam, and was tempted to conduct research on them as well. But Water Pythons were a full-time job, so Thomas and I ignored the smaller species. Sometimes we marked Slatey-Greys as well as pythons, but Keelbacks were beneath Thomas' dignity. Too small, too common, too drab. Drawing himself up to his full height (substantially greater than mine), Thomas would sneer "They are not *real* snakes" whenever I paused to examine a Keelback. These miserable creatures were beneath the dignity of a Viking.

My opportunity to study the colubrid denizens of Fogg Dam came about through an accident of cultural history. In 1995, I was invited to join a panel of the Australian Research Council (ARC), the main government agency for the funding of basic research. The panel's job was to evaluate grant applications by soliciting external reviews from experts in the field, and then to recommend which projects should be funded. I loathe committee work, but I passionately support the ARC's approach. They allocate funding based on the caliber of the research rather than the interests of their political masters. When I was asked to join the panel, I couldn't refuse. It was an opportunity to repay the ARC for their support of my own research.

In the 1990s, though, bureaucrats were busy codifying practices that previously had been handled by common sense. For example, the ARC drew up new rules to deal with "conflict of interest". Previously, if a panel member's own application was being discussed, he or she just left the room. But procedures became more formal. Any scientist (including a panel member) could submit two applications each year. With overall success rates of around 10%, nobody ever received two grants in the same year—- but it was sensible to buy two tickets in the lottery. Shortly after I joined the panel, the ARC's lawyers decreed that in order to avoid conflict of interest, grant applications by panel members had to be treated separately. They would be assessed by independent reviewers (like all applications) but wouldn't be added back into the main spreadsheet until after funding decisions were made about the rest of the grants. At that point, the panel member's applications would be interleaved with the others and funded at the same rate as the grants ranked on either side of them. It sounded simple and fair. There was no way for a panel member to influence the rating of his or her own grant, nor to affect how much funding it received.

But the system had a flaw. It didn't consider the (unlikely) possibility that both of a panel member's grant applications might be rated very highly. I submitted two applications that year, both for work at Fogg Dam. If one of them came through, I could afford to keep employing Thomas. But in fact, *both* of my applications had been highly ranked. If I hadn't been on the panel, I would have been given just one grant. But under the new "conflict of interest" rules, I had to be awarded both of them.

Having both applications succeed meant that I could expand the Fogg Dam work. Now that Thomas' salary was assured, who should I employ for the second grant? I advertised the job, and one of the applicants stood out. Greg Brown had just finished his Ph.D. on Watersnakes in Canada. He had strong references and his papers were top-class, so I offered him the three-year position. Greg came to Sydney first, so we could fill in all the necessary paperwork before he headed up to Fogg Dam. A quiet soul, he told me that the week in Sydney was a shocking experience; too many people, too much noise. When he recalls a football game I dragged him to, with thousands of screaming spectators, Greg shudders. But Fogg Dam, on the other hand, was all that an introverted Canadian could desire. He settled in immediately, and he's still there. More than 20 years later, it is difficult for me to imagine Fogg Dam without Greg. He is the heart and soul of my tropical research.

I wasn't sure what project Greg should do. My initial plan—to expand the Water Python research—was a non-starter. Greg and Thomas became good friends, but they were too independent to form a team. Smaller, quieter, and far less social than the Shouting Swede, Greg marches to a different drummer. He preferred the smaller, unloved, unappreciated snakes—especially the Keelbacks: the ones that Thomas sneered at. Slatey-Greys were big enough, mean enough, and likely enough to be recaptured, that Thomas was prepared to work on them. He'd started a mark-recapture project, but it was just a sidelight to the Water Python extravaganza. Greg took that over, with no fuss. But Keelbacks were a different matter. Thomas advised Greg (loudly, emphatically, and repeatedly) that working on such boring creatures would never yield any decent data. Night after night, as we relaxed with a beer or three after snake-collecting, Thomas launched into the same confident, well-argued, vociferous speech about the futility of doing research on those stupid little snakes. Thomas would begin by pointing out in a (relatively) gentle tone, with a look of regret in his eyes, that he really needed to explain something to Greg. Greg would nod and listen. Thomas would then point out that Keelbacks are not real snakes (too small, too numerous, too boring, no chance of recapturing marked snakes) and hence Greg should abandon all thought of studying them. Greg would thank Thomas for this suggestion. But realizing that he had failed to sway his Canadian colleague, Thomas would become more insistent. "Greggo", he would say, "you can't waste your time with Keelbacks. They are not *real* snakes". Again, Greg would thank Thomas for his suggestion. At the finale, Thomas would implore Greg to come to his senses, raising the decibel level ("*these are not real snakes!*") but without changing the message. And every night Greg would smile, agree with Thomas, and then ignore his advice.

For a while, though, it looked as if Thomas might be right. When Greg began marking and releasing Keelbacks, he ran into the same problem I had encountered with Water Pythons. He rarely recaptured any marked animals. And his project faced two additional obstacles. These little snakes are highly mobile (so they disperse away from the dam wall) and short-lived (so most marked animals die before Greg can recapture them). It would require a herculean effort to mark enough snakes to tell an interesting story. Gently, I tried to steer Greg toward other projects. He smiled, agreed with me, and kept on with his original plan.

Serenely determined, Greg was the right man for the job. He drove down to Fogg Dam at dusk every evening, walked along the dam wall shining his flashlight on the grassy verges, caught snakes, brought them back to the lab, then marked and released them the next day. Most people soon tire of the nocturnal insects that are attracted in their thousands by a light on the dam wall. Insect numbers vary from one night to the next, but a bad night is horrendous: beetles in your eyes and ears, mosquitoes attacking any exposed skin, bugs inside your shirt and underpants. It's difficult to look for snakes through a halo of insects in high-speed flight around your head. But even if you stoically ignore the bugs, it's boring to do the same thing in the same place every night. Any snake enthusiast could enjoy walking up and down the wall of Fogg Dam every night for a week. Very few could maintain their enthusiasm for a month. Remarkably, Greg has kept doing it almost every night for more than 20 years. When people ask him how he maintains that effort, he is puzzled. He looks forward to going out on the dam wall every night. In fact, as the years roll by, he enjoys it more than ever. If I force him into socializing with the other members of the lab over dinner, he tries to slip away for a quick trip to the dam wall beforehand. If he can't manage it, he sits in a corner of the room looking wistful and wondering what he's missing.

That original grant was for three years. Soon after it ended, Greg was awarded his own funding from the ARC. That kept him at Fogg Dam for another three years. Then I re-employed him on one of my grants. Then he won funding in his own right again. Then I got another grant to support him. And so on. It's difficult to imagine him leaving, especially since he is the proud owner of a very tame water buffalo named Mary. He talks about retiring to Canada one day, but it would be difficult to take his water buffalo with him. Easier, perhaps, to stay in the Northern Territory a bit longer.

In the course of his nightly strolls along the dam wall, Greg has caught and marked more than 5,500 Keelbacks and 1,500 Slatey-Grey Snakes. When he catches a pregnant female snake, he keeps her for a few days until she lays her eggs, then lets her go at the exact place he found her. He incubates the eggs, then marks and releases the babies (again, at the spot where he caught their mother). So far, he is proud foster-father to 8,000 Keelbacks and 1,000 Slatey-Grey Snakes. Add in recaptures of the same individuals, sometimes many times, and his total number of snake captures is up around 20,000. Greg's tally of marked and recaptured Keelbacks may well outnumber that of any other current study on snakes. The closest rivals might be Thomas' Water-Python studies (around 7,000 captures) and Xavier Bonnet's Sea Kraits in New Caledonia (10,500 captures of two species). In total captures, though, Greg still hasn't surpassed the Gold Standard of Snake Mark-Recapture Research: the pioneering 50-year study in Kansas by Henry Fitch (32,000 captures of 18 species). Not that Greg gives a damn. He just wants to find out what his snakes are doing.

Greg's massive sample sizes have shone a powerful light into the mysteries of Keelback society. In Greg's case, there was an extra dimension—the data on those baby snakes from eggs incubated in the laboratory. Female Keelbacks from all over the floodplain migrate to the dam wall during the dry season to lay their eggs in higher drier soil. As a result, many of the snakes are on the brink of laying eggs when Greg captures them. Within a day or two, a check of the cage reveals a slender mother and a clutch of white eggs. Like most snakes, female Keelbacks don't have anything more to do with their offspring after their eggs are laid. Greg releases the females but keeps the eggs.

With thousands of newly laid eggs to play with, Greg has been able to ask a wide range of questions. Some sites along the dam wall are warmer than others; some are wetter. Does that matter? Greg manipulated incubation conditions of his eggs, by changing temperature or moisture settings inside his laboratory incubators. Then, he could measure the effect of those treatments on the hatchlings. We had already conducted similar experiments on other kinds of reptiles, so it was no surprise when incubation conditions influenced the size, shape, and behavior of the baby Keelbacks. But Greg's huge sample sizes let us explore issues that had been inaccessible to earlier studies.

And even better, we could release those babies after they hatched, and monitor their subsequent lives. The earlier work had all been done in the lab, so we didn't know how those incubation-induced changes affected the subsequent life of a reptile. Greg's recaptures showed us how that variation played out in the real world, as the young snakes faced the perils of life on the dam wall. By monitoring thousands of young snakes, Greg could determine how nature (genetics) and nurture (the incubation environment) influence an individual's attributes at hatching; and then how the ruthless calculus of natural selection winnows that array of characteristics, creating winners and losers in the race to propagate one's genes. Those thousands of baby snakes gave us a direct view into how evolution shapes the ecological traits of Keelbacks. As Charles Darwin had first suggested in 1859, and George Williams cogently reinforced in his 1966 book *Adaptation and Natural Selection*, characteristics evolve because they help individual animals in their struggle to survive and reproduce. Greg's sustained efforts yielded a goldmine of data on exactly how a snake's success in life is influenced by its genes and its early environment.

Some of the stories from the Keelback research built on our earlier work. We already knew that annual variation in wet-season rainfall drove changes in prey, thus modifying the ecology of predators (fishes for Filesnakes; rats for Water Pythons). Greg found the same story for frogs versus Keelbacks. Good rains increased frog abundance the next year, and more frogs translated into faster growth and more frequent reproduction for Keelbacks. No surprise, but it was comforting to see the generality of our conclusions. I had framed those ideas about rainfall, prey, and predators for the initial grant application on Water Pythons. And a decade later, we had confirmed those concepts not only for Water Pythons, but for Filesnakes and Keelbacks as well.

As the research results accumulated, Greg and I wrote scientific papers about the results—just as Thomas and I were doing for the Water Pythons. The ideas behind those papers emerged from our conversations, and above all from our experiences on the dam wall. All three of us share a passionate desire to understand patterns in animal behavior. The data-gathering—the long and sweaty part of it all—was their job. I joined in, but my fieldwork contributions were paltry. Once the data were entered on spreadsheets, though, my enthusiasm for storytelling was given free rein. The combination of skills produced a stream of papers in high-impact journals, and the snakes of Fogg Dam became increasingly well known to other researchers.

Our growing fame was evident one year when the country's professional reptile researchers—the Australian Society of Herpetologists—held their annual scientific meeting near Darwin. Greg kept a low profile, because he prefers solitude. But Thomas and I offered to run evening tours along the Fogg Dam wall, an hour's drive from the conference venue. To my pleasure, almost all the conference attendees signed up. They had read our papers, seen the media coverage of our work, and were keen to see this Snake Shangri-La for themselves. It was surreal—walking along the dam wall, site of so many of my own research highlights, explaining the system to the luminaries in my field. We didn't need approval from our peers to validate the work at Fogg Dam, but it's nice to get that pat on the back sometimes.

But reality soon intervened. Every dry season, a large Freshwater Crocodile lays her eggs on the wall of Fogg Dam, and she vigorously attacks any intruder—be it a Heron, a Water Buffalo, or an eminent Professor. Her eggs hatched just before the conference, and she was guarding her babies in shallow water halfway along the dam wall, beneath a bird hide that extends out over the dam. Freshies are nowhere near as large, dangerous, or terrifying as Saltwater Crocodiles—indeed, most female Freshies are less than 2 meters (6 feet) long. But even at that size, a crocodile hurtling out of the water with her mouth open is spectacular, especially for people on their first trip to the tropics. So that night, I planned a Freshie display as the highlight of my guided tour. I packed the conference attendees (50 of them) into the bird hide, then clambered down the muddy bank beneath the hide to encourage the paranoid mother to show herself. I underestimated the fervor of this reptilian Mother of the Year, who immediately launched herself out of the water and up the bank toward me like a torpedo gone wild. And I also underestimated the difficulty of running uphill on slippery wet mud. My feet slid out from under me and I thrashed around

FIGURE 10.2 Monsoonal storms at the beginning of the Wet Season drop huge amounts of rainfall on Fogg Dam, inundating the floodplain. Photograph by Dudley Sergo.

frantically at the water's edge as Mum hurtled closer. Somehow or other, I made it up the bank before she could press home her charge. I suspect she wouldn't have bitten me anyway, even if she'd caught up to me. Instead, she would have stopped and laughed at me, like the 50 conference attendees were doing.

****

I kept traveling up to Fogg Dam whenever I could—if possible, for a week every month or two—so that Greg and I could walk along the dam wall in the evening and spend the day analyzing his expanding datasets. Thomas moved back to southern Australia with a new wife, a new life, and employment at a different university. He still came up to the Northern Territory for short periods of intense fieldwork, but Greg took over as the main scientific presence at the research station. For a few years, my trips were far more serene than before. The experience of walking along the dam wall at night changed from a rapid-fire high-volume exchange of ideas and opinions (the Thomas mode) to a quiet conversation about topics from snake ecology to Australian versus Canadian society (the Greg mode). Very different, but equally enjoyable.

And then the peace and quiet of the research station was shattered by a giant frog that had been marching its way across tropical Australia since long before I was born. Cane Toads are among the largest species of frogs in the world—a large toad can weigh more than 1 kg (2.2 lb). Brought to the Queensland coast in 1935 in a foolish attempt to control beetle pests in sugarcane plantations, the toads soon began spreading out in their new home. Their powerful poisons killed millions of native predators, like snakes, crocodiles, quolls, and bluetongue lizards. And the toads had another (although far less important) impact as well—they transformed my research career. With our background data on reptiles at Fogg Dam from Greg's nightly surveys, we were ideally placed to measure the impact of the toad invasion on native wildlife. And that's what we did. But it meant a huge expansion of our research program, as well as a shift in emphasis. Because Australians loathe and fear the invasive amphibians, research on toads was a far higher political priority than research on snakes. As a result, funding was easier to obtain. The ARC gave me a prestigious fellowship to study cane toads, with enough money to employ three postdoctoral fellows for five years. I've told the story of our toad research in my book *Cane Toad Wars.*

Overnight, that extra funding changed the research station from a sleepy backwater to a hive of activity. We expanded to take over the entire village, not just a single small demountable building. The Northern Territory government was looking for a new tenant, as the National Parks rangers moved out to their own newly constructed base, and the fellowship gave me enough cash to pay the rent—so we moved in. Three houses, four apartments, offices, a large laboratory—and all of it crammed full of new postdocs and research students, with their own toad-focused projects. The toad invasion would be upon us within a year, so everything was done at full speed.

Greg became the manager of the new research station, responsible for everything from fixing broken toilets to giving statistical advice to the students. And he took on toad research of his own, with population studies and nightly radio-tracking of the fast-moving toads in the invasion vanguard. Regretfully, I decided that the toad research would have to take priority. Much as I wanted to keep the snake research going, it was impossible. There weren't enough hours in the day for Greg to continue a full-time project on snakes, plus his new responsibilities as station manager and toad researcher. So, one night after we had walked the dam wall, I broached the subject. And got a puzzled stare in return. "No problem", said Greg, "I can easily do both projects. Walking the dam wall every night isn't work—it's my favorite form of relaxation". Realizing the futility of trying to change Greg's mind, I gave up. I figured that bitter experience would eventually show him the impossibility of the workload that he was planning to embrace.

I was, of course, completely wrong. Every time I came up from Sydney to visit Fogg Dam, I expected to see Greg looking a bit more tired, a bit more overextended. It never happened. He remained cheery, superbly organized, gathering more data in a week than most fieldworkers gather in a month. And after about a decade, I stopped worrying. He's just freakishly good at juggling multiple complex field programs.

With Greg's huge sample sizes, we could explore many issues that had long puzzled me, but which had remained untestable until the quiet Canadian amassed a vast amount of information. Some of those questions were really basic. For example, every snake-catcher knows that you can find a lot of animals one day and none the next. Fishermen have the same experience. Self-professed experts scratch their chins and say wise things like "no snakes tonight because it is too cold/too hot/too windy/too still/too humid/too dry". Or "no fish today because the tide is wrong/the water is dirty/the air pressure is falling". But nobody really knows, despite fisherman's almanacs that predict which days the fish will be biting (as accurate as flipping a coin, in my experience). But with Greg's phenomenal dataset—the same person walking along the same road, at the same time every night—surely we could tease apart the ways in which weather conditions affect the numbers and types of snakes that are active.

Sadly, we failed. Despite enormous amounts of data, we couldn't explain why more snakes cross the dam wall on some nights than others. Greg is a statistics nerd and he conducted all kinds of sophisticated mathematical tests. We looked for the effects of moon phases, temperature, humidity, wind, and various time delays (weather a day before, a week before, and so forth). Nothing worked. There were good nights (lots of snakes) and bad nights (no snakes), but nothing we could measure about the weather made much difference. Humbling. Snake activity isn't random, but it isn't predictable either—at least to us.

Although Keelbacks and Slatey-Grey Snakes are both colubrids, and live side by side, they are very different animals. The Keelbacks that Greg catches are part of a huge population, with snakes traveling long distances across the floodplain. They come back to the dam wall to lay their eggs in the higher drier soil, but for the rest of the year they go where the frogs are. It has taken extraordinary effort to work out what the snakes are doing. On any given night, only a tiny proportion of the local Keelbacks cross the dam wall. If Greg catches a snake and marks it, he'll probably never see it again.

Slatey-Grey Snakes are different. There are far fewer individuals, and the ones on the dam wall live there year-round. As a result, Greg often catches the same snake

FIGURE 10.3 Keelbacks mostly eat frogs, and are happy to take large meals (a). But these small snakes often become prey items themselves, for birds such as this Black-Necked Stork or Jabiru (b). Photographs by David Nelson and Dudley Sergo.

time after time, in the same place. We have really detailed information on individual Slatey-Greys, but we rarely catch a Keelback more than a few times during its life.

Another reason why we have so many recaptures of individual Slatey-Grey Snakes, and so few of individual Keelbacks, is a difference in lifespan. Keelbacks grow fast and die young. A male Keelback can mature at six to nine months of age, but rarely lives to see its second birthday. There are too many hungry mouths and beaks out there on the floodplain, ready to snap up a small harmless snake. But a Slatey-Grey Snake is a formidable proposition for any predator, and these snakes take a more cautious approach to life than Keelbacks do. Slatey-Greys grow slowly, mature at 2 years of age, and can live for at least nine years. The average Slatey-Grey Snake on the Fogg Dam wall thus has many opportunities to bite Greg during its lifetime. No red-blooded Slatey-Grey Snake would pass up any of those opportunities.

A short-lived study species like the Keelback is both a curse and a blessing for a researcher. A curse, because most of your marked snakes soon go to the Big Swamp in the Sky. From a scientific perspective, the researcher ends up with less data per unit effort. But the other side of the coin is that you can measure a snake's "lifetime reproductive success"—that is, you can assess what it does over its entire lifetime. That's the timescale that matters in terms of evolution: how many offspring are produced in the lifetime of an individual with Characteristic X compared to one with Characteristic Y? In Elephants or Albatrosses, where individuals can live for more than 50 years, even a long-term study covers only a fraction of each individual's life. Measuring "lifetime reproductive success" is out of the question. Even for some kinds of snakes, it's almost impossible; pampered captives of some species live for more than 30 years. With Keelbacks, though, a 2-year-old snake is a senior citizen. Greg's study has spanned many generations, so he can talk with confidence about how an experimental manipulation (like incubating an egg on dry versus moist soil) influences the animal's lifetime production of progeny.

And that rapid attainment of sexual maturity also allows us to compare a snake to its mother, its grandmother, and its great-grandmother. When Greg catches a pregnant female, incubates her eggs in the lab and then releases the babies after individually marking them, he's likely to see some of the daughters again. Many of the female Keelbacks that he catches on the dam wall, about to lay their eggs, are marked babies that have grown to adulthood. He certainly caught that animal's mother, and perhaps its grandmother as well. No other study on snakes has produced such data. It means we can estimate a Holy Grail of evolutionary biology—the degree to which offspring resemble their parents. We all know that the characteristics of an organism are influenced by both genes and environment. The incubation studies gave us a way to explore some of those environmental effects (what happens to a snake if it develops inside a moist versus dry nest?). Although it's easy to measure heritability (the degree of similarity between parents and offspring) in lab mice or dairy cows, it's a challenge to do it in wild animals. With Keelbacks, we can—because Greg walks along the dam wall night after night, month after month, year after year.

Like mother, like daughter? In what ways does a female Keelback resemble her mother? Mum left the nest-site as soon as she finished laying her eggs—so if a daughter is like her mother, it's not because of maternal care. Those two snakes never met each other. Any similarity between them is likely to be due to genetics. Greg accumulated data on 59 mother–daughter pairs for which he had data on egg-laying by the mother as well as by her adult daughter. Some characteristics weren't at all similar across generations, but two traits proved to be strongly heritable. One was egg size: some family lineages of Keelbacks produce big eggs, whereas others produce small ones. That was interesting, but not too surprising. But the other highly heritable trait was unexpected: daughters lay their eggs at the same place that their mothers did. If your mum laid her clutch (including the egg that produced you) at a spot halfway down the dam wall, near that big Pandanus palm, then that's where you will lay your own eggs. But if your mum laid her eggs in the paperbark-lined corridor along the eastern end of the dam wall, that's where Greg will find you when you are looking for a nest.

Ridiculous. Surely there can't be a gene for a specific egg-laying location? No, there isn't. That similarity between mother and daughter is due to environmental imprinting rather than to genes. Greg releases the babies a day or two after they hatch, at the same spot that he caught their mother two months earlier. And somehow or other, the tiny snakes remember that spot—the first piece of real estate they ever encounter—and they return when they are ready to lay their own eggs. This is a good tactic, because if that was a successful place to lay eggs a year ago (and it must have been, because here you are, about to have eggs of your own), it's sensible to use it again. So, daughters use the same nest-site that their mums did. The unsuccessful nests (wrong soil conditions, too easy for predators to find, etc.) didn't produce any surviving daughters, so those sites don't get used in the following year. Passing down rules of behavior from mother to daughter, it's eerily reminiscent of human culture.

Can a baby snake really learn the location of that site? Yes it can. Every once in a while, even Greg makes a mistake. He releases a baby at the wrong place, not where he caught her mum. And sure enough, he finds that baby snake a year later (as she's about to lay her own eggs) at the place she was released as a baby, not at the spot where he should have released her. Young Keelbacks aren't the only creatures that learn the location of the place they were born and return as breeding adults. In Canadian Salmon, subtle chemical differences between streams enable an adult fish to find its way home. We have no idea what cues the young Keelbacks use, but they are astonishingly good at finding the old family hacienda where they began their own existence.

Keelback growth and reproduction depends upon frog abundance, and so vary among years, whereas the broad diet of the Slatey-Grey buffers it from that variation. One prey type goes up as another one goes down. If the frogs are scarce this year, a Slatey-Grey dines on rats or lizard eggs instead. If the floodplain goes under water, the Slatey-Grey heads up into the trees. As an ecological generalist, the Slatey-Grey keeps ticking along, flexibly exploiting local opportunities. And, of course, relishing the chance to desanguinate any human being foolish enough to interrupt its leisurely lifestyle.

Because Greg catches the same Slatey-Grey Snakes again and again, year after year, his data for that species are very different from those he gathers on Keelbacks. Some questions (like lifetime reproductive success and heritability) are harder to investigate in a long-lived species. But at the same time, other kinds of questions become accessible for study. To succeed in research, you need to match your projects to your study species. Our work on Slatey-Grey genetics is a good example. I'm not a geneticist, and neither is Greg—but modern genetical methods are incredibly powerful. So, I thought, one day I will find a geneticist to work with us on Slatey-Greys—and in the meantime, we'll take a scale-clip sample every time we catch a snake. It doesn't hurt the snake, and a single clip contains enough DNA to run many analyses. The alcohol-filled tubes with those scale-clips steadily accumulated in the chaotic room that Greg calls his office.

In 2007, I was approached by a dapper young biologist looking for a project on snake genetics. Sylvain Dubey had recently finished his Ph.D. in Switzerland and was enthusiastic about spending time in Australia. Always well-dressed and well-groomed, Sylvain looked like a fashion model among the rough-and-ready fieldworkers at Fogg Dam. But he enjoyed fieldwork as much as anyone; and importantly, he also enjoyed running genetic analyses in the laboratory. Those genetic tests enabled Sylvain to unravel some intimate aspects of the Slatey-Grey's sex life. By then, Greg had tissue samples from every Slatey-Grey Snake that had lived on the dam wall over several years, as well as from 200 babies that had hatched in the lab. So—we had the DNA of offspring and of mothers, as well as of all the adult male Slatey-Greys on the dam wall—the potential fathers. We could run paternity analyses, to see which male was the father of which baby.

Greg sorted all his samples, and Sylvain extracted the DNA and examined it. And sure enough, he could identify which males were the fathers. A single clutch of eggs was usually fathered by at least two or three males—the females were obviously willing to tango with more than one male. But there was a strong overall pattern: the largest males were the most successful in passing on their genes. On average, a large male sired six times as many offspring as the smaller guy who lived just down the road.

That's the pattern we had predicted. Male Slatey-Grey Snakes grow much bigger than females, and they fight vigorously with each other (although in such grumpy snakes, it is hard to tell whether a battle is motivated by sex, cannibalism, or bloody-mindedness). So, we'd expect the bigger blokes to have an advantage in fathering offspring. In a different species, we might have been able to work out this kind of process by directly observing the animals as they fought, courted and mated. But that was impossible with the Thug of the Floodplain. Slatey-Greys are secretive, nocturnal and pugnacious, and live in complex habitats where snake hanky-panky is invisible to an observer. Fortunately for us, the collaboration between an ecologist (Greg) and a geneticist (Sylvain) gave us a way to peer into the Slatey-Grey's private life. In the end, the high-tech approach gave us the answers we were after. And quite a bit more besides.

Sylvain also found that within any small section of the dam wall, female Slatey-Greys were more closely related to each other than were the males. That's a clear signature of sex-biased dispersal—the females stay where they hatch (and so, the ladies end up living close to their sisters, cousins, etc.), whereas the lads take off for parts unknown, and broaden the genetic composition of the local serpents wherever they settle. Like mating success, dispersal is hard to measure in the field—but it can be assessed quite easily with genetic data. Those scale-clips told us a great deal about the lives of Slatey-Greys.

The strangest story to emerge from the research on Slatey-Greys involves the way that an individual's sex is determined. Like humans, Slatey-Grey Snakes have sex chromosomes: males have two Z chromosome in every cell (ZZ), whereas females have one copy of the Z chromosome and one of the W chromosome (ZW). Surely that's the end of the story? A snake develops as a male if it lacks the female sex chromosome (W), and it develops as a female if it has that female sex chromosome. But our laboratory incubation studies revealed something much more interesting. If we incubated eggs in dry soil, most of the babies were male, especially from smaller-than-average eggs. My statistical tests showed that this sex bias was too great to be due to chance. What was going on? We saw the bias even in clutches where all of the eggs survived to hatch, so it couldn't be differential mortality. Somehow or other, embryos that are genetically female switch their gender if they encounter dry conditions. The end result is although most male Slatey-Greys are ZZ, some of them are ZW. That latter group, although possessing "female" sex chromosomes, have testes and penises rather than ovaries and uteri—weird.

We're still not sure *how* the young snakes make this transgender switch, but Greg's mark-recapture studies have shown us *why* they do it. The reason that it pays off in terms of evolutionary fitness begins with the fact that female embryos don't deal with dry incubation conditions as well as male embryos do. A son from dry conditions hatches into a reasonably sized young snake, but a daughter from dry conditions is tiny: her egg dries out so much that she isn't able to use up all of the yolk. She has to leave some of it behind in the eggshell, instead of using it to build her body. The field data show that hatching as a small snake carries a major penalty.

FIGURE 10.4 By incubating eggs and releasing the baby snakes, Greg Brown has shown that Slatey-Grey Snakes have evolved an unusually sophisticated way to determine the sex of their offspring. Photograph by Greg Brown.

Small hatchlings grow slowly and are unlikely to survive. As a result, a Slatey-Grey embryo that develops in a dry nest benefits by becoming a male, even if it is genetically female. By turning into a male, the snake attains a larger size at hatching and so survives to adulthood. In sex-determining systems as in so many other aspects, snakes have evolved amazing flexibility in response to unpredictable environmental conditions.

More generally, our work on Keelbacks and Slatey-Greys exemplifies how snake research has evolved over the course of my career. It's now far more sophisticated than it was a few decades ago. During my Ph.D., I walked around swamps, caught snakes, and dissected them. I constructed crude radio transmitters and put them inside snakes, but the units were large, the battery life was short, and the numbers of snakes I could follow were abysmally low. These days, it's a different ballgame. Now I also pursue projects with sophisticated conceptual themes and methods. Thirty years ago, all we could ask was "what are the snakes doing?" then devise explanations after we had the data. Today, we know enough to pose the questions beforehand. And we are answering those questions with decades of data, and thousands of individual animals followed through time; we are monitoring prey species as well as snakes; and we are integrating our results with other kinds of information, ranging from climatic variation to environmental mapping and molecular genetics. The original toolkit of methods has been massively expanded. Technology has given us tiny transmitters, longer-lasting and more powerful than the early models; and as a result of all these changes, the datasets on my computer are a thousand times larger than in the Good Old Days. Fortunately, though, the computers are quicker and the software is easier to use.

Instead of being the lone snake ecologist, watching snakes in the bushland and in the laboratory, now I collaborate with smart young people whose expertise enhances the quality of everything I do. When I flick through the pages of a high-impact scientific journal, I'm no longer surprised to see a paper on snake ecology. Snakes have emerged from the scientific shadows, to bask in the sunlight as legitimate subjects for ecological research. If I ever meet that English postdoc again—the one who tried to talk me out of snake research—I'll tell him about how wrong he was, all those years ago. His views about snakes as simple primitive creatures, and ecosystems functioning in less sophisticated ways in Australia than in other countries, were an accurate reflection of opinions that were widely held at the time. But those opinions were based on guesswork, because we simply didn't have the hard evidence available. Now we do. The results are in—and continuing to come in, at an increasing pace—and everything we learn tells us that snakes and Australian ecosystems differ in important ways from other organisms and other ecological systems. But in no sense are the serpents or the Australian bush simpler, or less sophisticated, or inferior, to their equivalents in other evolutionary lineages and other places. Purely and simply, they are different. And in retrospect, having my assumptions challenged so forcefully, so early in my career, was a valuable gift. It made me think long and hard about what kinds of evidence were needed to confront those opinions, and it motivated me to go out and do the studies that were needed to gather that evidence. If I ever run into that Englishmen again, I might even buy him a beer.

11 Our Evolving Relationship with Serpents

Attitudes toward snakes have changed dramatically over the course of my career. These days, I rarely hear anyone say that "the only good snake is a dead snake". And the view of Australian wildlife as primitive and inferior has gone. Nowadays we celebrate the Australian fauna as a spectacular success story, adapted to a harsh continent, playing the game by rules very different from those that apply in the climatically benevolent lands of the Northern Hemisphere.

The early European settlers of Australia were terrified by the harsh new landscape and by the venomous snakes that dwelt in it. In writings from colonial times, snakes are the Devil. Those attitudes were deeply entrenched, and snakes were still feared and detested when I was growing up in the 1950s and 1960s. But as society changed, opinions about snakes changed as well. Australia was inundated with immigrants from Asia and television shows from the United States. A new generation grew their hair longer, protested against the Vietnam War, and challenged the wisdom of their elders. Society was in flux, and attitudes to Australia—and even to its least popular inhabitants, the snakes—were caught up in a cultural war. My parents' generation saw themselves as expatriate English folk, but my generation is proudly Australian. We learned about the role of snakes in Aboriginal dreamtime mythology, as creative not destructive beings, and began to look more kindly upon our true-blue local reptiles.

In our understanding of the ecology of snakes, as in so many topics, the paradigms are now very different from the ones that held sway earlier. Snakes have evolved from Everyman's Enemy into being seen as part of the wildlife, and a valuable component of the distinctive ecosystems of Australia.

That change in attitudes was driven by several factors, not just the avalanche of information. As a scientist who has contributed to our understanding about Australian snakes, I would love to believe that science played a seminal role in changing public attitudes: brave scientists providing cold hard facts to sweep away centuries of superstition. But in truth, public opinion depends on how people feel, not what they learn; many people still reject ideas that are backed by overwhelming evidence, like evolution or climate change.

So although scientists can claim a small slice of credit, we were only "supporting actors" in modifying community attitudes toward snakes. Cultural dimensions were more important. Increasing urbanization reduced encounters between people and wild snakes, but encouraged a rapid growth in reptiles as domestic pets. Apartment living eradicates the chicken-coop, but there's still enough room for a snake-cage.

DOI: 10.1201/b22815-11

And pampered pet snakes are superb ambassadors. It's difficult to believe in stereotype about "aggressive snakes" when your next-door neighbors share their house with a well-behaved python named Monty.

The natural-history documentaries of my childhood were all about large mammals. But from the late 1960s onward, grainy depictions of life in the Arctic or Serengeti were superseded by high-quality TV shows on more diverse topics. Hungry for new material and backed by new technology, those shows began to feature a wide range of species. It was cheaper to film snakes in the local bushland than to send a crew to Alaska to film Grizzly Bears. Information flooded into our living rooms, eroding the status of the neighborhood "experts". When I was a child, Old Glenda down the road told me that geckos were "stone adders" and could inflict a deadly sting with their tails. But the TV told a different story. And with the advent of the internet, information became even more accessible. Local myths were drowned out by the digital revolution. And importantly, many of those TV shows treated reptiles as interesting and attractive. A few "documentaries" still feature muscular men exaggerating the deadliness and aggression of a snake as they torment it to perform for the camera. Macho posturing elevated to an art form. But in other shows, the narrator talks about snakes as animals not demons; an unintended public relations campaign on behalf of the cold-blooded Australians.

The change in Australian attitudes to snakes was more complicated than a simple transition from "snakes are horrible" to "snakes are OK". Snakes didn't just move upward in the popularity stakes. We also began to view them in a different light. One dramatic change that resonates with me, because I underwent it myself, is the shift away from thinking of snakes as "lower" organisms. The books I read as a child described reptiles as the remnants of evolution's first stumbling attempts to create a "higher vertebrate".

Much as I adored reptiles, I accepted that view. Snakes were "primitive". A "cold-blooded" (ectothermic) animal was less advanced than a "warm-blooded" (endothermic) one. But we now understand that the idea of a "primitive" species is nonsense. Some species evolved earlier than others, but the fact that they are still around today is a hallmark of success not fallibility. There's nothing about the process of evolution that leads recently evolved species to be "better" or more complicated than older taxa. Natural selection just adapts organisms to local conditions. Knowing that a group of animals evolved recently (like primates) doesn't imply that they are any more complex than a group that arose earlier (like snakes). Evolution often favors simplicity. For example, the first snakes evolved from lizards, losing their limbs (and hence becoming "simpler" in shape) in the process. And non-venomous snakes evolved from venomous ancestors, as well as the other way around.

That process of shedding intellectual blinkers involved other kinds of animals as well. I had been taught that Australia's distinctive wildlife (whether a Platypus, a kangaroo, or a Death Adder) were "oddities"—inferior beings, surviving only because an accident of geological history kept them apart from the "advanced" European and American species. If they ever met, the Aussies would be outcompeted or consumed. I felt sorry for Wallabies and Wombats, and (as an Australian) was somehow ashamed of their weirdness.

An inferiority complex colored the public's attitudes toward anything Antipodean: the reproductive anatomy of marsupials, the colors and leaf shapes of eucalypts, the poetry of Australian bush balladeers. A tiny outpost of England, close to the teeming hordes of Asia, we clung tenaciously to our British heritage. At school, I learned about English kings not about Aboriginal culture. I was taught that Captain Cook "discovered" Australia in 1770, neglecting the people who had arrived more than 60,000 years earlier. My biology lessons were based on examples from European animals and plants. The education system made young Australians ashamed of their homeland. We were second-class Europeans, in a third-class continent, inhabited by fourth-class wildlife.

Looking back, a stronger and more vibrant perspective on Australian ecosystems was already available—but it was in the pages of poetry books not biology texts. One of my favorite stanzas comes from a poem by Banjo Patterson in "The Animals Noah Forgot", written in 1933:

Land of plenty or land of want, where the grey Companions dance,
Feast or famine, or hope or fear, and in all things land of chance,
Where Nature pampers and Nature slays, in her ruthless red, romance.

Those three lines capture the essence of everything that Andrewartha and Birch said about chaotic weather-driven ecosystems, and that Thomas Madsen and I learned from the Water Pythons. And just to tie it all together, in the dry season, the floodplain below Fogg Dam is home to dozens of the weird and wonderful dancing waterbirds that Patterson called "grey Companions" (now called Brolgas or Australian Cranes).

Poets may have grasped the harsh reality of Australian ecology long before scientists managed to do so, but whatever its faults (and they are many), science learns from its mistakes. Evidence trumps opinion. If the information doesn't fit your ideas, you change your ideas. That's very different from other mainstream approaches. In politics, religion, and support for sporting teams, people select a position and refuse to budge from it. It's difficult to convince a conspiracy theorist to abandon his favorite myth when new evidence becomes available.

Fortunately, Australian views about our place in the world were easier to change than ideas about the Fake Moon Landing. As a groundswell of cultural change swept through society, embracing the heretical notion that "different" does not mean "inferior", pioneering Aussie scientists showed us that kangaroos are better suited to our ecosystems than sheep could ever be; and that Aussie birds have more complex social lives than any feathered denizens of the woods of Sussex. By the time my own research revealed a sophisticated diversity of lifestyles in Australian snakes, I was following a well-trodden path. Australian wildlife is unique because Australia imposes novel challenges, not because the Land of Oz is a lifeboat of inferior beings.

What is so special about Australian ecosystems? As Andrewartha and Birch realized in 1954, Australian ecosystems are dynamic. Not balanced, not stable, not in equilibrium. They bounce from one state to another with the fervor of a hungry tree snake chasing an escaping frog. A few years of heaven, punctuated by decades of hell. It should have been an easy lesson to learn because the harshness of the bush was a central theme of early Australian literature. But ecologists weren't the only slow learners. In the 1800s, farmers saw the Australian inland during good years, after heavy rain, and it took them a century to understand that drought was the usual condition. A green landscape was a rarity. Australia isn't the same as Ireland. Even today, politicians dream of diverting the coastal rivers to create a fertile mini-Europe in Australia's dead heart. Old ideas die hard.

But for most Australians, yesterday's inferiority complex has given way to a more confident stance. In the field of conservation biology, decades of innovation have given Antipodean biologists a high profile on the world stage. A uniquely Aussie approach has evolved, with a pragmatic perspective on understanding how our ecosystems operate. We no longer borrow ideas developed in other landscapes, on other kinds of animals, and then twist them around to fit the Australian fauna. We look at what *our* animals are doing and ask why. That may lead us back into orthodoxy or may take us in new directions. But abandoning that Northern-Hemisphere perspective has enabled us to see the Australian wildlife in new ways, and we have learned a great deal.

One of those lessons is that if we want to conserve the unique ecosystems of Australia, we can't afford to ignore the role played by reptiles—including snakes. Step aside, Kookaburra. Move over, Koala. There is room on the public podium for a wider range of Australian animals—not just the furred and feathered creatures that top the popularity polls. To maintain Aussie ecosystems, we need to conserve all kinds of organisms.

In days gone by, Australians defined "wildlife" in very restricted ways. We doted on "cute" or "useful" animals, and we waged war on "ugly" and "dangerous" beasts. In New South Wales, native mammals and birds were protected in the 1800s, but reptiles didn't join them until 1974 (and frogs, not until 1992). Once viewed as evil predators that massacre nice furry animals, snakes are now seen as part of the ecosystem. In most western countries, snakes are legally protected. Even the brutal "rattlesnake round-ups" of the American west are evolving into information sessions not displays of machismo. When I see a snake crossing the road in front of a car, I'm no longer surprised if the vehicle slows down and allows the snake to pass in safety.

The snake is no longer seen as a remnant of an earlier phase of vertebrate evolution, an archaic bystander watching on as "proper" animals run the world. Decades of research have refigured the snake as a sophisticated and flexible predator, critically important in ecosystems worldwide. And the exact nature of the snake's role varies enormously. Each species is unique; and within each population, snakes of different sizes or sexes exploit different ecological niches. And lastly, individual animals adjust their lifestyles to the local environment. Snakes are the ultimate opportunists.

The Broadheaded Snake and the Slatey-Grey Snake will never win an award for Miss Congeniality. The Arafura Filesnake will never wear the crown of Beauty Queen. The Red-Bellied Blacksnake will never be everyone's favorite suburban neighbor. But if we stand back and take a good hard look, there's a lot to admire—and even, to love—about the serpents that enrich Australia's unique ecosystems. Without the snakes, it wouldn't be Australia.

And our debt to snakes may extend even further. More than any other kind of animal, snakes have shaped the course of human evolution.

FIGURE 11.1 Increasingly, children see snakes as creatures of interest rather than as terrible monsters. Photograph by Terri Shine.

The chief proponent of that view—indeed, its creator—is Lynne Isbell, a professor at the University of Davis in California. During her field studies on monkeys, Lynne noticed that primates are incredibly good at detecting snakes. Even a tiny patch of scales, or a loop of sinuous body half-hidden within a patch of grass, is enough to arouse alarm. Snakes eat monkeys, so it pays to be on the lookout. But as Lynne looked more deeply into the timing and trajectory of primate evolution, she began to suspect that snakes have played a major role. In short, she developed the "Snake Detection Hypothesis".

The first clue was that the primates (monkeys, apes, and so forth) with the most acute vision are those that evolved side by side with dangerous snakes. Throughout their evolutionary history, African and Asian monkeys have lived under constant threat of being crushed by a python or (more recently) fatally struck by a venomous snake. Those monkeys have terrific eyesight. But South American primates only encountered Rattlesnakes after North and South America joined to form a single landmass, about three million years ago. The monkeys from that part of the world have far less acute vision. And venomous snakes never reached Madagascar, so a Lemur that encounters a snake is curious not frightened. And Lemurs have terrible vision compared to the other primates.

Why should evolutionary coexistence with snakes improve eyesight in primates? Because if snakes are a threat, an ability to rapidly detect a well-hidden snake pays off. A monkey that didn't see the lurking python was killed, whereas a monkey that noticed a small patch of scales between the bushes avoided that fate. And so the monkeys that survived were the ones that did two things very well. First, they could detect a snake even if it was well-hidden. And second, they could instantly react to that situation.

That second shift—the ability to skedaddle as soon as we see a snake—rewired our brains. Everyone knows the sensation of panic, when an unexpected threat looms into view. It's like a lightning bolt through your brain. And monkeys experience that same sensation, based on measuring their brain activity, heart rate, and skin conductance—classic indicators of the fear response. Even a photograph of a snake can send a monkey into overdrive. Primates not only are incredibly good at noticing the presence of a snake, but they also go ape when they see it. For most primates, there's something especially scary about snakes.

A picture of a flower leaves us unmoved, a picture of a tiger can stimulate the fear response, but a picture of a snake jolts us into panic. That response is extraordinarily rapid. Lynne Isbell thinks we have special wiring in our brains to enable those super-fast responses. In essence, a neural shortcut bypasses the higher-order brain centers, instead sending a link straight from the eyes to the part of the brain that controls arousal. This special "low road" abandons the usual information-processing pathway ("send the data to the forebrain, let it decide what to do, then do it") in favor of a simpler direct connection ("Snake detected!! Run away!!"). It's a quick and dirty system, but it saved a lot of monkeys. And so, when humans evolved, we inherited a special "snake detection" module deep within our brains.

For a modern-day human being living in a city, there's nothing useful about that brainware. Today, the threats to a city-dweller come from vehicles, guns, overeating, and faulty electrical circuits. But a phobia about snakes is entrenched in our psyches. For some people, it's debilitating—a rope lying on the floor can induce a panic attack. Gradual exposure can help, and therapists have good success at overcoming snake phobia. But for most of us, an unexpected encounter with a snake at close quarters evokes panic.

That's true even for snake professionals. I've spent much of my life close to snakes, and the intensity of that reflex—that feeling of shock on sighting a long scaly body—has been eroded by thousands of hours of close contact. If I'm working with snakes, I feel nothing but pleasure in finding a serpent. But if I'm strolling through the bush, thinking about other things, a snake materializing on the path can cause that adrenalin surge. It's brief, but that ancient pathway ("Snake! Beware!") still lurks somewhere inside my brain.

So, snakes shaped our brains. We should be grateful. Snakes are the reason why you can read this book—why primates can see tiny squiggles on a page as letters, in a way impossible for the blurry-eyed majority of living creatures. And that power to see the intricate details of the world around us gives us the joy of seeing the faces of people we love, and the kaleidoscope of colors across the sky at sunset. In a very real sense, snakes made us who we are today. Let's return the favor by offering them some respect. Let's admire them, and try to save their shrinking populations on our imperiled planet.

FIGURE 11.2 Snakes arouse a special kind of fear in humans, reflecting the importance of our interactions with them over evolutionary time. I feel awe and wonder when I see a Red-Bellied Blacksnake, but many people have a less positive reaction. Photograph by Rob Valentic.

Acknowledgments

I owe an enormous debt not only to my family, friends, and fellow snake enthusiasts (pet-keepers as well as researchers), but also to the institutions that have housed me (first the University of Sydney, and now Macquarie University), and my primary source of funding (the Australian Research Council). I've been able to spend my career doing things that fascinate me, and that couldn't have happened without their support.

On a personal level, my students and fellow researchers have been a joy to work with. I have avoided inserting lots of names in the text, to keep things simple, but I am grateful to each and every one of you. My snake-biology postdocs include some of the oddest, hardest-working, and most inspiring human beings I have ever encountered. I owe special thanks to Thomas Madsen for enabling my leap into tropical biology, to Mats Olsson for countless collaborations and enduring friendship, to Greg Brown for quietly assembling a truly extraordinary research program, to Jonno Webb for seeing opportunities hidden to everyone else, and to Ben Phillips for a dash of genius that lifted our research to a new level. My students taught me much more than I ever taught them, and my research assistants toiled hard, overcame challenges, and educated me in the hard slog of field biology. My debt to them will be obvious to anyone who reads this book. Special thanks to Melanie Elphick who has been running my laboratory for more than 20 years, doing her best in every task, and supporting everyone in the group.

Several friends donated photographs or provided comments as the book was being written. For their fantastic photographs, I thank Rob Valentic, Kris Bell, Steve Wilson, Ru Somaweera, David Nelson, Sylvain Dubey, Terri Shine, Greg Brown, Stephen Mahoney, Scott Eipper, Matt Greenlees, Dudley Sergo, John Weigel, Peter Harlow, Mark Hutchinson, Etienne Littlefair, Geoff Boorman, Dane Trembath, and Shannon Kaiser. For comments, I'm especially grateful to Simon Baeckens, Mark Fitzgerald, Sean Graham, Crystal Kelehear, David Pearson, Glenn Shea, Georgia Ward-Fear, Jim Bull, and Ben Phillips. Chuck Crumly from CRC Press has guided the publication of this book with enthusiasm and professionalism.

Most of all, thanks to my wife Terri and my sons Mac and Ben. I'm a snake-a-holic, and it hasn't been easy for them to live with a zealot who kept disappearing into the wilderness to chase venomous animals. My family has not only allowed me to indulge that obsession; they've also actively helped out in many projects. For that, as well as for so many other things, I am deeply grateful.

Bibliography

GENERAL BOOKS

Cogger, H. 2014. *Reptiles and Amphibians of Australia*. CSIRO Publishing, Collingwood, Victoria, Australia.

Greene, H. W. 1997. *Snakes: The Evolution of Mystery in Nature*. University of California Press, Berkeley, CA.

Shine, R. 1991. *Australian Snakes. A Natural History*. Reed New Holland, Sydney, New South Wales. Reprinted 1993, 1994, 1998, 1999, 2001, 2003, 2005, 2007, 2009, 2016.

Shine, R. 2018. *Cane Toad Wars*. University of California Press, Berkeley, CA.

PREFACE

Gilbert, P. A. 1935. The black snake (*Pseudechis porphyriacus* Shaw). *Proceedings of the Royal Zoological Society of New South Wales* 1934–35:35–37.

Lawson, H. 1894. Bush cats. In *Short Stories in Prose and Verse*. L. Lawson, Sydney, New South Wales.

CHAPTER 1. BOYHOOD AND ADOLESCENCE

McPhee, D. R. 1959. *Some Common Snakes and Lizards of Australia*. Jacaranda Press, Brisbane, Queensland.

Strahan, M. J. 1965a. *The Snake Catchers*. A.H. and A.W. Reed, Sydney, New South Wales.

Strahan, M. J. 1965b. *Robert's Reptiles*. A.H. and A.W. Reed, Sydney, New South Wales.

CHAPTER 2. SERPENTS IN THE SHEEP PADDOCK

Cogger, H. G. 1967. *Australian Reptiles in Colour*. A. H. and A. W. Reed, Sydney.

Dawkins, R. 1976. *The Selfish Gene*. Oxford University Press, New York.

Heatwole, H. and R. Shine. 1976. Mosquitoes feeding on ectothermic vertebrates: A review and new data. *Australian Zoologist* 19:69–75.

Midnight Oil. 1990. *King of the Mountain*. Track 6 on Blue Sky Mining. Columbia Records, New York.

Pope, C. H. 1958. *Snakes Alive and How They Live*. Viking Press, New York.

Shine, R. 1977a. Habitats, diets and sympatry in snakes: A study from Australia. *Canadian Journal of Zoology* 55:1118–1128.

Shine, R. 1977b. Reproduction in Australian elapid snakes. I. Testicular cycles and mating seasons. *Australian Journal of Zoology* 25:647–653.

Shine, R. 1977c. Reproduction in Australian elapid snakes. II. Female reproductive cycles. *Australian Journal of Zoology* 25:655–666.

Shine, R. 1978. Growth rates and sexual maturation in six species of Australian elapid snakes. *Herpetologica* 34:73–79.

Shine, R. 1979. Activity patterns in Australian elapid snakes (Squamata: Serpentes: Elapidae). *Herpetologica* 35:1–11.

Shine, R. and J. J. Bull. 1977. Skewed sex ratios in snakes. *Copeia* 1977:228–234.

Williams, G. C. 1966. *Adaptation and Natural Selection*. Princeton University Press, Princeton, NJ.

CHAPTER 3. PEERING INTO THE LOVE LIVES OF BLACKSNAKES

Andrewartha, H. G. and L. C. Birch. 1954. *The Distribution and Abundance of Animals*. University of Chicago Press, Chicago, IL.

Bull, J. J. and R. Shine. 1979. Iteroparous animals that skip opportunities for reproduction. *American Naturalist* 114:296–303.

Greer, G. 1970. *The Female Eunuch*. MacGibbon and Kee, London.

Harlow, P. S. and R. Shine. 1988. Voluntary ingestion of radiotransmitters by snakes in the field. *Herpetological Review* 19:57.

Harlow, P. S. and R. Shine. 1992. Food habits and reproductive biology of the Pacific island boas (*Candoia*). *Journal of Herpetology* 26:60–66.

Mell, R. 1929. *Beitrage zur Fauna sinica. IV. Grundziige einer Okologie der chinesischen Reptilien und einer herpetologischen Tiergeographe Chinas*. Walter de Gruyter, Berlin.

Sergeev, A. M. 1940. Researches in the viviparity of reptiles. *Moscow Society of Naturalists* Jubilee Issue:1–34.

Shine, C., G. Ross, P. Harlow, and R. Shine. 1989. High body temperatures in an Australian frog, *Litoria caerulea*. *Victorian Naturalist* 106:138–139.

Shine, R. 1978. Sexual size dimorphism and male combat in snakes. *Oecologia* 33:269–278.

Shine, R. 1980. "Costs" of reproduction in reptiles. *Oecologia* 46:92–100.

Shine, R. 1983a. Food habits and reproductive biology of Australian elapid snakes of the genus *Denisonia*. *Journal of Herpetology* 17:171–175.

Shine, R. 1983b. Reptilian viviparity in cold climates: Testing the assumptions of an evolutionary hypothesis. *Oecologia* 57:397–405.

Shine, R. 1987. Intraspecific variation in thermoregulation, movements and habitat use by Australian blacksnakes, *Pseudechis porphyriacus* (Elapidae). *Journal of Herpetology* 21:165–177.

Shine, R. 1997. Not as black as it's painted. *Australian Geographic* 46:77–87.

Shine, R. and J. J. Bull. 1979. The evolution of live-bearing in lizards and snakes. *American Naturalist* 113:905–923.

Shine, R., G. C. Grigg, T. G. Shine, and P. Harlow. 1981. Mating and male combat in the Australian blacksnake, *Pseudechis porphyriacus* (Serpentes, Elapidae). *Journal of Herpetology* 15:101–107.

Shine, R. and R. Lambeck. 1990. Seasonal shifts in the thermoregulatory behavior of Australian blacksnakes, *Pseudechis porphyriacus*. *Journal of Thermal Biology* 15:301–305.

Weekes, H. C. 1933. On the distribution, habitat and reproductive habits of certain European and Australian snakes and lizards, with particular regard to their adoption of viviparity. *Proceedings of the Linnean Society of New South Wales* 8:270–274.

CHAPTER 4. LONG-DEAD SNAKES AT THE MUSEUM

Charles, N., R. Field, and R. Shine. 1985. Notes on the reproductive biology of Australian pythons, genera *Aspidites, Liasis* and *Morelia*. *Herpetological Review* 16:45–48.

Charles, N., A. Watts, and R. Shine. 1983. Captive reproduction in an Australian elapid snake, *Pseudechis colletti*. *Herpetological Review* 14:16–18.

Charles, N., P. Whitaker, and R. Shine. 1980. Oviparity and captive breeding in the spotted blacksnake, *Pseudechis guttatus* (Serpentes: Elapidae). *Australian Zoologist* 20:361–364.

Cogger, H. 2014. *Reptiles and Amphibians of Australia*. CSIRO Publishing, Collingwood, Victoria.

Dubey, S., J. S. Keogh, and R. Shine. 2010. Plio-pleistocene diversification and connectivity between mainland and Tasmanian populations of Australian snakes (*Drysdalia*, Elapidae, Serpentes). *Molecular Phylogenetics and Evolution* 56:1119–1125.

Fukada, H. 1992. *Snake Life History in Kyoto*. Impact Shuppankai, Tokyo.
How, R. A. and R. Shine. 1999. Ecological traits and conservation biology of five fossorial "sand-swimming" snake species (*Simoselaps*: Elapidae) in southwestern Australia. *Journal of Zoology (London)* 249:269–282.
Markwell, K. and Cushing, N. 2010. *Snake Bitten: Eric Worrell and the Australian Reptile Park*. UNSW Press, Sydney.
Mengden, G. A., R. Shine, and C. Moritz. 1986. Phylogenetic relationships within the Australasian venomous snakes of the genus *Pseudechis*. *Herpetologica* 42:215–229.
Scanlon, J. and R. Shine. 1988. Dentition and diet in snakes: Adaptations to oophagy in the Australian elapid genus *Simoselaps*. *Journal of Zoology (London)* 216:519–528.
Shea, G. M. and J. D. Scanlon. 2007. Revision of the small tropical whipsnakes previously referred to *Demansia olivacea* (Gray, 1842) and *Demansia torquata* (Gunther, 1862) (Squamata: Elapidae). *Records of the Australian Museum* 59:117–142.
Shine, R. 1980a. Reproduction, feeding and growth in the Australian burrowing snake *Vermicella annulata*. *Journal of Herpetology* 14:71–77.
Shine, R. 1980b. Comparative ecology of three Australian snake species of the genus *Cacophis* (Serpentes: Elapidae). *Copeia* 1980:831–838.
Shine, R. 1980c. Ecology of eastern Australian whipsnakes of the genus *Demansia*. *Journal of Herpetology* 14:381–389.
Shine, R. 1980d. Ecology of the Australian death adder, *Acanthophis antarcticus* (Elapidae): Evidence for convergence with the Viperidae. *Herpetologica* 36:281–289.
Shine, R. 1981a. Venomous snakes in cold climates: Ecology of the Australian genus *Drysdalia* (Serpentes: Elapidae). *Copeia* 1981:14–25.
Shine, R. 1981b. Ecology of the Australian elapid snakes of the genera *Furina* and *Glyphodon*. *Journal of Herpetology* 15:219–224.
Shine, R. 1982. Ecology of an Australian elapid snake, *Echiopsis curta*. *Journal of Herpetology* 16:388–393.
Shine, R. 1983a. Arboreality in snakes: Ecology of the Australian elapid genus *Hoplocephalus*. *Copeia* 1983:198–205.
Shine, R. 1983b. Food habits and reproductive biology of Australian elapid snakes of the genus *Denisonia*. *Journal of Herpetology* 17:171–175.
Shine, R. 1984a. Ecology of small fossorial Australian snakes of the genera *Neelaps* and *Simoselaps* (Serpentes, Elapidae). *University of Kansas Museum of Natural History, Special Publication* 10:173–183.
Shine, R. 1984b. Reproductive biology and food habits of the Australian elapid snakes of the genus *Cryptophis*. *Journal of Herpetology* 18:33–39.
Shine, R. 1986. Natural history of two monotypic snake genera of southwestern Australia, *Elapognathus* and *Rhinoplocephalus* (Elapidae). *Journal of Herpetology* 20:436–439.
Shine, R. 1987a. Ecological ramifications of prey size: Food habits and reproductive biology of Australian copperhead snakes (*Austrelaps*, Elapidae). *Journal of Herpetology* 21:21–28.
Shine, R. 1987b. Food habits and reproductive biology of Australian snakes of the genus *Hemiaspis* (Elapidae). *Journal of Herpetology* 21:71–74.
Shine, R. 1987c. Ecological comparisons of island and mainland populations of Australian tigersnakes (*Notechis*, Elapidae). *Herpetologica* 43:233–240.
Shine, R. 1987d. Reproductive mode may determine geographic distributions in Australian venomous snakes (*Pseudechis*, Elapidae). *Oecologia* 71:608–612.
Shine, R. 1987e. The evolution of viviparity: Ecological correlates of reproductive mode within a genus of Australian snakes (*Pseudechis*, Elapidae). *Copeia* 1987:551–563.
Shine, R. 1988. Food habits and reproductive biology of small Australian snakes of the genera *Unechis* and *Suta* (Serpentes, Elapidae). *Journal of Herpetology* 22:307–315.
Shine, R. 1989. Constraints, allometry and adaptation: Food habits and reproductive biology of Australian brownsnakes (*Pseudonaja*, Elapidae). *Herpetologica* 45:195–207.

Shine, R. 1991. Strangers in a strange land: Ecology of the Australian colubrid snakes. *Copeia* 1991:120–131.
Shine, R. and S. Allen. 1980. Ritual combat in the Australian copperhead, *Austrelaps superbus* (Serpentes, Elapidae). *Victorian Naturalist* 97:188–190.
Shine, R. and N. Charles. 1982. Ecology of the Australian elapid snake *Tropidechis carinatus*. *Journal of Herpetology* 16:383–387.
Shine, R. and J. Covacevich. 1983. Ecology of highly venomous snakes: The Australian genus *Oxyuranus* (Elapidae). *Journal of Herpetology* 17:60–69.
Shine, R. and J. S. Keogh. 1996. Food habits and reproductive biology of the endemic Melanesian elapids: Are tropical snakes really different? *Journal of Herpetology* 30:238–247.
Shine, R. and D. J. Slip. 1990. Biological aspects of the adaptive radiation of Australasian pythons (Serpentes: Boidae). *Herpetologica* 46:283–290.
Shine, R., C. L. Spencer, and J. S. Keogh. 2014. Morphology, reproduction and diet in Australian and Papuan Death Adders (*Acanthophis*, Elapidae). *PLoS One* 9:e94216.
Shine, R. and J. Webb. 1990. Natural history of Australian typhlopid snakes. *Journal of Herpetology* 24:357–363.
Strahan, R. 1979. *Rare and Curious Specimens: An Illustrated History of the Australian Museum, 1827–1979*. Australian Museum Publications, Sydney.
Webb, J. K. and R. Shine. 1992. To find an ant: Trail-following behaviour in the eastern Australian blindsnake *Rhamphotyphlops nigrescens*. *Animal Behaviour* 43:941–948.
Webb, J. K. and R. Shine. 1993a. Prey-size selection, gape limitation and predator vulnerability in Australian blindsnakes (Typhlopidae). *Animal Behaviour* 45:1117–1126.
Webb, J. K. and R. Shine. 1993b. Dietary habits of Australian blindsnakes. *Copeia* 1993:762–770.
White, B. S., J. S. Keogh, and R. Shine. 1995. Reproductive output in two species of small elapid snakes. *Herpetofauna* 25:20–22.
Worrell, E. 1964. *Reptiles of Australia*. Angus & Robertson, Sydney.

CHAPTER 5. A PLETHORA OF PYTHONS

Ayers, D. Y. and R. Shine. 1997. Thermal influences on foraging ability: Body size, posture and cooling rate of an ambush predator, the python *Morelia spilota*. *Functional Ecology* 11:342–347.
Eyre, E. J. 1841. *Journals of Expeditions of Discovery into Central Australia, and Overland from Adelaide to King George's Sound, in the Years: 1840–1: Sent by the Colonists of South Australia, with the Sanction and Support of the Government; Including an Account of the Manners and Customs of the Aborigines and the State of Their Relations with Europeans*. T and W. Boone, London.
Fearn, S., B. Robinson, J. Sambono, and R. Shine. 2001. Pythons in the pergola: The ecology of "nuisance" carpet pythons (*Morelia spilota*) from suburban habitats in south-eastern Queensland. *Wildlife Research* 28:573–579.
Fearn, S., L. Schwarzkopf, and R. Shine. 2005. Giant snakes in tropical forests: A field study of the Australian scrub python, *Morelia kinghorni*. *Wildlife Research* 32:193–201.
Fitzgerald, M. and R. Shine. 2018. Mate-guarding in free-ranging Carpet Pythons (*Morelia spilota*). *Australian Zoologist* 39:434–439.
Pearson, D. J. and R. Shine. 2002. Expulsion of interperitoneally-implanted radiotransmitters by Australian pythons. *Herpetological Review* 33:261–263.
Pearson, D., R. Shine, and R. How. 2002. Sex-specific niche partitioning and sexual size dimorphism in Australian pythons (*Morelia spilota imbricata*). *Biological Journal of the Linnean Society* 77:113–125.

Pearson, D., R. Shine, and A. Williams. 2002. Geographic variation in sexual size dimorphism within a single snake species (*Morelia spilota*, Pythonidae). *Oecologia* 131: 418–426.

Pearson, D., R. Shine, and A. Williams. 2003. Thermal biology of large snakes in cool climates: A radiotelemetric study of carpet pythons (*Morelia spilota imbricata*) in south-western Australia. *Journal of Thermal Biology* 28:117–131.

Pearson, D. J., R. Shine, and A. Williams. 2005. The spatial ecology of a threatened python (*Morelia spilota imbricata*) and the effects of anthropogenic habitat change. *Austral Ecology* 30:261–274.

Porter, R., J. Weigel, and R. Shine. 2012. Natural history of the rough-scaled python, *Morelia carinata*. *Australian Zoologist* 36:137–142.

Shine, C., N. Shine, R. Shine, and D. Slip. 1988. Use of subcaudal scale anomalies as an aid in recognising individual snakes. *Herpetological Review* 19:79.

Shine, R. 1989. Ecological causes for the evolution of sexual dimorphism: A review of the evidence. *Quarterly Review of Biology* 64:419–464.

Shine, R. 1991a. Intersexual dietary divergence and the evolution of sexual dimorphism in snakes. *American Naturalist* 138:103–122.

Shine, R. 1991b. *Australian Snakes: A Natural History*. A. H. and A. W. Reed, Sydney.

Shine, R. 1992. Ecological studies on Australian pythons. Pages 29–40 in *Proceedings of the Fifteenth International Herpetological Symposium on Captive Propagation and Husbandry* (M. J. Uricheck, ed.). International Herpetological Symposium, Seattle, WA.

Shine, R. 1994. The biology and management of diamond pythons (*Morelia s. spilota*) and carpet pythons (*M. s. variegata*). Species Management Report Number 15, New South Wales National Parks and Wildlife Service (50 pp).

Shine, R. and M. Fitzgerald. 1995. Variation in mating systems and sexual size dimorphism between populations of the Australian python *Morelia spilota* (Serpentes: Pythonidae). *Oecologia* 103:490–498.

Shine, R. and M. Fitzgerald. 1996. Large snakes in a mosaic rural landscape: The ecology of carpet pythons, *Morelia spilota* (Serpentes: Pythonidae) in coastal eastern Australia. *Biological Conservation* 76:113–122.

Shine, R. and J. Koenig. 2001. Snakes in the garden: An analysis of reptiles "rescued" by community-based wildlife carers. *Biological Conservation* 102:271–283.

Simon, P., R. Whittaker, and R. Shine. 1999. *Morelia spilota* (Australian carpet python). Caudal luring. *Herpetological Review* 30:102–103.

Slip, D. J. and R. Shine. 1988a. Feeding habits of the diamond python, *Morelia s. spilota*: Ambush predation by a boid snake. *Journal of Herpetology* 22:323–330.

Slip, D. J. and R. Shine. 1988b. Thermophilic response to feeding of the diamond python, *Morelia s. spilota* (Serpentes, Boidae). *Comparative Biochemistry and Physiology A* 89:645–650.

Slip, D. J. and R. Shine. 1988c. Thermoregulation of free-ranging diamond pythons, *Morelia spilota* (Serpentes, Boidae). *Copeia* 1988:984–995.

Slip, D. J. and R. Shine. 1988d. Reptilian endothermy: A field study of thermoregulation by brooding diamond pythons. *Journal of Zoology (London)* 216:367–378.

Slip, D. J. and R. Shine. 1988e. The reproductive biology and mating system of diamond pythons, *Morelia spilota* (Serpentes, Boidae). *Herpetologica* 44:396–404.

Slip, D. J. and R. Shine. 1988f. Habitat use, movements and activity patterns of free-ranging diamond pythons, *Morelia s. spilota* (Serpentes: Boidae): A radiotelemetric study. *Australian Wildlife Research* 15:515–531.

Sues, L. and R. Shine. 1999. *Morelia amethistina* (Australian scrub python). Male-male combat. *Herpetological Review* 30:102.

CHAPTER 6. BETWEEN A ROCK AND A HARD PLACE

Brischoux, F., L. Pizzatto, and R. Shine. 2010. Insights into the adaptive significance of vertical pupil shape in snakes. *Journal of Evolutionary Biology* 23:1878–1885.
Croak, B. M., M. S. Crowther, J. K. Webb, and R. Shine. 2013. Movements and habitat use of an endangered snake, *Hoplocephalus bungaroides* (Elapidae): Implications for conservation. *PLoS One* 8:e61711.
Croak, B. M., D. A. Pike, J. K. Webb, and R. Shine. 2008. Three-dimensional crevice structure affects retreat-site selection by reptiles. *Animal Behaviour* 76:1875–1884.
Croak, B. M., D. A. Pike, J. K. Webb, and R. Shine. 2010. Using artificial rocks to restore non-renewable shelter sites in human degraded systems: Colonization by fauna. *Restoration Ecology* 18:428–438.
Croak, B. M., D. A. Pike, J. K. Webb, and R. Shine. 2012. Habitat selection in a rocky landscape: Experimentally decoupling the influence of retreat-site attributes from that of landscape features. *PLoS One* 7:e37982.
Croak, B. M., J. K. Webb, and R. Shine. 2013. The benefits of habitat restoration for rock-dwelling geckos (*Oedura lesueurii*). *Journal of Applied Ecology* 50:432–439.
Darwin, C. 1839. *The Voyage of the Beagle*. Wordsworth Editions, London.
Du, W., J. K. Webb, and R. Shine. 2009. Heat, sight and scent: Multiple cues influence foraging site selection by an ambush-foraging snake (*Hoplocephalus bungaroides*, Elapidae). *Current Zoology* 55:266–271.
Dubey, S., B. M. Croak, D. A. Pike, J. K. Webb, and R. Shine. 2012. Phylogeography and dispersal in a rock-dwelling gecko (*Oedura lesueurii*), and its potential implications for conservation of an endangered snake (*Hoplocephalus bungaroides*). *BMC Evolutionary Biology* 12:67.
Dubey, S., J. Sumner, D. A. Pike, J. S. Keogh, J. K. Webb, and R. Shine. 2011. Genetic connectivity among populations of an endangered snake species from southeastern Australia (*Hoplocephalus bungaroides*, Elapidae). *Ecology and Evolution* 1:218–227.
Fitzgerald, M., B. Lazell, and R. Shine. 2010. Ecology and conservation of the pale-headed snake, *Hoplocephalus bitorquatus*. *Australian Zoologist* 35:283–290.
Fitzgerald, M., R. Shine, and F. Lemckert. 2002a. A radiotelemetric study of habitat use by the arboreal snake *Hoplocephalus stephensii* (Elapidae) in eastern Australia. *Copeia* 2002:321–332.
Fitzgerald, M., R. Shine, and F. Lemckert. 2002b. Spatial ecology of arboreal snakes (*Hoplocephalus stephensii*, Elapidae) in an eastern Australian forest. *Austral Ecology* 27:537–545.
Fitzgerald, M., R. Shine, and F. Lemckert. 2003. A reluctant heliotherm: Thermal ecology of the arboreal snake *Hoplocephalus stephensii* (Elapidae) in dense forest. *Journal of Thermal Biology* 28:515–524.
Fitzgerald, M., R. Shine, and F. Lemckert. 2004. Life history attributes of a threatened Australian snake (Stephen's Banded Snake, *Hoplocephalus stephensii*, Elapidae). *Biological Conservation* 119:121–128.
Fitzgerald, M., R. Shine, F. Lemckert, and A. Towerton. 2005. Habitat requirements of the threatened snake species *Hoplocephalus stephensii* (Elapidae) in eastern Australia. *Austral Ecology* 30:465–474.
Keogh, J. S., I. A. W. Scott, M. Fitzgerald, and R. Shine. 2003. Molecular phylogeny of the Australian venomous snake genus Hoplocephalus (Serpentes, Elapidae) and conservation genetics of the threatened *H. stephensii*. *Conservation Genetics* 4:57–65.
Llewelyn, J., J. K. Webb, and R. Shine. 2010. Flexible defense: Context-dependent antipredator responses of two species of Australian elapid snakes. *Herpetologica* 66:1–11.
Penman, T. D., D. A. Pike, J. K. Webb, and R. Shine. 2010. Predicting the impact of climate change on Australia's most endangered snake, *Hoplocephalus bungaroides*. *Diversity and Distributions* 16:109–118.

Pike, D. A., B. M. Croak, J. K. Webb, and R. Shine. 2010. Subtle—but easily reversible—anthropogenic disturbance seriously degrades habitat quality for rock-dwelling reptiles. *Animal Conservation* 13:411–418.

Pike, D. A., J. K. Webb, and R. Shine. 2011a. Removing forest canopy cover restores a reptile assemblage. *Ecological Applications* 21:274–280.

Pike, D. A., J. K. Webb, and R. Shine. 2011b. Chainsawing for conservation: Ecologically informed tree removal for habitat management. *Ecological Management and Restoration* 12:110–118.

Pizzatto, L., S. M. Almeida-Santos, and R. Shine. 2007. Life-history adaptations to arboreality in snakes. *Ecology* 88:359–366.

Pough, F. H. 1980. The advantages of ectothermy for tetrapods. *American Naturalist* 115:92–112.

Pringle, R., M. Syfert, J. K. Webb, and R. Shine. 2009. Quantifying historical changes in habitat availability for endangered species: Vegetation structure and broad-headed snakes in Australia. *Journal of Applied Ecology* 46:544–553.

Pringle, R. M., J. K. Webb, and R. Shine. 2003. Canopy structure, microclimate, and habitat selection by a nocturnal ectotherm. *Ecology* 84:2668–2679.

Reed, R. N. and R. Shine. 2002. Lying in wait for extinction? Ecological correlates of conservation status among Australian elapid snakes. *Conservation Biology* 16:451–461.

Shine, R. 1990. The broad-headed snake. *Australian Natural History* 23:442.

Shine, R. and M. Fitzgerald. 1989. Conservation and reproduction of an endangered species: The broad-headed snake, *Hoplocephalus bungaroides* (Elapidae). *Australian Zoologist* 25:65–67.

Shine, R., J. K. Webb, M. Fitzgerald, and J. Sumner. 1998. The impact of bush-rock removal on an endangered snake species, *Hoplocephalus bungaroides* (Serpentes: Elapidae). *Wildlife Research* 25:285–295.

Sumner, J., J. K. Webb, R. Shine, and J. S. Keogh. 2010. Molecular and morphological assessment of Australia's most endangered snake, *Hoplocephalus bungaroides*, reveals two evolutionarily significant units for conservation. *Conservation Genetics* 11:747–758.

Webb, J. K., B. W. Brook, and R. Shine. 2002a. Collectors endanger Australia's most threatened snake, the broad-headed snake *Hoplocephalus bungaroides*. *Oryx* 36:170–187.

Webb, J. K., B. W. Brook, and R. Shine. 2002b. What makes a species vulnerable to extinction? Comparative life-history traits of two sympatric snakes. *Ecological Research* 17:59–67.

Webb, J. K., B. W. Brook, and R. Shine. 2003. Does foraging mode influence life history traits? A comparative study of growth, maturation and survival of two species of sympatric snakes from southeastern Australia. *Austral Ecology* 28:601–610.

Webb, J. K., W-G. Du, D. A. Pike, and R. Shine. 2009. Chemical cues from both dangerous and non-dangerous snakes elicit antipredator behaviours from a nocturnal lizard. *Animal Behaviour* 77:1471–1478.

Webb, J. K., W-G. Du, D. A. Pike, and R. Shine. 2010. Generalization of predator recognition: Velvet geckos display anti-predator behaviours in response to chemicals from non-dangerous elapid snakes. *Current Zoology* 56:337–342.

Webb, J. K., D. A. Pike, and R. Shine. 2010. Olfactory recognition of predators by nocturnal lizards: Safety outweighs thermal benefits. *Behavioral Ecology* 21:72–77.

Webb, J. K., R. Pringle, and R. Shine. 2004. How do nocturnal snakes select diurnal retreat sites? *Copeia* 2004:919–925.

Webb, J. K., R. Pringle, and R. Shine. 2005. Canopy removal restores habitat quality for an endangered snake in a fire suppressed landscape. *Copeia* 2005:894–900.

Webb, J. K., R. Pringle, and R. Shine. 2009. Intraguild predation, thermoregulation, and microhabitat selection by snakes. *Behavioral Ecology* 20:271–277.

Webb, J. K. and R. Shine. 1997a. Out on a limb: Conservation implications of tree hollow use by a threatened snake species (*Hoplocephalus bungaroides*: Serpentes, Elapidae). *Biological Conservation* 81:21–33.

Webb, J. K. and R. Shine. 1997b. A field study of spatial ecology and movements of a threatened snake species, *Hoplocephalus bungaroides*. *Biological Conservation* 82:203–217.

Webb, J. K. and R. Shine. 1998a. Ecological characteristics of a threatened snake species, *Hoplocephalus bungaroides* (Serpentes, Elapidae). *Animal Conservation* 1:185–193.

Webb, J. K. and R. Shine. 1998b. Thermoregulation by a nocturnal elapid snake (*Hoplocephalus bungaroides*) in south-eastern Australia. *Physiological Zoology* 71:680–692.

Webb, J. K. and R. Shine. 1998c. Using thermal ecology to predict retreat-site selection by an endangered snake species. *Biological Conservation* 86:233–242.

Webb, J. K. and R. Shine. 2000. Paving the way for habitat restoration: Can artificial rocks restore degraded habitats for endangered reptiles? *Biological Conservation* 92:93–99.

Webb, J. K. and R. Shine. 2008. Differential effects of an intense wildfire on survival of two species of sympatric snakes. *Journal of Wildlife Management* 72:1394–1398.

CHAPTER 7. SNAKES IN NEED OF A DEFAMATION LAWYER

Aubret, F., X. Bonnet, D. Pearson, and R. Shine. 2005. How can blind tigersnakes (*Notechis scutatus*) forage successfully? *Australian Journal of Zoology* 53:283–288.

Aubret, F., X. Bonnet, R. Shine, and S. Mamelaut. 2005. Swimming and pregnancy in tiger snakes, *Notechis scutatus*. *Amphibia-Reptilia* 26:396–400.

Aubret, F., R. J. Michniewicz, and R. Shine. 2011. Correlated geographic variation in predation risk and antipredator behaviour within a wide-ranging snake species (*Notechis scutatus*, Elapidae). *Austral Ecology* 36:446–452.

Aubret, F. and R. Shine. 2007. Rapid prey-induced shift in body size in an isolated snake population (*Notechis scutatus*, Elapidae). *Austral Ecology* 32:889–899.

Aubret, F. and R. Shine. 2009a. Causes and consequences of aggregation by neonatal tiger snakes (*Notechis scutatus*, Elapidae). *Austral Ecology* 34:210–217.

Aubret, F. and R. Shine. 2009b. Genetic assimilation and the postcolonization erosion of phenotypic plasticity in island tiger snakes. *Current Biology* 19:1932–1936.

Aubret, F. and R. Shine. 2010a. Thermal plasticity in young snakes: How will climate change affect the thermoregulatory tactics of ectotherms? *Journal of Experimental Biology* 213:242–248.

Aubret, F. and R. Shine. 2010b. Fitness costs may explain the post-colonisation erosion of phenotypic plasticity. *Journal of Experimental Biology* 213:735–739.

Aubret, F., R. Shine, and X. Bonnet. 2004. Adaptive developmental plasticity in snakes. *Nature* 431:261–262.

Bonnet, X., D. Bradshaw, R. Shine, and D. Pearson. 1999. Why do snakes have eyes? The (non-) effect of blindness in island tigersnakes. *Behavioral Ecology and Sociobiology* 46:267–272.

Bonnet, X., R. Shine, G. Naulleau, and C. Thiburce. 2001. Plastic vipers: Genetic and environmental influences on the size and shape of Gaboon vipers, *Bitis gabonica*. *Journal of Zoology (London)* 255:341–351.

Cann, J. 2014. *Historical Snakeys*. ECO Publishing, Rodeo, New Mexico.

Keogh, J. S., I. A. Scott, and C. Hayes. 2005. Rapid and repeated origin of insular gigantism and dwarfism in Australian tiger snakes. *Evolution* 59:226–233.

Ladyman, M., E. Seubert, and D. Bradshaw. 2020. The origin of tiger snakes on Carnac Island. *Journal of the Royal Society of Western Australia* 103:39–42.

Mirtschin, P. J., R. Shine, T. J. Nias, N. L. Dunstan, B. J. Hough, and M. Mirtschin. 2002. Influences on venom yield in Australian tigersnakes (*Notechis scutatus*) and brownsnakes (*Pseudonaja textilis*: Elapidae, Serpentes). *Toxicon* 40:1581–1592.

Palika, L. 1998. *The Complete Idiot's Guide to Reptiles and Amphibians*. Alpha Books, New York.

Pearson, D., R. Shine, X. Bonnet, A. Williams, B. Jennings, and O. Lourdais. 2001. Ecological notes on crowned snakes, *Elapognathus coronatus*, from the Archipelago of the Recherche in southwestern Australia. *Australian Zoologist* 31:610–617.
Schwaner, T. D. 1985. Population structure of black tiger snakes, *Notechis ater niger*, on offshore islands of South Australia. Pages 35–46 in *Biology of Australasian Frogs and Reptiles* (G. C. Grigg, R. Shine, H. Ehmann, eds.). Surrey-Beatty, Sydney.
Schwaner, T. D. 1989. A field study of thermoregulation in black tiger snakes (*Notechis ater niger*: Elapidae) on the Franklin Islands, South Australia. *Herpetologica* 45:393–401.
Schwaner, T. D. and S. D. Sarre. 1990. Body size and sexual dimorphism in mainland and island tiger snakes. *Journal of Herpetology* 24:320–322.
Waddington, C. H. 1953. Genetic assimilation of an acquired character. *Evolution* 7:118–126.
Whitaker, P. B., K. Ellis, and R. Shine. 2000. The defensive strike of the eastern brownsnake (*Pseudonaja textilis*, Elapidae). *Functional Ecology* 14:25–31.
Whitaker, P. B. and R. Shine. 1999a. When, where and why do people encounter Australian brownsnakes (*Pseudonaja textilis*, Elapidae)? *Wildlife Research* 26:675–688.
Whitaker, P. B. and R. Shine. 1999b. Responses of free-ranging brownsnakes (*Pseudonaja textilis*, Elapidae) to encounters with humans. *Wildlife Research* 26:689–704.
Whitaker, P. B. and R. Shine. 2000. Sources of mortality of large elapid snakes in an agricultural landscape. *Journal of Herpetology* 34:121–128.
Whitaker, P. B. and R. Shine. 2002. Thermal biology and activity patterns of the eastern brownsnake (*Pseudonaja textilis*): A radiotelemetric study. *Herpetologica* 58:436–462.
Whitaker, P. B. and R. Shine. 2003. A radiotelemetric study of movements and shelter-site selection by free-ranging brownsnakes (*Pseudonaja textilis*, Elapidae). *Herpetological Monographs* 17:130–144.

CHAPTER 8. ROUGH CHARACTERS IN THE BILLABONG

Camilleri, C. and R. Shine. 1990. Sexual dimorphism and dietary divergence: Differences in trophic morphology between male and female snakes. *Copeia* 1990:649–658.
Houston, D. L. and R. Shine. 1993. Sexual dimorphism and niche divergence: Feeding habits of the Arafura filesnake. *Journal of Animal Ecology* 62:737–749.
Houston, D. L. and R. Shine. 1994a. Low growth rates and delayed maturation in Arafura filesnakes (Serpentes: Acrochordidae) in tropical Australia. *Copeia* 1994:726–731.
Houston, D. L. and R. Shine. 1994b. Movements and activity patterns of Arafura filesnakes (Serpentes: Acrochordidae) in tropical Australia. *Herpetologica* 50:349–357.
Houston, D. L. and R. Shine. 1994c. Population demography of Arafura filesnakes (Serpentes: Acrochordidae) in tropical Australia. *Journal of Herpetology* 28:273–280.
Idriess, I. 1935. *Man Tracks*. Angus and Robertson, Sydney.
Idriess, I. 1941. *Nemarluk: King of the Wilds*. Angus and Robertson, Sydney.
Idriess, I. 1946. *In Crocodile Land*. Angus and Robertson, Sydney.
Madsen, T. and R. Shine. 2000. Rain, fish and snakes: Climatically-driven population dynamics of Arafura filesnakes in tropical Australia. *Oecologia* 124:208–215.
Madsen, T. and R. Shine. 2001a. Conflicting conclusions from long-term *versus* short-term studies on growth and reproduction of a tropical snake. *Herpetologica* 57:147–156.
Madsen, T. and R. Shine. 2001b. Do snakes shrink? *Oikos* 92:187–188.
Shine, R. 1986a. Diets and abundances of aquatic and semi-aquatic reptiles of the Alligator Rivers Region. *Supervising Scientist for the Alligator Rivers Region, Technical Memorandum* 16:1–54.
Shine, R. 1986b. Predation upon filesnakes (*Acrochordus arafurae*) by aboriginal hunters: Selectivity with respect to size, sex and reproductive condition. *Copeia* 1986:238–239.

Shine, R. 1986c. Ecology of a low-energy specialist: Food habits and reproductive biology of the Arafura filesnake (Acrochordidae). *Copeia* 1986:424–437.

Shine, R. 1986d. Sexual differences in morphology and niche utilization in an aquatic snake, *Acrochordus arafurae*. *Oecologia* 69:260–267.

Shine, R. 1988. Constraints on reproductive investment: A comparison between aquatic and terrestrial snakes. *Evolution* 42:17–27.

Shine, R. 1994. Rough character in the billabong. *Australian Geographic* 35:106–117.

Shine, R., P. Harlow, J. S. Keogh, and Boeadi. 1995. Biology and commercial utilization of acrochordid snakes, with special reference to karung (*Acrochordus javanicus*). *Journal of Herpetology* 29:352–360.

Shine, R. and R. Lambeck. 1985. A radiotelemetric study of movements, thermoregulation and habitat utilization of Arafura filesnakes (Serpentes, Acrochordidae). *Herpetologica* 41:351–361.

Ujvari, B., S. Andersson, G. P. Brown, R. Shine, and T. Madsen. 2010. Climate-driven impacts of prey abundance on the population structure of a tropical aquatic predator. *Oikos* 119:188–196.

Vincent, S. E., R. Shine, and G. P. Brown. 2005. Does foraging mode influence sensory modalities for prey detection? A comparison between male and female filesnakes (*Acrochordus arafurae*, Acrochordidae). *Animal Behaviour* 70:715–721.

CHAPTER 9. SNAKES, RATS, AND RAINFALL

Flannery, T. 1994. *The Future Eaters*. Reed Books, Chatswood.

Frith, H. J. and S. J. J. F. Davies. 1961. Ecology of the Magpie Goose, *Anseranas semipalmata* Latham (Anatidae). *CSIRO Wildlife Research* 6:91–141.

Madsen, T. and R. Shine. 1996a. Determinants of reproductive output in female water pythons (*Liasis fuscus*: Pythonidae). *Herpetologica* 52:146–159.

Madsen, T. and R. Shine. 1996b. Seasonal migration of predators and prey: Pythons and rats in tropical Australia. *Ecology* 77:149–156.

Madsen, T. and R. Shine. 1998a. Spatial subdivision within a population of tropical pythons (*Liasis fuscus*) in a superficially homogeneous habitat. *Australian Journal of Ecology* 23:340–348.

Madsen, T. and R. Shine. 1998b. Quantity or quality? Determinants of maternal reproductive success in tropical pythons (*Liasis fuscus*). *Proceedings of the Royal Society B* 265:1521–1525.

Madsen, T. and R. Shine. 1999a. The adjustment of reproductive threshold to prey abundance in a capital breeder. *Journal of Animal Ecology* 68:571–580.

Madsen, T, and R. Shine. 1999b. Rainfall and rats: Climatically-driven dynamics of a tropical rodent population. *Australian Journal of Ecology* 24:80–89.

Madsen, T. and R. Shine. 1999c. Life-history consequences of nest-site variation in tropical pythons (*Liasis fuscus*). *Ecology* 80:989–997.

Madsen, T. and R. Shine. 2000a. Energy *versus* risk: Costs of reproduction in free-ranging pythons in tropical Australia. *Austral Ecology* 25:670–675.

Madsen, T. and R. Shine. 2000b. Silver spoons and snake sizes: Prey availability early in life influences long-term growth rates of free-ranging pythons. *Journal of Animal Ecology* 69:952–958.

Madsen, T. and R. Shine. 2002. Short and chubby or long and thin? Food intake, growth and body condition in free-ranging pythons. *Austral Ecology* 27:672–680.

Madsen, T., B. Ujvari, M. M. Olsson, and R. Shine. 2005. Paternal alleles enhance female reproductive success in tropical pythons. *Molecular Ecology* 14:1783–1787.

Madsen, T., B. Ujvari, R. Shine, W. Buttemer, and M. Olsson. 2006. Size matters: Extraordinary rodent abundance on an Australian tropical floodplain. *Austral Ecology* 31:361–365.
Madsen, T., B. Ujvari, R. Shine, and M. Olsson. 2006. Rain, rats and pythons: Climate-driven population dynamics of predators and prey in tropical Australia. *Austral Ecology* 31:30–37.
Pizzatto, L., T. Madsen, G. P. Brown, and R. Shine. 2009. Spatial ecology of hatchling water pythons (*Liasis fuscus*) in tropical Australia. *Journal of Tropical Ecology* 25:181–191.
Redhead, T. D. 1979. On the demography of *Rattus sordidus colletti* in monsoonal Australia. *Australian Journal of Ecology* 4:115–136.
Shine, R. and T. Madsen. 1996. Is thermoregulation unimportant for most reptiles? An example using water pythons (*Liasis fuscus*) in tropical Australia. *Physiological Zoology* 69:252–269.
Shine, R. and T. Madsen. 1997. Prey abundance and predator reproduction: Rats and pythons on a tropical Australian floodplain. *Ecology* 78:1078–1086.
Shine, R., T. R. L. Madsen, M. J. Elphick, and P. S. Harlow. 1997. The influence of nest temperatures and maternal brooding on hatchling phenotypes of water pythons. *Ecology* 78:1713–1721.
Stahlschmidt, Z., R. Shine, and D. DeNardo. 2012a. The consequences of alternative parental care tactics in free-ranging pythons (*Liasis fuscus*) in tropical Australia. *Functional Ecology* 26:812–821.
Stahlschmidt, Z., R. Shine, and D. DeNardo. 2012b. Temporal and spatial complexity of maternal thermoregulation in tropical pythons. *Physiological and Biochemical Zoology* 85:219–230.
Ujvari, B., G. P. Brown, R. Shine, and T. Madsen. 2016. Floods and famine: Climate-induced collapse of a tropical predator-prey community. *Functional Ecology* 30:453–458.
Ujvari, B., R. Shine, and T. Madsen. 2011. How well do predators adjust to climate-mediated shifts in prey spatial distribution? A field study on Australian water pythons (*Liasis fuscus*). *Ecology* 92:777–783.
Wittzell, H., T. Madsen, T. Westerdahl, R. Shine, and T. von Schantz. 1999. MHC variation in birds and reptiles. *Genetica* 104:301–309.

CHAPTER 10. SCIENCE ON THE FLOODPLAIN

Brown, G. P., T. Madsen, S. Dubey, and R. Shine. 2017. The causes and ecological correlates of head scale asymmetry and fragmentation in a tropical snake. *Scientific Reports* 7:11363.
Brown, G. P., T. Madsen, and R. Shine. 2017. Resource availability and sexual size dimorphism: Differential effects of prey abundance on the growth rates of tropical snakes. *Functional Ecology* 31:1592–1599.
Brown, G. P., C. M. Shilton, and R. Shine. 2006. Do parasites matter? Assessing the fitness consequences of haemogregarine infection in snakes. *Canadian Journal of Zoology* 84:668–676.
Brown, G. P. and R. Shine. 2002a. Reproductive ecology of a tropical natricine snake, *Tropidonophis mairii* (Colubridae). *Journal of Zoology (London)* 258:63–72.
Brown, G. P. and R. Shine. 2002b. The influence of weather conditions on activity of tropical snakes. *Austral Ecology* 27:596–605.
Brown, G. P. and R. Shine. 2004a. Maternal nest-site choice and offspring fitness in a tropical snake (*Tropidonophis mairii*, Colubridae). *Ecology* 85:1627–1634.
Brown, G. P. and R. Shine. 2004b. Effects of reproduction on the antipredator tactics of snakes (*Tropidonophis mairii*, Colubridae). *Behavioral Ecology and Sociobiology* 56:257–262.

Brown, G. P. and R. Shine. 2005a. Do changing moisture levels during incubation influence phenotypic traits of hatchling snakes (*Tropidonophis mairii*)? *Physiological and Biochemical Zoology* 78:524–530.

Brown, G. P. and R. Shine. 2005b. Female phenotype, life-history, and reproductive success in free-ranging snakes (*Tropidonophis mairii*). *Ecology* 86:2763–2770.

Brown, G. P. and R. Shine. 2005c. Nesting snakes (*Tropidonophis mairii*, Colubridae) selectively oviposit in sites that provide evidence of previous successful hatching. *Canadian Journal of Zoology* 83:1134–1137.

Brown, G. P. and R. Shine. 2006a. Why do most tropical animals reproduce seasonally? Testing alternative hypotheses on the snake *Tropidonophis mairii* (Colubridae). *Ecology* 87:133–143.

Brown, G. P. and R. Shine. 2006b. Effects of nest temperature and moisture on phenotypic traits of hatchling snakes (*Tropidonophis mairii*, Colubridae) from tropical Australia. *Biological Journal of the Linnean Society* 89:159–168.

Brown, G. P. and R. Shine. 2007a. Repeatability and heritability of reproductive traits in free-ranging snakes. *Journal of Evolutionary Biology* 20:588–596.

Brown, G. P. and R. Shine. 2007b. Like mother, like daughter: Inheritance of nest-site location in snakes. *Biology Letters* 3:131–133.

Brown, G. P. and R. Shine. 2007c. Rain, prey and predators: Climatically-driven shifts in frog abundance modify reproductive allometry in a tropical snake. *Oecologia* 154:361–368.

Brown, G. P. and R. Shine. 2016. Maternal body size influences offspring immune defence in an oviparous snake. *Royal Society Open Science* 3:160041.

Brown, G. P. and R. Shine. 2018. Immune configuration in hatchling snakes is affected by incubation moisture, and is linked to subsequent growth and survival in the field. *Journal of Experimental Zoology A* 329:222–229.

Brown, G. P., R. Shine, and T. Madsen. 2002. Responses of three sympatric snake species to tropical seasonality in northern Australia. *Journal of Tropical Ecology* 18:549–568.

Brown, G. P., R. Shine, and T. Madsen. 2005. Spatial ecology of slatey-grey snakes (*Stegonotus cucullatus*, Colubridae) on a tropical Australian floodplain. *Journal of Tropical Ecology* 21:605–612.

Dubey, S., G. P. Brown, T. Madsen, and R. Shine. 2008a. Characterization of tri- and tetranucleotide microsatellite loci for the slatey-grey snake (*Stegonotus cucullatus*, Colubridae). *Molecular Ecology Resources* 8:431–433.

Dubey, S., G. P. Brown, T. Madsen, and R. Shine. 2008b. Male-biased dispersal in a tropical Australian snake (*Stegonotus cucullatus*, Colubridae). *Molecular Ecology* 17:3506–3514.

Dubey, S., G. P. Brown, T. Madsen, and R. Shine. 2009. Sexual selection favours large body size in males of a tropical snake (*Stegonotus cucullatus*, Colubridae). *Animal Behaviour* 77:177–182.

Lindström, T., B. L. Phillips, G. P. Brown, and R. Shine. 2015. Identifying the time scale of synchronous movement: A study on tropical snakes. *Movement Ecology* 3:12.

Mayer, M., G. P. Brown, B. Zimmerman, and R. Shine. 2015. High infection intensities, but negligible fitness costs, suggest tolerance towards gastrointestinal nematodes in a tropical snake. *Austral Ecology* 40:683–692.

Sergo, D. and R. Shine. 2015. Snakes for lunch: Bird predation on reptiles in a tropical floodplain. *Australian Zoologist* 37:311–320.

Shine, R. and G. P. Brown. 2002. Effects of seasonally varying hydric conditions on hatchling phenotypes of keelback snakes (*Tropidonophis mairii*, Colubridae) from the Australian wet-dry tropics. *Biological Journal of the Linnean Society* 76:339–347.

Shine, R., G. P. Brown, and M. J. Elphick. 2004. Field experiments on foraging in free-ranging water snakes *Enhydris polylepis* (Homalopsinae). *Animal Behaviour* 68:1313–1324.

Webb, J. K., G. P. Brown, and R. Shine. 2001. Body size, locomotor speed and antipredator behaviour in a tropical snake (*Tropidonophis mairii*, Colubridae): The influence of incubation environments and genetic factors. *Functional Ecology* 15:561–568.

CHAPTER 11. OUR EVOLVING RELATIONSHIP WITH SERPENTS

Isbell, L. A. 2006. Snakes as agents of evolutionary change in primate brains. *Journal of Human Evolution* 51:1–35.

Isbell, L. A. 2009. *The Fruit, the Tree, and the Serpent: Why We See So Well.* Harvard University Press, Cambridge, MA.

Patterson, A. B. 1933. *The Animals Noah Forgot.* Endeavour Press, Sydney.

Van Le, Q., L. A. Isbell, J. Matsumoto, M. Nguyen, E. Hori, R. S. Maior, C. Tomaz, A. H. Tran, T. Ono, and H. Nishijo. 2013. Pulvinar neurons reveal neurobiological evidence of past selection for rapid detection of snakes. *Proceedings of the National Academy of Sciences USA* 110:19000–19005.

Index

A

Aboriginal culture, 141, 177, 180, 188, 192, 203, 247, 249
Aboriginal hunters, 185, 192, 193, 199
academic employment, 60, 63, 112, 238
Acrochordidae, 181
activity patterns, 65
Adder, Death (*Acanthophis spp.*), 55, 70, 92, 95–97, 106, 114, 153, 157, 188, 229, 248
Adder, Floodplain Death (*Acanthophis praelongus*), 97
Addicks Dam, Houston, Texas, 54
Adelaide River, Northern Territory, 204, 217, 223, 225
age at sexual maturity, 47, 241
agile wallaby (*Macropus agilis*), 21, 25, 180, 223
agriculture, 37, 148, 172
alcohol, 11, 20, 37, 73, 81–84, 88, 179, 201, 219, 243
Alice Springs, Northern Territory, 132, 220
Amazon Tree Boa (*Corallus hortulanus*), 92
ambush predation, 92, 96, 97, 106, 107, 118, 125, 130, 145–147, 184, 186, 198, 199
American Coachwhip (*Masticophis flagellum*), 158
American Copperhead (*Agkistrodon contortrix*), 92
American Gartersnakes (*Thamnophis* spp.), 201, 229, 231
American Racer (*Coluber constrictor*), 158
American Rattlesnakes (*Crotalus* spp.), 11, 201, 250, 252
American Watersnakes (*Nerodia* spp.), 43, 54, 157, 158, 184, 232
amphibian abundance, 71, 172, 179
ancestral form, 95, 165
Andrewartha, Herbert, 61, 79, 207, 215–218, 249
animal behavior, 2, 22, 41, 44, 47, 55, 65, 66, 71, 74–77, 94, 96, 110, 113, 144, 151, 158, 169, 175, 176, 204, 234, 235, 242
animal tracks, 54, 142
annual variation, 200, 206, 235
ant, nest and larvae, 91, 93, 94
ant, scent trail, 93, 262
antivenom, 171, 172
Ants, Bulldog (*Myrmecia*), 93
Aquarius Arts Festival, 115
aquatic snakes, 181, 195, 198
Arafura Filesnake (*Acrochordus arafurae*), 177, 180, 202–207, 213, 229, 235, 250
Armidale, New South Wales, 19, 23–44, 48–71, 84, 86, 87, 93, 157, 178
Arnhem Highway, Northern Territory, 191
artificial rocks, 138, 139
Asia, 21, 27, 95, 97, 181, 229, 231, 247, 249, 252
attacks, unprovoked, 162, 163
attitudes, change in, vii, 62, 63, 247–249
Aubret, Fabien, 170, 266
Australian Academy of Science, 82, 83
Australian Copperhead (*Austrelaps labialis*), 10, 27, 29, 30–31, 38, 47, 48, 95, 101, 119, 135
Australian ecosystems, dynamic, 79, 201, 207, 216, 249
Australian fauna, 10, 82, 85, 88, 95, 104, 122, 123, 141, 177, 196, 227, 247, 250
Australian Magpie (*Gymnorhina tibicen*), 32, 35
Australian Museum, 37, 81, 84, 86–89, 99, 168, 189, 195, 206
Australian Reptile Park, 91, 138, 139, 171
Australian Research Council, 139, 212, 213, 231
Australia Society of Herpetologists, 235

B

Banded Watersnake (*Nerodia fasciata*), 155, 157
Bandy-Bandy (*Vermicella annulata*), 91, 96
Bangalow, New South Wales, 118
Bardick (*Echiopsis curta*), 114
Barramundi (*Lates calcarifer*), 187, 191, 198, 199
Barwick, Dick, 20, 22, 62
basking, 10, 19, 23, 47, 59, 63, 65, 71, 76–79, 95, 106, 110, 111, 130, 132, 146, 172, 187, 204, 221, 245
Beatrice Hill, Northern Territory, 223
behavioral thermoregulation, 129, 221
Billabong, v, 173, 186, 188–194, 196, 197, 199, 200, 201, 213, 229
biodiversity, 73, 177, 206, 220, 225
Biology Department, University of Sydney, 60, 66, 67
Birch, Charles, 60–62, 67, 79, 207, 215–218, 249
bird-eaters, 172
Black Flying Fox (Fruit-Bat) (*Pteropus alecto*), 187
Black Mamba (*Dendroaspis polylepis*), 27, 96, 114, 227
Black-Necked Stork (also Jabiru) (*Ephippiorhynchus asiaticus*), 240
Black-Tailed Rattlesnake (*Crotalus molossus*), 41
blindness in snakes, 171, 172
Blindsnake (*Anilios nigrescens*), 91, 93, 94, 112, 129, 177
Bluetongue Skink (*Tiliqua scincoides*), 3, 6, 9, 39, 238

Boa, Amazon Tree (*Corallus hortulanus*), 92
Boa, New Guinea Ground (*Candoia aspera*), 92
Boa, Pacific Tree (*Candoia bibroni*), 92, 132, 260
body temperature regulation, 221
Bonnet, Xavier, 83, 169–171, 234
boom and bust, 1, 158, 216, 218
Branch, Bill, 89
Brazil, 92, 158, 222
Brindabella Range, Australian Capital Territory, 63–65
Brisbane, Queensland, 10, 177
Broadheaded Snake (*Hoplocephalus bungaroides*), 131, 133–148, 172, 176, 201, 220, 250
Brolga (*Antigone rubicunda*), 249
brooding, 111, 214
Brown, Greg, 156, 175, 176, 228, 229, 232–235, 238–244, 257
Brownsnake, Eastern (*Pseudonaja textilis*), 1, 2, 4, 11, 13–15, 20, 27, 29, 38, 48, 70, 71, 89, 96, 111, 119, 151, 153, 157–164, 172, 229
Brownsnake, Western (*Pseudonaja nuchalis*), 114
Brown Tree Snake (*Boiga irregularis*), 7, 89, 90, 132
Brushtail Possum (*Trichosurus vulpecula*), 9, 106, 107, 111, 112, 119, 124, 125, 216
Bubalahlah Creek, New South Wales, 79
Buffalo, Water (*Bubalus bubalis*), 187, 188, 199, 200, 204, 206, 233, 235
"build-up" season, 203
Bull, Jim, 40–48, 54, 56, 57, 59, 172
Bulldog Ants (*Myrmecia*), 93
Burger, Joanna, Rutgers University, New Jersey, 222
Burrows, shearwater, 168
Burton's Legless Lizard (*Lialis burtonis*), 114
Bushmaster (*Lachesis muta*), 114
Bushrat (*Rattus fuscipes*), 146
Bushrock removal, 136
"bush tucker", 180, 196

C

Cairns, Queensland, 112
camouflage, 1, 92, 96, 122, 158, 164, 199
Canada, 6, 232, 233
Canadian Pike (*Esox lucius*), 43
Canberra, ACT, 1, 9, 10, 13, 16, 17, 20–23, 25, 33, 62, 63, 78, 126, 172, 225
Cane Toad (*Rhinella marina*), 92, 238
Cape York, Queensland, 114, 173, 178
Cappo, Mike, 227
car accident, 34, 35, 61
Carnac Island, Western Australia, 167–172
Carpet Python (also Diamond Python) (*Morelia spilota*), 16, 33, 101, 110–126, 135, 145, 146, 176
Carpet Python, Centralian (*Morelia bredli*), 132
Carpet Python, Southwest (*Morelia spilota imbricata*), 165, 197
Centralian Carpet Python (*Morelia bredli*), 132
Charnov, Ric, 52, 54, 62
Cliffs, habitat, 6, 25, 106, 127, 129, 130–132, 135, 139, 145, 155, 177
climate change, 38, 141, 204, 205, 216, 225, 247
climatic fluctuations and extremes, 204, 215, 223, 225, 245
clutch size, 111, 172, 224
Coachwhip, American (*Masticophis flagellum*), 158
Coastal Taipan (*Oxyuranus scutellatus*), 11, 20, 93, 96, 98, 114, 151, 153, 157
Cobra, King (*Ophiophagus hannah*), 27, 133, 134
Coffs Harbour, New South Wales, 33
Cogger, Hal, 27, 86, 87, 189, 206
cold-blooded, vii, 17, 23, 62, 111, 128, 162, 215, 216, 248
cold climate, 27, 59, 63, 201, 221
cold climate theory, 63
Collett's Snake (*Pseudechis colletti*), 58
colonization events, 95, 165, 167, 170
colubrid snakes, 91, 92, 135, 158, 177, 229, 231, 239
combat, male-male, 33, 53, 55, 56, 71, 77, 95, 96, 119, 120–124
Commonwealth Scientific Research Organisation (CSIRO), 21, 73, 206
conference, 10, 17, 42, 63, 89, 94, 157, 221, 222, 235, 238
conservation, python, 111, 122, 125
conservation biology, 73, 128, 140, 144, 149, 194, 216, 250
conservation threats and planning, 38, 99, 176, 205, 219
constrict prey, 115, 185
Cooloola Tree Frog (*Litoria cooloolensis*), 88
Coomonderry Swamp, New South Wales, 79
Copperhead, American (*Agkistrodon contortrix*), 92
Copperhead, Australian (*Austrelaps labialis*), 10, 27, 29, 30, 38, 47, 48, 95, 101, 119, 135
Coral Snakes (*Micrurus* spp.), 27, 96
Corroboree Frog (*Pseudophryne pengilleyi*), 64
"costs" of reproduction, 65
Cottonmouth Moccasin (*Agkistrodon piscivorus*), 54, 157, 158
Covacevich, Jeanette, 87
Crimson Rosella (*Platycercus elegans*), 79
critical habitat, 140, 146
Croak, Ben, 139
Crocodile, Freshwater (*Crocodylus johnstoni*), 91, 174, 175, 235
Crocodile, Saltwater (*Crocodylus porosus*), 155, 173, 174, 175, 180, 187, 189, 194, 196, 199, 204, 217, 235
crocodile conservation, 175
Crowned Snakes (*Cacophis* spp.), 86, 87
CSIRO (Commonwealth Scientific Research Organisation), 21, 73, 206

Cunningham's Skink (*Egernia cunninghami*), 22, 23
Curl Snake (*Suta suta*), 89
cyclones, 206, 224

D

Dangars Falls, New South Wales, 25, 35
Darwin, Charles, 41, 43, 44, 61, 123, 129, 130, 165, 169
Darwin, Northern Territory, 88, 91, 112, 151, 152, 173–179, 203–206, 219, 220, 228, 235
data analysis, 48, 82, 88, 201, 223, 235, 239, 243, 245
data collection, 19, 37, 42, 53, 55, 74, 78, 79, 82–85, 87, 90, 93, 100, 104, 119, 175, 200, 207, 213, 235, 239, 242, 245
datasets, 45, 48, 79, 82, 95, 112, 129, 164, 198, 213, 219, 213, 234, 238, 239, 245
Dawkins, Richard, 44
deadly snakes, 3, 11, 14, 20, 24, 29, 32, 47, 93, 98, 102, 114, 119, 134, 135, 151–171, 217, 228, 229
Death Adder (*Acanthophis antarcticus*), 55, 70, 92, 95–97, 106, 114, 153, 157, 188, 229, 248
De Bavay, John, 25
deserts, 41, 49, 54, 91, 97, 98, 102, 135, 151, 218, 220
Desert Tortoise (*Gopherus agassizii*), 54
DeVis' Banded Snake (*Denisonia devisi*), 70
Diamond Python (also Carpet Python) (*Morelia spilota*), 9, 10, 101–125, 132, 141, 186, 203
digestive efficiency, 125, 196
Dingo (*Canis lupus*), 24, 64, 142, 143, 201
diversity, vii, 27, 41, 95, 124, 222, 249
DjaDja Billabong, Kakadu, 196, 197, 200, 213
DNA, 88, 243
Dragon, Komodo (*Varanus komodoensis*), 142
Dreamtime, Aboriginal, 203, 247
drought, 52, 61, 76, 78, 79, 143, 148, 151, 205, 215, 223, 225, 250
dry season, 175, 179, 199, 203, 206, 216, 223, 234, 235, 249
Dryandra Forest, Western Australia, 122, 124
Du, Weiguo, Chinese Scholar, 143
Dubbo, New South Wales, 70
Dubey, Sylvain, 26, 94, 103, 134, 142, 243, 257
Dugite (*Pseudonaja affinis*), 114, 165
Dungowan, New South Wales, 19, 20
Dusky Rat (*Rattus colletti*), 210, 216, 217, 223
dynamic ecosystem, 79, 201, 207, 216, 249

E

Eastern Brownsnake (*Pseudonaja textilis*), 1, 2, 4, 11, 13–15, 20, 27, 29, 38, 48, 70, 71, 89, 96, 111, 119, 151–164, 172, 229
Echidna (*Tachyglossus aculeatus*), 57
ecological flexibility, 126, 144, 145, 169, 172, 201, 214, 224, 225, 245
ecological niche, 92, 128, 146, 186, 198, 229, 250
ecological specialization, 81, 124, 132, 145, 146, 198
ectothermic biology, 62, 111, 128, 130, 148, 215, 216
ectotherms, 111, 132, 149, 155, 176, 204, 205, 215, 218, 225
ectothermy, advantages of, 111, 128, 176, 186, 201, 248
egg-eaters, 81, 82
egg incubation, 63, 90, 110, 111, 214, 234, 241, 244
egg-laying, 57, 63, 65, 87, 203, 242
elapid snakes, 27, 29, 56, 91, 92, 95–98, 112, 114, 135, 153, 158, 176, 177, 229
embryonic development, 45, 46, 57, 59, 63, 65, 86, 94, 110, 111, 132, 152, 181, 195, 244, 245
endangered snakes, 89, 129, 138, 140, 141, 144–146, 148, 156, 172, 176
endotherms, 62, 111, 128, 149, 155, 215, 216, 248
energy allocation, 25, 27, 52, 55, 111, 125, 128, 130, 148, 149, 155, 162, 185, 186, 196, 198, 201, 214–218
environmental vandalism, 136, 138, 139, 149
Esperance, Western Australia, 123, 165
Evolution of live-bearing (viviparity), 57, 59, 63, 65, 95, 152
evolutionary biology, 24, 40, 43, 44, 45, 46, 52, 57, 83, 169, 197, 241
evolutionary ecology, 17, 41, 42, 46, 51, 126, 234
evolutionary fitness, 55, 168, 198, 241, 244
evolutionary processes, 40, 98, 170, 215, 248
evolution on islands, 123, 165, 170
extinction, 64, 89, 112, 129, 135, 141, 145, 148, 224
Eyre, Edward John, 114, 262

F

fake rocks, 138–141
farmers, 37, 38, 71, 73, 115, 146, 158, 164, 172, 219, 250
fear of snakes, 3, 14, 33, 85, 151, 228, 229, 247, 252, 254
fellowship, 48, 60, 63, 65, 67, 144, 171, 238
field studies, 79, 89, 252
fieldwork, 21, 22, 25, 26, 54, 59, 60, 52, 84, 89, 98, 102, 112, 114, 118, 138, 142, 144, 151, 162, 171, 176–179, 196, 200, 220, 221, 224, 228, 231, 235, 238, 243
Filesnake, Arafura (*Acrochordus arafurae*), 177, 180–202, 204, 205, 207, 213, 229, 235, 250
firearms, 54, 143
"firestick farming", 141, 180
Fish, Canadian Pike (*Esox lucius*), 43
Fish, Gudgeon (*Oxyeleotris* spp.), 198

Fish, Rainbowfish (*Melanotaenia* spp.), 198
Fitch, Henry, 234
Fitzgerald, Mark, 115, 118–120, 126, 146–148, 257
Flannery, Tim, 216
Flat Spiders (*Hemicloea* spp.), 132, 133, 139
flooding, 1, 61, 70, 71, 93, 175, 179, 187, 196, 212, 217, 223, 225, 248
floodplain, v, 173, 180, 191, 196, 199, 201, 204, 206, 207, 212, 215–217, 220, 222–225, 227, 230, 231, 234, 236, 239, 241–243, 249
Floodplain Death Adder (*Acanthophis praelongus*), 97
Florida, 157
Fogg Dam, 107, 175, 201, 203–243, 249
foraging, 77, 92, 96, 106, 113, 118, 143, 145, 175, 199, 207, 231
forest clearing, regrowth, fire, 112, 122, 141, 143
Forests New South Wales, 139
forked tongue, 93, 106, 172, 185
formalin, 33, 84, 88, 89
Fox (*Vulpes vulpes*), 74, 119, 124, 140, 141, 168, 172
France, 100, 158, 169, 216
Fraser Island, Queensland, 88
Freshwater Crocodile (*Crocodylus johnstoni*), 91, 174, 175, 235
Frillneck Lizard (*Chlamydosaurus kingii*), 178, 187
Frog, Cooloola Tree (*Litoria cooloolensis*), 88
Frog, Corroboree (*Pseudophryne pengilleyi*), 64
Frog, Green Tree (*Litoria caerulea*), 78
Frog, Wallum Sedge (*Litoria olongburensis*), 88
Fruit-Bat, Black Flying Fox (*Pteropus alecto*), 187
Fukada, Hajime, 83
funding, 51, 99, 104, 176, 180, 181, 196, 205, 206, 213, 214, 223, 231–233, 238, 257
funnel-traps, 181, 194
fyke net, 194, 197, 213

G

Gaboon Viper (*Bitis gabonica*), 169, 170
Gagadju people, Kakadu, 180, 200
Galapagos Islands, 111, 123, 165
Galapagos Land Iguana (*Conolophus subcristatus*), 111
gape limitation, 199, 262
Garden Island, Western Australia, 123, 124
Gartersnakes, American (*Thamnophis* spp.), 201, 229, 231
Gecko, Velvet (*Oedura lesueurii*), 136, 139
genes, 44–46, 55, 76, 119, 120, 145, 165, 169, 170, 177, 197, 218, 234, 235, 241–245
genetic analyses, 45, 60, 88, 97, 145, 165, 167, 229, 231, 243, 245
genetic assimilation, 169–171
Gila Monster (*Heloderma suspectum*), 54
Glauert, Ludwig, museum curator, 114
Gold Coast, Queensland, 177
Golden Crowned Snake (*Cacophis squamulosus*), 86, 87
Golden Tree Snake (also Green Tree Snake) (*Dendrelaphis punctulatus*), 187
Golden Water Skink (*Eulamprus quoyii*), 25, 26
Goldsbrough, Claire, 132
Goose, Magpie (*Anseranas semipalmata*), 206
Goulburn, New South Wales, 78
government authorities, 104, 152, 167, 190
Gow, Graeme, 88
grant applications, 213, 231, 232
Greater Gliding Possum (*Petauroides volans*), 22
Great Tit (*Parus major*), 61, 225
Green Tree Frog (*Litoria caerulea*), 78
Green Tree Python (*Morelia viridis*), 92
Green Tree Snake (also Golden Tree Snake) (*Dendrelaphis punctulatus*), 6, 7, 9, 10, 135, 162
Greer, Germaine, 62
Grey Snake (*Hemiaspis damelii*), 70
Griffith, Bobbi, mother-in-law, 157
Griffith (nee), Terri, wife, 59, 60, 63–65, 73, 79, 80, 82, 83, 87, 100, 122, 131, 137, 139, 140, 208, 214, 224, 251, 257
Grigg, Gordon, 66
Gudgeon (*Oxyeleotris* spp.), 198
Gwardar (*Pseudonaja nuchalis*), 114

H

habitat, loss, restoration, 172
Harlow, Peter, 51, 58, 72–75, 82, 89, 121, 200, 207, 220, 257
hatchlings, 91, 218, 234, 245
Heatwole, Hal, 19, 23, 24, 39, 41, 47, 62, 173, 178
helicopter surveys, 138, 217
Herdsman Lake, Perth, 170
heritability, 44, 241, 242, 270
high-impact journals, 99, 213, 235, 245
hippies, 52, 115, 119
Honors research, Australian National University, 22, 25, 62
Hore, Russell, 207
House-Mouse (*Mus domesticus*), 74–76, 98, 107, 158, 159, 162, 172, 207
Houston, Darryl, 196–198, 213
How, Ric, 81
humidity, 177, 179, 203, 204, 239
Humpty Doo, Northern Territory, 175, 220
Huxley, Thomas, 61

I

Ice Age, 95, 97, 165
Idriess, Ion, 177, 188, 191

Iguana, Galapagos Land (*Conolophus subcristatus*), 111
incubation moisture, 234, 241, 244
incubation period, 90, 111
incubation temperature, 63, 65, 110, 214, 234
India, 133, 177
Indigenous Australians, 64, 180, 188
Indigenous hunters, 141, 180, 188, 192–194, 199
Indigenous people, Wik-Mungkan, Cape York, 114
Indonesia, 95, 142, 231
inherited characters, 45, 169, 252
Inland Taipan (*Oxyuranus microlepidotus*), 98
insects, 38, 91, 92, 96, 180, 207, 233
interbreeding, 115, 118, 165
interest in snakes, 2, 6, 7, 10, 22, 27, 48, 81, 86, 95, 123, 127, 157, 158, 180, 219, 222, 242, 248–251
invasive species, 64, 92, 238
Isbell, Lynne, University of Davis, California, 252
island evolution, 95, 123, 124, 165, 166, 167–172, 197

J

Jabiru (also Black-Necked Stork) (*Ephippiorhynchus asiaticus*), 240
Jabiru, Kakadu, 180, 181, 186, 192, 196, 201, 206, 227
James, Craig, 188
Jamison Valley, Blue Mountains, New South Wales, 79, 130
Japan, 83, 92
journals, scientific, 41, 42, 56, 63, 65, 83, 99, 113, 125, 126, 129, 170, 212, 213, 219, 235, 245

K

Kakadu National Park, Northern Territory, 152, 179–181, 187–190, 192, 196, 199, 201, 205, 206
Kansas, 234
Kata Tjuta, Northern Territory, 132
Keelback (*Tropidonophis mairii*), 230–235, 239–242, 245
Keogh, Scott, 145, 167
Kimberley region, Western Australia, 114, 177, 179
King Brown Snake (also Mulga Snake) (*Pseudechis australis*), 33, 70, 88, 95, 151, 152, 154, 156, 155, 217, 229
King Cobra (*Ophiophagus hannah*), 27, 133, 134
Koala (*Phascolarctos cinereus*), 45, 79, 129, 250
Komodo Dragon (*Varanus komodoensis*), 142
Krefft, Gerard, 136

L

Lace Monitor (*Varanus varius*), 111, 118, 141, 142
Lake Cronin Snake (*Paroplocephalus atriceps*), 135
Lake George, Canberra, 172
Lamarck, French biologist, 169, 170
Lambeck, Rob, 181, 188, 190, 207
Land Mullet (*Bellatorias major*), 8
Lawson, Henry, vii
Lazell, Brian, 147
Leeton, New South Wales, 158, 162, 164
Legler, John, 39, 40, 48, 49, 63, 181
Lesser Black Whipsnake (*Demansia vestigiata*), 56
life history, snake, 27, 79, 83
lifetime reproductive success, 241, 242
Lifou, New Caledonia, 132
Lindner, Dave, 189, 190, 194, 197
litter size, 41, 42, 44–47, 96, 145, 147, 195, 216
Little Bellingen River, 33
live-bearing (viviparity), 57–59, 63, 65, 86, 87, 95, 96, 152
Lizard, Burton's Legless (*Lialis burtonis*), 114
Lizard, Frillneck (*Chlamydosaurus kingii*), 178, 187
Lizard, Gila Monster (*Heloderma suspectum*), 54
Lizard, Land Mullet (*Bellatorias major*), 8
lizards, as prey, 1, 26, 39, 84, 91, 92, 95–97, 106, 110, 124, 136, 145, 148, 157, 168, 169, 223, 242
Lizard, Tuatara (*Sphenodon punctatus*), 60
lizard body temperature, 65, 132
lizard-catching, 3, 7, 19, 35, 177, 181, 184
lizard eggs, 57, 64, 81, 111, 227, 231
lizard research, 22, 24, 25, 27, 59, 63, 180, 181, 221
Llangothlin Lagoon, New South Wales, 29
long-term study, 74, 102, 143, 205, 206, 223, 241
low metabolic rate, 185, 186, 204, 215

M

Macleay's Watersnake (*Pseudoferania polylepis*), 175
Macquarie Marshes, 51, 67, 70–73, 78, 79, 109, 147, 207, 223
Madsen, Monika, 213, 219
Madsen, Thomas, 201, 212–214, 216–219, 221, 222–224, 228, 231–235, 238, 249, 257
Magela Creek, Kakadu, 181, 184, 187–189, 192
Magpie, Australian (*Gymnorhina tibicen*), 32, 35
Magpie Goose (*Anseranas semipalmata*), 206
male-male combat, 55, 56, 71, 96, 119, 124
manuscript reviews, 42, 56, 57, 59, 129
mark-recapture, 232, 234, 244
Mary River, Northern Territory, 155, 191, 197
mating aggregation, 109
Maynard Smith, John, 83
McKinley, Jim, 54
Mell, Rudolf, 59
Melrose, William, 167
Mengden, Greg, 41, 115
metabolic rate, 149, 185, 186, 198, 204, 215
Metallic Skink (*Niveoscincus metallicus*), 85
Middle Point Village, Northern Territory, 212, 213
Midnight Oil, 38

Mitchell's Water Monitor (*Varanus mitchelli*), 187
Moccasin, Cottonmouth (*Agkistrodon piscivorus*), 54, 157, 158
molecular genetics, 60, 245
Molonglo River, Australian Capital Territory, 1, 13
Monitor, Lace (*Varanus varius*), 111, 118, 141, 142
Monitor, Mitchell's Water (*Varanus mitchelli*), 187
Monitor, Yellow-Spotted (*Varanus panoptes*), 180, 181
monsoons, 175, 179, 187, 199, 201, 203, 215–218, 220, 223, 227, 236
mortality rate, 42, 45, 110, 145
mosquitoes, 47, 175, 200, 219, 233
Mother of Ducks Lagoon, New South Wales, 29
mothers, reptile, 45, 46, 57, 65, 111, 203, 204, 214, 215, 234, 235, 241, 242, 243
Mouse, House (*Mus domesticus*), 74–76, 98, 107, 158, 159, 162, 172, 207
Mulga Snake (also King Brown Snake) (*Pseudechis australis*), 151
Mullumbimby, New South Wales, 114, 115, 118, 119, 146
Murrumbidgee River, 10
museum, Australian, 37, 81, 84, 86–89, 99, 168, 189, 195, 206
museum, Northern Territory, 88
museum, Queensland, 87, 88
museum, South Australian, 168
museum, Western Australian, 81, 114
museum curator, reptile, 81, 86–88, 114, 206
museum dissections, 37, 92–95, 100, 106, 112, 118, 147, 151, 152

N

Namoi River, 147
Nancy, pet chicken, 83
National Parks and Wildlife Service (NPWS), 104, 112, 136, 138
natural history, snakes, 26, 27, 48, 60, 92, 99, 113, 206, 215
natural selection, 41, 43–45, 55, 119, 128, 234, 248
Nelson, David, 56, 186, 210, 230, 240, 257
Nemarluk, Aboriginal warrior, 178
nesting behaviour, 110
nest-site, 111, 241, 242
New England Highlands, 23, 27, 33, 49
New Guinea, 92, 181, 231
New Guinea Ground Boa (*Candoia aspera*), 92
New Zealand, 14, 20, 60, 130, 228
Newton, Isaac, 223
niche partitioning, 199
Nimbin, New South Wales, 115
nocturnal, 54, 64, 92, 233, 243
non-venomous snakes, 3, 158, 229, 230, 248
Noorlangie, Kakadu National Park, 180
North America, 44, 90, 92, 177, 184, 200, 201, 216, 221
Northern Corroboree Frog (*Pseudophryne pengilleyi*), 64
Northern Hemisphere, ecosystem stability, 61, 149, 215, 216, 247, 250
Northern Territory, 88, 98, 107, 151, 152, 155, 173–238
Northern Territory Museum, 88
Numbat (*Myrmecobius fasciatus*), 122

O

offspring, reptile, 45, 55, 57, 63, 65, 77, 78, 80, 87, 96, 110, 111, 145, 169, 201, 203, 214, 216, 218, 234, 241, 243, 244
Ogilby, Douglas, reptile curator, 88
Oklahoma, 157
Ontario, Canada, 43
oviparity, egg-laying, 57, 65, 96

P

Pacific Tree Boa (*Candoia bibroni*), 92, 132
Pale-Headed Snake (*Hoplocephalus bitorquatus*), 147, 148
Panic, 109, 252, 253
Papua New Guinea, 97, 231
parental attitudes, 2, 6, 7, 9, 13, 14, 247
parental care, in reptiles, 203, 214
Patonga homestead, Kakadu, 190, 192
Patterson, Banjo, 249
Pearl Beach, New South Wales, 101, 102
Pearson, David, 122–126, 197, 257
Perth, Western Australia, 81, 112, 122, 123, 125, 167, 170
Ph.D. thesis, 24, 48
phobia, snake, 252, 253
Pike, Canadian (*Esox lucius*), 43
Pike, David, 141
Pilliga Scrub, New South Wales, 147
pioneers, research, ix, 11, 27, 51, 57, 59, 60, 61, 79, 115, 168, 169, 172, 207, 222, 234, 249
plasticity, phenotypic, 169, 170, 202
Platypus (*Ornithorhynchus anatinus*), 57, 248
poaching, 122, 190, 191
poetry, 14, 118, 249
pollution, 37, 38, 196, 204
Pope, Clifford, American biologist, 29
"popular book", 90, 113
population-level study, 45, 129, 176, 177, 197, 229
Possum, Greater Gliding (*Petauroides volans*), 22
postdoctoral fellowship, 48, 60, 63, 171
Pough, Harvey, 128, 176, 186, 201, 207
Press, Tony, 155, 156
Pretoria, South Africa, 89

prey abundance, 71, 206, 212, 217, 219, 235, 242, 245
prey size, 39, 85, 96, 124, 168–171, 197, 199
primates, vii, 248, 252, 253
primitive, snakes seen as, vii, 17, 23, 62, 111, 128, 245, 248
protected fauna, 104, 109, 250
publications, ix, 41, 42, 63, 66, 67, 82, 86, 99, 128, 212
public opinion, 88, 180, 247–249
Python, Carpet (also Diamond Python) (*Morelia spilota*), 16, 33, 101, 110, 112, 114–116, 118–126, 135, 145, 146, 176
Python, Centralian Carpet (*Morelia bredli*), 132
Python, Diamond (also Carpet Python) (*Morelia spilota*), 9, 10, 101–106, 109, 110–113, 115, 118–120, 122–125, 132, 141, 186, 203
Python, Green Tree (*Morelia viridis*), 92
Python, Scrub (*Simalia amethistina*), 96, 112
Python, Southwest Carpet (*Morelia spilota imbricata*), 197
Python, Water (*Liasis fuscus*), 91, 105, 107, 108, 203–225, 229, 231–235, 249
python teeth, recurved, 9, 101, 107, 108, 119

Q

Quambone, New South Wales, 70, 80
Queensland, vii, 101, 151, 173, 175, 177–179, 196, 231, 238
Queensland Museum, 87, 88

R

Racer, American (*Coluber constrictor*), 158
radio transmitters, 51–79, 104, 109, 125, 135, 146, 158, 164, 186–188, 203, 207, 212, 214, 245
radio-tracking, 71, 74, 77, 78, 104, 119, 124, 125, 129, 146, 158, 159, 186, 189, 217, 238
rainbow serpent, 203
Rainbowfish (*Melanotaenia* spp.), 198
rainfall, monsoonal, 223, 236
rainfall, seasonal variation, 79, 179, 200, 201, 205, 206, 215, 217–219, 235
rainforest, vii, 33, 102, 119, 132, 146, 178–180, 231
rangers, park, 33, 34, 37, 138, 139, 152, 153, 155, 184, 188, 238
Rat, Bushrat (*Rattus fuscipes*), 146
Rat, Dusky (*Rattus colletti*), 210, 216, 217, 223
rat abundance, 217, 218
rates of reproduction, 146, 176
Rattlesnake, Black-Tailed (*Crotalus molossus*), 41
Rattlesnakes, American (*Crotalus* spp.), 11, 201, 250, 252
Recherche Archipelago, Western Australia, 123, 124, 165
Red-Bellied Blacksnake (*Pseudechis porphyriacus*), 3, 12, 25, 27, 53–81, 92, 95, 151, 152, 250, 254
Reed, Bob, 145, 265
Renfree, Marilyn, 22
reproductive cycles, 40, 87, 207
reproductive mode, 57, 87
research assistants, 21, 51, 60, 74, 82, 83, 89, 100, 102, 175, 181, 220, 221, 257
research funding, 51, 99, 104, 176, 180, 181, 196, 205, 206, 213, 214, 223, 231–233, 238, 257
research station, 101, 227, 238, 239
research supervisor, 19, 23–27, 37, 39, 47, 60, 62, 93, 102, 115, 122, 173, 178
retreat sites, 96, 130
ritual combat, 33, 71
River, Adelaide, 204, 217, 223, 225
road-kill, 33, 71, 84, 142
rock crevice, 7, 106, 130, 131, 132, 135, 136, 138, 139, 140, 144, 148
rock-thieves, 136, 138, 139
Rosella, Crimson (*Platycercus elegans*), 79
Ross, Geoff, 51, 73, 82
Rottenest Island Shingleback (*Tiliqua rugosa konowi*), 165
Rough-Scaled Snake (*Tropidechis carinatus*), 135
Ryukyu Island Pit Viper (*Ovophis okinavensis*), 92

S

Saint George, Utah, 54
salamanders, 55
Salt Lake City, Utah, 54, 59, 62
Saltwater Crocodile (*Crocodylus porosus*), 155, 173, 174, 175, 180, 187, 189, 194, 196, 199, 204, 217, 235
sample size, 70, 90, 109, 125, 164, 176, 198, 234, 239
sandstone cliffs, escarpments, 6, 106, 127, 129, 130, 131
scale-clips, 207, 211, 212, 243
scent trail, 93, 106, 109, 119, 172, 199
Schwaner, Terry, 168
scientific career, vii, viii, 17, 18, 23, 36, 39–42, 48, 49, 52, 57, 60, 62, 63, 66, 67, 79, 82, 94, 98, 126, 151, 155, 176, 181, 189, 191, 221, 222, 228, 238, 245, 247, 257
Scrub Python (*Simalia amethistina*), 96, 112
seabird colony, 168–170
Sea Kraits (*Laticauda* spp.), 95, 234
sea-level change, 97, 165–167
seasonal migration, predator and prey, 217
Sergeev, Alexei, 59
sex ratios, 29, 42–46, 48, 49, 54
sexual size dimorphism, 48
Shark of the Jungle (*Loveridgelaps elapoides*), 114

Shea, Glenn, 97, 98, 133, 247
Shine, Ben, son, 83, 100, 122, 224, 257
Shine, Cooper, family dog, 82, 83
Shine, James Macquarie (Mac), son, 80, 82, 83, 87, 100, 122, 224, 257
Shine, John, brother, 7, 9, 60
Shine, Judith, sister, 7
Shine, Molly, mother, 6, 9, 49
Shine, Patrick, father, 6, 9
Shine Terri (nee Griffiths), wife, 59, 60, 63–65, 73, 79, 80, 82, 83, 87, 100, 122, 131, 137, 139, 140, 208, 214, 224, 251, 257
Shine's Whipsnake (*Demansia shinei*), 98, 99
Shoalhaven Heads, New South Wales, 79
Sidewinder Rattlesnake (*Crotalus cerastes*), 54
Skink, Bluetongue (*Tiliqua scincoides*), 3, 6, 9, 39, 238
Skink, Cunningham's (*Egernia cunninghami*), 22, 23
Skink, Golden Water (*Eulamprus quoyii*), 25, 26
Skink, Metallic (*Niveoscincus metallicus*), 85
Skink, Rottenest Island Shingleback (*Tiliqua rugosa konowi*), 165
Skink, Tussock Grass (*Pseudemoia pagenstecheri*), 66
Slatey-Grey Snake (*Stegonotus australis*), 227–232, 234, 239–245, 250
Slip, David (Dave), 104, 105, 115, 122, 125, 126, 197, 203, 207
Slowinski, Joe, 171
Small-Eyed Snake (*Cryptophis nigrescens*), 127, 132, 143
Smith, Sil, farmer, 172
Snake, Collett's (*Pseudechis colletti*), 58
snake, mothers, 45, 46, 57, 111, 203, 204, 214, 215, 234, 241–243
snake anatomy, 39
snake "attacks", 13, 36, 38, 76, 134, 158, 162–164, 227
snakebite, 1, 7, 14, 20, 21, 27, 32, 33, 85, 89, 93, 98, 102, 107, 123, 133, 135, 143, 153, 156, 157, 163, 167, 185, 227–230, 241
snakebite cure, antidote, 167
snake-catching, 7, 13, 29, 33, 41, 93, 192, 217
snake conservation, ix, 128, 129, 140, 144, 145, 148
snake courtship and mating, 26, 47, 51, 53, 55, 56, 76, 77, 79, 109, 112, 113, 115, 122–124, 197, 243
snake creche, 77, 78
"Snake Detection Hypothesis", 252
snake diet, 27, 38, 85, 91, 92, 95, 157, 168, 170, 172, 207, 231, 242
snake dissection, 37, 92–95, 100, 106, 112, 118, 147, 151, 152
snake embryos, 45, 46, 57, 59, 86, 94, 110, 111, 132, 152, 181, 195, 244, 245
snake enthusiast, 8–11, 17, 25, 83, 86, 93, 104, 127, 128, 132, 155, 158, 233, 257
snake evolution, 17, 27, 39, 42, 45, 46, 48, 49, 55–57, 59, 94–96, 98, 115, 124, 126, 132, 145, 151, 152, 165, 167–170, 185, 195, 198, 201, 229, 234, 241, 244, 248, 254
snake genetics, 45, 46, 55, 76, 97, 145, 165, 167, 169, 171, 177, 218, 229, 231, 234, 241–245
snake habitat, 27, 38, 49, 51, 70, 74, 81, 99, 102, 106, 122, 124, 126, 132, 136, 138–141, 144–148, 151, 158, 160, 172, 176, 180, 208, 243
snake-keeping, 14, 103, 104, 107, 130, 133, 147
snake life history, 27, 79, 83
snake myths, 38, 62, 71
snake personality, 76, 124, 133, 162
snake prey, 9, 26, 38, 39, 71, 85, 87, 89, 91, 96, 97, 106, 107, 124, 130, 132, 144, 145, 165, 168–172, 185, 196–199, 205, 206, 210, 212, 217, 219, 223, 231, 235, 242, 245
snakes as pets, 3, 90, 104, 112, 120, 247
snake sex lives, 51, 118, 122, 197
snake sex ratios, 42, 45, 46, 48, 49, 54
snake stomach contents, 81, 195
snake-tongs, 11, 28, 54, 89, 153
South Africa, 89, 158
South America, 27, 92, 229, 252
Southwest Carpet Python (*Morelia spilota imbricata*), 197
Spiders, Flat (*Hemicloea* spp.), 132, 133, 139
spotlighting for snakes, 54, 123, 213, 219
Stephens' Banded Snake (*Hoplocephalus stephensii*), 115, 146, 147, 148, 179
Stork, Black-Necked (*Ephippiorhynchus asiaticus*), 240
striking pose, 133, 162, 164, 198
suburban reptiles, 3, 6, 7, 86, 90, 92, 102, 104, 106, 109, 112, 120, 125, 133, 136, 203, 250
Sumner, Jo, 145
sun-exposed rocks, 130, 131, 132, 141, 148
surgical procedure, transmitters, 44, 74, 106, 109, 187
Swamps, snake habitat, 29, 38, 51, 54, 68, 70, 71, 76, 78, 79, 167, 168, 172, 204, 215, 216, 231, 245
Swamp Snake (*Hemiaspis signata*), 96
Swan, Gerry, 104, 106
Sweden, 212, 213, 217, 232
Switzerland, 180, 243
Sydney Harbour Bridge, 101, 102
Sydney suburbs, snakes in, 6, 7, 9, 86, 90, 93, 101, 102, 104, 105, 112, 113, 120, 124, 125, 132–136, 203
Sydney University, 51, 59–63, 66, 67, 84, 86, 93, 112, 213, 257

T

Taipan, Coastal (*Oxyuranus scutellatus*), 11, 20, 93, 96, 98, 114, 151, 153, 157
Taipan, Inland (*Oxyuranus microlepidotus*), 98
Tasmania, 92, 157, 165, 167, 168, 176
telemetry, 76, 78, 102, 112, 119, 159, 187, 188, 204, 217
termites, 91, 93, 111
Texas, 41, 54
thermal biology, 65, 78, 130, 132, 136, 138, 201, 221
threat display, of snakes, 13, 133, 164
Tianjara Falls, New South Wales, 127, 132, 144
Tigersnake (*Notechis scutatus*), 10, 20, 27–29, 32, 36, 38, 41, 44–49, 54, 93, 95, 96, 102, 114, 119, 135, 153, 157, 165, 166–172, 231
tissue samples, snake, 167, 243
Tit, Great (*Parus major*), 61, 225
Toad, Cane (*Rhinella marina*), 92, 238
tongue-flicking, 32, 52, 77, 109, 184
"Top End", Northern Territory, 151, 179, 180, 192, 196, 201, 219
Tortoise, Desert (*Gopherus agassizii*), 54
Townsville, Queensland, 112, 179
traditional knowledge, Aboriginal, 141, 180, 188, 192, 200
Trans-Pecos Ratsnake (*Bogertophis subocularis*), 41
transmitters, radio, 51, 52, 74, 76, 79, 104, 109, 125, 135, 146, 158, 164, 186–188, 203, 207, 212, 214, 245
tree-climbing snakes, 135, 227
Tree Frog, Cooloola (*Litoria cooloolensis*), 88
tree-hollows, 112, 146, 148, 216
Trefoil Island, Tasmania, 167
tropical ecology, 174–227, 229, 232, 238, 257
Tuatara (*Sphenodon punctatus*), 60
Tussock Grass Skink (*Pseudemoia pagenstecheri*), 66
Tyndale-Biscoe, Hugh, 22
Typhlopidae, blindsnakes, 93, 94

U

Ujvari, Bea, 201
University of New England, Armidale, New South Wales, 23, 196
University of Sydney, 51, 59–63, 66, 67, 84, 86, 93, 112, 213, 257
University of Utah, 48, 52, 56, 60, 63, 67
Utah, 48, 49, 53, 54, 59, 63
unpredictable Australian environment, 79, 144, 149, 201, 204, 207, 215, 216, 218, 245
Uralla Lagoon, New South Wales, 27, 28
uranium mining, 180, 196
urbanization, 3, 37, 83, 106, 109, 112, 145, 148, 247

V

Vane, Rocky (aka Lindsay Vane), showman, 167, 170
Velvet Gecko (*Oedura lesueurii*), 136, 139
venom, vii, 1, 3, 4, 13, 20, 33, 71, 93, 96, 98, 101, 133, 135, 143, 158, 162, 164, 172, 227
venomous snakes, 2, 3, 7, 10, 11, 13, 14, 20, 22, 27, 33, 39, 67, 78, 79, 83, 86, 91, 93, 95, 102, 119, 143, 148, 151–153, 155, 157, 159, 163, 229, 247, 248, 252, 257
Viper, Gaboon (*Bitis gabonica*), 169, 170
viviparity, live-bearing, 57–59, 65, 95, 96

W

Waddington, C. H., British biologist, 169–171
Wallaby, Agile (*Macropus agilis*), 21, 25, 180, 223
Wallaby, Rock (*Petrogale* spp.), 140
Wallaby, Tammar (*Macropus eugenii*), 124, 125, 198
Wallum Sedge Frog (*Litoria olongburensis*), 88
warm-blooded, endotherms, 62, 106, 111, 136, 148, 202, 216, 248
Water Buffalo (*Bubalus bubalis*), 187, 188, 199, 200, 204, 206, 233, 235
Water Python (*Liasis fuscus*), 91, 105, 107, 108, 203, 204, 206, 207, 210–215, 217–219, 221–225, 229, 231–235, 249
Watersnake, Banded (*Nerodia fasciata*), 155, 157
Watersnakes, American (*Nerodia* spp.), 43, 54, 157, 158, 184, 232
Webb, Grahame, 175, 178
Webb, Jonathan (Jonno), 89, 93, 94, 112, 129, 133, 135, 138, 140–148, 257
Weekes, Claire, 59
Weigel, John, Australian Reptile Park, 138, 171, 257
Western Australia, 81, 98, 114, 122, 123, 125, 126, 165, 167, 197
Western Australian Museum, 81, 114
Western Brownsnake (*Pseudonaja nuchalis*), 114
"wet-dry" tropics, 151, 179, 196, 220
wet season, 175, 179, 196, 199, 201, 203, 206, 212, 215, 217, 223, 235, 236
Whipsnake, Lesser Black (*Demansia vestigiata*), 56
Whipsnake, Shine's (*Demansia shinei*), 98, 99
Whipsnake, Yellow-Faced (*Demansia psammophis*), 7, 19–21
Whitaker, Pat, 157–159, 162–164
White-Lipped Snake (*Drysdalia coronoides*), 85

wildfire, 143, 144
Williams, George C., evolutionary biologist, 43, 52, 57, 128, 234
Williams Island, South Australia, 167, 171
Wombat (*Vombatus ursinus*), 64, 248
Wombey brothers (John and Russell), 10, 11, 14
Worrell, Eric, 91
Woylie (*Bettongia penicillate*), 122, 124
Wright, Roy (Wrighty), 190, 191

Y

Yarrawonga Zoo, Northern Territory, 152
Yellow-Faced Whipsnake (*Demansia psammophis*), 7, 19–21
Yellow-Spotted Monitor (*Varanus panoptes*), 180, 181

Z

zoos, 2, 73, 139, 152, 153, 157, 169, 173, 203

For Product Safety Concerns and Information please contact our EU
representative GPSR@taylorandfrancis.com
Taylor & Francis Verlag GmbH, Kaufingerstraße 24, 80331 München, Germany

www.ingramcontent.com/pod-product-compliance
Ingram Content Group UK Ltd.
Pitfield, Milton Keynes, MK11 3LW, UK
UKHW020931180425
457613UK00012B/317

* 9 7 8 1 0 3 2 2 3 2 9 2 8 *